Discovering Nothing

David L. Nicandri

Discovering Nothing

In Pursuit of an Elusive Northwest Passage

UBCPress · Vancouver

Originally printed in Canada
Print and bound by CPI Group (UK) Ltd, Croydon, CR0 4YY

UBC Press is a Benetech Global Certified Accessible™ publisher. The epub version of this book meets stringent accessibility standards, ensuring it is available to people with diverse needs.

Library and Archives Canada Cataloguing in Publication
Title: Discovering nothing : in pursuit of an elusive Northwest Passage / David L. Nicandri.
Names: Nicandri, David L., 1948- author.
Description: Includes bibliographical references and index.
Identifiers: Canadiana (print) 20240310993 | Canadiana (ebook) 20240311094 | ISBN 9780774868877 (hardcover) | ISBN 9780774868891 (PDF) | ISBN 9780774868907 (EPUB)
Subjects: LCSH: Northwest Passage—History. | LCSH: Northwest Passage—Discovery and exploration—British. | LCSH: Explorers—Great Britain—History.
Classification: LCC G640 .N53 2024 | DDC 910.9163/27—dc23

Canada Council for the Arts | Conseil des arts du Canada | Canada
British Columbia Arts Council | British Columbia

UBC Press gratefully acknowledges the financial support for our publishing program of the Government of Canada, the Canada Council for the Arts, and the British Columbia Arts Council.

This book has been published with the help of a grant from the Canadian Federation for the Humanities and Social Sciences, through the Scholarly Book Awards, using funds provided by the Social Sciences and Humanities Research Council of Canada.

UBC Press is situated on the traditional, ancestral, and unceded territory of the xʷməθkʷəy̓əm (Musqueam) people. This land has always been a place of learning for the xʷməθkʷəy̓əm, who have passed on their culture, history, and traditions for millennia, from one generation to the next.

UBC Press
The University of British Columbia
www.ubcpress.ca

For Robin Inglis

Contents

Illustrations

Cast of Major Characters

Government-Sponsored Maritime Explorers

Juan Francisco Bodega y Quadra, Spanish explorer sent to Nootka Sound in 1792 to negotiate the terms of the Nootka Sound Convention with his British counterpart, George Vancouver, intended to defuse international tensions in the region.

James Cook, British explorer whose third voyage, 1776–80, was the first to test the existence of the Pacific portal to the Northwest Passage.

Dionisio Alcalá Galiano, Spanish explorer and the principal navigator on the Malaspina expedition to the Pacific. In 1792, his ship, the *Sutil,* was detached, along with the *Mexicana* under Cayetano Valdés, to survey the inland waters of the Northwest Coast in search of the Northwest Passage.

Alejandro Malaspina, Spanish explorer whose circumnavigation of the globe, 1789–94, was his country's attempt at emulating the model of scientific discovery popularized by Cook.

Esteban José Martínez, Spanish explorer who, in response to repeated incursions from British maritime fur traders along the Northwest Coast claimed by Spain, took actions at Nootka Sound in 1789 that almost led to war, indirectly precipitating what became Vancouver's voyage.

José María Narváez, Spanish explorer who, in 1789, commanded the first European ship to sail deep into the waters of the Strait of Juan de Fuca.

Cayetano Valdés, Spanish explorer who, together with Galiano, encountered Vancouver's expedition and worked in tandem with the British in the summer of 1792. His *Mexicana* and Galiano's *Sutil* were the first European craft to circumnavigate Vancouver Island.

George Vancouver, British explorer who was dispatched to the Northwest Coast to negotiate the terms of the Nootka Convention with his Spanish counterpart, Bodega y Quadra, and led a cartographic survey of the Northwest Coast, 1792–94.

Charles Wilkes, American explorer who conducted an expedition to the Pacific, 1838–42, focusing on Antarctica. His side trip to the Northwest Coast in 1841 solidified American resolve for a division of the Oregon Country to include the deep waters of Puget Sound.

Exploration Sponsors and Northwest Passage Advocates

Joseph Banks, naturalist on Cook's first voyage. As president of the Royal Society, he later promoted John Ledyard's efforts and exploration of the Northwest Coast and the continental interior of Africa.

Alexander Dalrymple, geographer for the East India Company and the British Admiralty. Formerly a contender for the command of what became Cook's first voyage, he was the leading proponent of a revised version of the Northwest Passage in the aftermath of Cook's third voyage.

Sandford Fleming, Scottish-Canadian engineer and early proponent of an all-Canadian transcontinental railroad. He surveyed the first proposed route for the Canadian Pacific Railway (CPR) and later invented the notion of global standard time.

Thomas Jefferson, third president of the United States and member of the American Philosophical Society, offices that he used to promote exploration of the American West, including what became the Lewis and Clark Expedition.

William Cornelius Van Horne, Fleming's successor as general manager of the CPR who completed the railway. He named the line's western terminus after Vancouver in recognition of the railroad's potential to realize the promise of the Northwest Passage that the maritime explorer had sought earlier.

Asa Whitney, American merchant who became an early and effective proponent of an American transcontinental railroad.

Maritime Fur Traders and Explorers

Charles William Barkley, English ship captain. In 1787, while searching for pelts, he entered a large opening on the Northwest Coast and "rediscovered" what he surmised was the legendary Strait of Juan de Fuca.

James Colnett, veteran of Cook's second voyage who joined the search for furs in the wake of Cook's third voyage. He became the central figure in the sovereignty dispute with Martínez at Nootka Sound.

George Dixon, veteran of Cook's third voyage who sailed with Nathaniel Portlock for a royally chartered trading venture. The account of their 1786–87 expedition, which found a more complex coastline than Cook had intimated, helped repopularize the notion of a mid-latitude Northwest Passage.

Charles Duncan, an associate of Colnett's who sketched the first chart of the Strait of Juan de Fuca, published by Dalrymple.

Robert Gray, an American ship captain generally credited with the discovery in 1792 of the Columbia River, named after his ship. He was the first American sailor to command a circumnavigation of the globe.

James Hanna, the first European, in 1785, to sail to the Northwest Coast in search of furs after Cook's third-voyage account had publicized that prospect.

John Meares, a partner in several merchant ventures who publicized the difficulties of British traders with Spanish officials at Nootka Sound, almost leading to war. His 1790 account helped resurrect the promise of a Northwest Passage and laid the groundwork for Vancouver's expedition.

Nathaniel Portlock, a veteran of Cook's third voyage, like his partner Dixon.

Terrestrial Explorers

WILLIAM CLARK, US Army explorer. He was the principal geographical problem solver during his journey with Meriwether Lewis (1803–06). His reputation never recovered from the cartographic trap inadvertently left for him by the Vancouver expedition.

JOHN C. FRÉMONT, US Army explorer. His first two expeditions on the Oregon Trail to the Pacific Northwest and California were emblematic of American belief in Manifest Destiny in the 1840s.

SAMUEL HEARNE, Hudson's Bay Company fur trade explorer who led an expedition north and west of said bay to the mouth of the Coppermine River on the Arctic coast in the early 1770s. News of his trip guided Cook's third voyage to the subarctic latitudes since it seemed to preclude any possibility of a mid-latitude Northwest Passage.

ALEXANDER HENRY (the elder), partner in several fur trade enterprises, friend of Banks, and student of Cook. He touted the prospect of an overland version of a trade network to China, inspiring others to pursue this vision.

JOHN LEDYARD, corporal of marines on Cook's third voyage. After failing to organize a fur trade venture to the Northwest Coast, he sought the sponsorship of Banks and Jefferson for terrestrial circumambulation and later an expedition to the interior of Africa.

MERIWETHER LEWIS, a protege of Jefferson who led the first American overland expedition to the Pacific with Clark, whom he invited to serve as co-commander. His mission was a response to the mercantile challenges Mackenzie laid out in his account of exploration.

ALEXANDER MACKENZIE, a protege of Pond who led successive expeditions to the Arctic (1789) on the river that bears his name, and to Pacific tidewater (1793). The account of his explorations prompted Jefferson to launch what became known as the Lewis and Clark Expedition.

GEORGE MCCLELLAN, a captain in the US Army who was second in command to Isaac Stevens during the northern railroad survey of 1853. His diffident comportment during the survey perfectly prefigured his performance as a Civil War general.

Peter Pond, a fur trade explorer for the North West Company who pieced together many elements of a network of rivers and lakes in the Canadian Northwest that he thought could serve as an overland version of the Northwest Passage.

Isaac Stevens, the Washington Territory's first governor. He was tapped by US Secretary of War Jefferson Davis to lead a survey for a northern transcontinental railroad from the Great Lakes to Puget Sound. Uniquely among his colleagues who surveyed more southerly routes, Stevens touted his line as a latter-day version of the Northwest Passage.

Preface

THIS BOOK BUILDS UPON the history of the search for the Northwest Passage from the Pacific Basin that I began in *Captain Cook Rediscovered: Voyaging to the Icy Latitudes,* published in 2020. In that book, I challenged the orthodox understanding of Cook's third voyage by arguing that the great navigator believed he had finished his "tropical" career well before the conclusion of the second voyage in 1775. In his next (and final) expedition, Cook conducted planned stops in New Zealand and Tahiti for staging purposes only; there was no exploratory purpose in doing so. Inadvertently, he was forced to replenish his supplies in Tonga and Hawai'i, the latter perhaps his greatest geographical find, on the way to the actual discovery zone for the third voyage – the Northwest Coast of America. After transecting the Bering Strait and heading into the Arctic Ocean in August 1778, Cook was stymied by the ice off the northwest corner of Alaska, a land whose subcontinental dimensions he was the first to chart. That environmental encounter effectively ended Cook's career as an explorer. He died seven months later in Hawai'i when the expedition was overwintering in advance of a planned second assault on the Northwest Passage's Pacific portal.

Historians found, perhaps even invented, many shortcomings in Cook's third-voyage deportment, but most of these presumed deficiencies were the result of a fundamental misreading of what Cook and his employer, the British Admiralty, were trying to accomplish. Cook had tired of sorting out the insular particularity of the Pacific Basin during the second voyage, and said as much. Gradually he turned his gaze to geographical issues of a continental dimension. The second-voyage search for *Terra Australis Incognita,* the hypothetical southern continent mirroring the size and antipodal placement of Eurasia, was followed by the quest to determine if there was a northern passage through or above North America linking the Pacific and Atlantic Oceans. The ordering of that link is probative because Cook was the first to attempt the Northwest Passage via its Pacific portal, with the exception of Francis Drake's half-hearted attempt at the end of his famous piratical voyage two hundred years earlier. Between Drake and Cook, every effort by navigators hailing from several nations in Western Europe, but primarily the English, had logically attempted to find this imagined passage from its supposed Atlantic gateway, the one closest to their home ports. The names of these explorers – Hudson, Baffin, Frobisher, and Davis – populate the map of Atlantic Canada. The mistake historians have made is failing to see that Cook's third voyage was *not* a continuation of his earlier explorations in the Pacific, but rather the first step in Great Britain's assault on the Northwest Passage, an effort that the Admiralty was able to sustain continuously after the defeat of Napoleon.

The ice Cook encountered north of Alaska prevented him from discovering that which he sought. This book completes the story of the quest Cook initiated. Spoiler alert: Those who followed him to the Northeast Pacific, or overland to the Pacific Northwest quadrant of North America, didn't find it either. Of necessity, the great navigator's immediate nautical successors, like maritime fur trader John Meares or a more scientific voyager like George Vancouver, pursued the new template of west-to-east exploration that Cook modelled. However, Cook's theoretical perception of the Pacific coastline of North America above latitude 45°N also influenced terrestrial explorers. John Ledyard's somewhat quixotic attempt to circle the globe, also west to east, was first in line. Ledyard prefigured an evolving discovery agenda in which the Northwest Coast remained a focal point, although the overland expeditions conducted or sponsored by such practical and theoretical geographers as Peter Pond, Alexander Mackenzie, or Thomas Jefferson proceeded in the more traditional mode of east to west.

It is important to note here that this book is not a history of the Northwest Passage in its entirety. There are many such treatments, all of which privilege the approach from the Atlantic before and after Cook, relegating attempts at discerning the Pacific gateway to an interpretive backwater. The chapters in this volume, though arranged generally in chronological order, proceed episodically by following a specific thematic approach. The narrative emphasizes efforts to discover the passage's western gateway, work that Cook pioneered, but also the evolution of its cartographic conception and commercial potential over time. This is a wide spectrum. For example, it includes the overt, if still all too little appreciated, recalibrations of the notion of the Northwest Passage found in the writings of theoretical geographer and Cook rival Alexander Dalrymple.

More subtly, this book treats aspects of terrestrial exploration in Canada and the United States that have never been previously perceived as being a part of the Northwest Passage nexus. This includes the expeditions of Sandford Fleming and Robert Palliser to the north, or the one led by Isaac I. Stevens, the Washington Territory's first governor. In 1853, Stevens commenced a survey for a potential railroad from Minnesota to Puget Sound. More importantly, unlike his US Army peers investigating three east-west routes further south, his reports framed the value of his corridor in the long tradition of the Northwest Passage ideal as the shortcut to Asia. Fleming's and Palliser's efforts were consciously modelled upon the *Pacific Railroad Reports,* to which Stevens was the most prolific contributor.

By the same token, this book is not intended as a comprehensive history of Euro-American exploration of the continent's North Pacific Slope either. Not every explorer subsequent to Cook's third voyage, including some important ones, worked within the Northwest Passage framework. Take Charles Wilkes, for example. Wilkes was unquestionably a first-rank explorer in Northwest history, even if measured only in the number of place names he imprinted on the landscape south of the 49th parallel. His Jacksonian-era voyage in some respects geographically overlapped the discovery agenda of Cook's third voyage as well as George Vancouver's, but unlike those British expeditions, his mission's main goal lay in Antarctic waters. In his 1841 Pacific Northwest adjunct to that southern voyage, Wilkes was guided by a purely instrumentalist outlook, not one grounded in geographical theory. Specifically, he emphasized the navigability of the Columbia River and the suitability of harbours in the Salish Sea for the benefit of American policymakers eager for guidance on where to divide the Oregon Country, as the region was by then generally called. Curiously for a navy man, among his most ambitious efforts were the several terrestrial surveys he launched into the Hudson's Bay Company fur-trading districts in the Willamette Valley and what is now eastern Washington, in furtherance of an American Manifest Destiny agenda. In this respect, Wilkes had more in common with the contemporaneous Oregon Trail surveys of US Army explorer and national expansionist John C. Frémont than any explorer working within the Northwest Passage idiom. This includes their common American successor in the field, Stevens, who, as I demonstrate in these pages, was clearly working within the same theoretical framework as Cook, Mackenzie, and Lewis and Clark, albeit with an Industrial Revolution twist.

Heretofore the literature on the subject has taken a narrow view of what constitutes the Northwest Passage; the concept has usually been limited to the classic saltwater route connecting two oceans – the one Cook looked for. This book argues that there was actually a *series* of Northwest Passages, the shifting nature of which was concentrated on the Pacific side of the North American continent. The idea of a shortcut between the two oceans was so durable and compelling that when one version failed to materialize another was proffered hypothetically, or attempted through different means.

The Sequence of Northern Exploration

The evolution of the Northwest Passage began even before the ink was dry on the Admiralty's post-voyage account of Cook's final voyage, published in 1784. John Ledyard, an American-born corporal in the Royal Marines, had sailed with Cook without particular distinction. But Ledyard seems to have been especially captivated by the relative narrowness of the Bering Strait, separating the eastern cape of Asia from the westernmost point on the North American mainland, an aspect of geographical reality established cartographically by Cook. Ledyard published one of the handful of unauthorized accounts of the third voyage that were rushed into print to capitalize on the market appeal of all things Cook, even before the posthumous publication of the commander's own narrative. Having participated in a worldwide circuit with Cook, and having witnessed first-hand that the Arctic ice pack appeared to be a permanent obstacle to the Northwest Passage ideal, Ledyard generated a companion idea as audacious as his mentor's serial circumnavigations – global circumambulation, that is, "walking" around the world.

This book begins with the Ledyard story. Though schooled in the concept of identifying the Northwest Passage as a nautical enterprise, Ledyard would prove to be a pivotal figure in the evolution of its cartographic conception. One of the principal consequences of Cook's third voyage was the inadvertent discovery, after the great navigator's death, that the sea otter pelts his crew acquired at Nootka Sound for warmth, in anticipation of their voyage into the Arctic's icy latitudes, were worth a small fortune in Asian ports. Cashing in on the resultant Northwest Coast maritime fur trade boom was Ledyard's first goal. When, after the US War of Independence, he failed to find American sponsors for a commercial voyage or to catch on as a supercargo on expeditions departing Great Britain, Ledyard settled on the scheme of travelling around the world. Anticipating Jules Verne's fictional Phileas Fogg in *Around the World in Eighty Days* (1873), his venture necessarily projected a series of conveyances, such as coach rides and oceanic passages. Ledyard truly expected to walk only across North America, and even then just its western part.

As detailed in this book's Prologue, Ledyard's planning brought him into contact with Joseph Banks, a naturalist on Cook's first voyage and later Royal Society president and exploration sponsor, as well as Thomas Jefferson, Benjamin Franklin's successor at the royal court in Paris. Ledyard captured Jefferson's imagination as the prototype of the adventuresome and inquisitive American explorer, later realized in the person of Meriwether Lewis. Even though he was ultimately unsuccessful in his global quest, Ledyard prefigured the transition of scientific discovery from the geography bordering the world's oceans to its continental interiors. After all, as several Spanish, French, and British navigators complained, after Cook, what truly noteworthy accomplishments were left to rack up at sea? Thus, the terrestrial interiors of Africa and the Americas became the new frontier. Emblematic of this conversion were Alexander von Humboldt's Spanish-American expedition (1799–1804) and the explorations of Mungo Park. The latter's *Travels in the Interior of Africa* (1799) became a travel literature bestseller.

In this book's area of focus – the Northwest Passage nexus – the comparable continental interior voyagers were Peter Pond and Alexander Mackenzie. Their stories are recounted episodically in Chapters 1 to 3. As terrestrial fur trade explorers, they extended westward from Montreal toward the Arctic and Pacific in the wake of the Cook third-voyage peltry epiphany. Like Ledyard, Pond and Mackenzie would toy with the idea of a global expedition, but in reverse – stretching to Atlantic Europe by way of the Russian Far East after reaching the western limits of North America. The American venture Lewis led was in large measure a response to theirs. As for Ledyard, he would eventually die during the course of an attempt to plumb the depths of the African interior.

As we shall see in Chapter 5, what became known as the Vancouver expedition actually had its roots in African exploration too. That is only one aspect of an origin story that is far more complicated than generally conceded. Vancouver's maritime mission (1792–94) is often mistakenly posited as an attempt to clean up Cook's supposedly sloppy work in 1778, a regional extension of the long-dominant tired third-voyager hypothesis that I critiqued in *Captain Cook Rediscovered.* As I explain at length in the body of this book, Vancouver's survey bore only a surface resemblance to Cook's mission. Ironically, Vancouver was himself often wrong in his calculations of latitude and longitude and more general geographical deductions. Rivers were a particular challenge: Vancouver might never have found the mouth of the Columbia River if American fur trader Robert Gray had not pointed out its existence, and he completely missed the outfalls of the Fraser and Stikine Rivers.

But this may be explained by the fact that Vancouver's guidance emphasized scouring the Northwest Coast of America looking for a western analogue to Hudson Bay, which, as it was imagined, shortened the carrying distance between the Pacific Ocean and the Atlantic Basin. This idea of a North American "Mediterranean," accessible from the Pacific, had a long history. The *Mer de l'Ouest* had been a favourite notion, pre-Cook, of French cartographers. Cook and Samuel Hearne, a Hudson's Bay Company fur trade explorer, combined efforts to seemingly destroy the idea once and for all in the 1770s. As I explained in *Captain Cook Rediscovered,* Cook's guidance from the Admiralty, issued in 1776, directed that he avoid exploring for a passage in the temperate latitudes of the Northwest Coast. This was because Hearne had proved, by walking from the bay's shore to the mouth of the Coppermine River on the Arctic coast in 1771, that there could be no Northwest Passage in the subarctic latitudes. However, the maritime fur traders who followed Cook's track revived the old idea of this great inland western sea after finding a far more complex Northwest Coast shoreline – one populated with many indentations and openings in the temperate latitudes – than Cook's charts had intimated.

Another echo from Cook's time, in the person of Alexander Dalrymple, is fully introduced in Chapter 2. Dalrymple is a frequently misunderstood character in Cook literature. He had once been a legitimate candidate to lead what became Cook's first voyage to the South Pacific on the transit of Venus mission. Dalrymple was later caricatured by historians as someone who sought revenge against his supposed nemesis by suggesting that Cook's third voyage along the Northwest Coast had not been pursued diligently. Though Dalrymple revelled in geographical speculation (which, while frequently off-base, did nonetheless have the productive effect of spurring voyages of discovery), he was too accomplished a geographer to believe that Cook had somehow missed a mid-continental opening that led all the way to the Atlantic. After all, it was Dalrymple who had studied Hearne's pivotal account and corrected the fur trader's estimate for the latitude of the Coppermine's outflow into the Arctic.

In a sense, Dalrymple would reprise the role of an earlier hypothetical geographer, Daines Barrington, whose theories of an ice-free Arctic prompted Cook's search for the Northwest Passage from the Pacific side of the continent in the summer of 1778. Informed by the exploits and theories of the post-Cook maritime and terrestrial fur traders, Dalrymple repropagated the old speculative French theory of the *Mer de l'Ouest* via his famous pamphlet, the *Plan for Promoting the Fur-Trade and Securing It to This Country* (1789). Dalrymple's new, second-generation Northwest Passage did not offer a direct route from one ocean to the other. Instead it conveyed the purposely vague concept of a malleable "communication" that would shorten the distance across North America west of Hudson Bay that goods and people would need to traverse to get from the Atlantic to the Pacific, and vice versa. Vancouver was dispatched to the Northwest Coast to assess the practicability of this successor theory, as covered in Chapters 5 to 7.

It is important to note here in passing, however, that the principal purpose of Vancouver's voyage was diplomatic in nature, not geographical. John Meares precipitated an international crisis in 1790 when he returned to London with news that a Spanish captain had captured an English fur trader at Nootka Sound and otherwise discouraged British maritime trade activity in adjacent waters. This was an international incident of the first order, almost bringing Spain and Great Britain to war. Such a threat forced Spain to capitulate, an aspect of which entailed the formal return of British property at Nootka Sound to a designated representative of the United Kingdom at the site of the offending expropriation – in the event, Captain George Vancouver. It was only because an official voyage was already being dispatched to the Northwest Coast that, in response to Dalrymple's theory and allied speculations by Meares criticizing Cook's supposed lack of thoroughness, geography was added to the mission. However, nothing had changed the force and effect of what I have elsewhere described as the Hearne dictum. Even with the decision to dispatch Vancouver, the British Admiralty was still fully confident in the authoritativeness of Cook's geostrategic assessment that the Northwest Coast was not the home of a temperate-latitude passage to the Atlantic. In the end, all Vancouver ended up doing was to transform Cook's relatively straight-edged Northwest Coastline into the labyrinthine maze that was nonetheless bound by a wall of passage-blocking mountains to the east.

But Cook actually did significantly mislead geographers and fellow explorers in one respect. Not being any more of a river man than Vancouver later proved to be, he mistakenly concluded that what the Admiralty would, after his death, call "Cook's River" had its sources deep in the North American interior. Cook was confused by the glacial melt flowing down neighbouring mountain ranges that had the effect of diluting the saline content in a body of water that Vancouver, in his final season of exploration (1794), would properly and permanently rename Cook Inlet. But the legacy of Cook's original cartographic image of a large river reaching the Pacific in the heart of what his expedition concurrently discovered to be a rich sea otter fur zone was of enduring significance to the history of the Northwest Passage.

The relatively short time Vancouver spent tracking down this illusory concept was the least of it. The account of his multi-year voyage, published in 1798, banished the concept from further consideration, but following the publication of Cook's third-voyage report in 1784, two terrestrial explorers, Pond and then Mackenzie, spent a decade pursuing the chimera of this mythical river's headwaters. In one sense, like Cook on his third voyage, they would discover nothing, for, as Vancouver would later prove and document, there was no continent-penetrating "Cook's River." But in the geographical tangle that was the northwest quadrant of the continent, Pond and Mackenzie instead laid out the principal forms of, and interfaces between, the Mackenzie, Yukon, Stikine, and Fraser River basins. Along the way, they extended the fur trade network of the North West Company from its headquarters in Montreal to the Pacific.

The conceptual breakthrough and early geographical insight was Pond's. Like Isaac Stevens south of the border a century later, Pond has never really received the credit he deserves for originating a theoretical reframing of the Northwest Passage in the wake of Cook's third voyage. As detailed in Chapter 3, Pond's managerial difficulties with his partners and other traders forced him into premature retirement. His successor and protege, Mackenzie, pursued the vision with two magnificent voyages of exploration in 1789 and 1793, but they took him not to Cook Inlet but, in the first instance, to the Arctic Ocean via the river that bears the Scotsman's name, and then to Pacific tidewater on the north central British Columbia coast. In his post-expeditionary account published in 1801, *Voyages from Montreal,* Mackenzie imagined that he had pieced together a workable Northwest Passage of interlocking lakes and rivers. His scheme replaced both the classic saline solution that had foiled Cook, and Dalrymple's non-existent inland western sea communicating in some fashion with Hudson Bay, whose dubiety Vancouver proved.

The events covered in the first seven chapters of this book overlap in time and play out in vastly divergent spaces (Great Britain, Montreal, the Northwest Coast) and modes (some practical exploration, others of a literary nature). Accordingly, readers are strongly encouraged to refer to the exploration chronology provided as an Appendix while working through that part of the book. To cite two intertwined examples, on July 14, 1789, coincidentally the same day the Bastille fell in Paris, Mackenzie reached the Arctic Ocean (not "Cook's River" off the Pacific as expected); while simultaneously James Colnett, a former voyager with Cook turned maritime fur trader and a commercial ally of Meares, sailed out of Nootka Sound, on what would later be perceived as Vancouver Island, as a Spanish prisoner. The imperial fracas this and related events precipitated led to the reframing of what became Vancouver's expedition away from its originally intended destination off Africa toward the Northwest Coast of America instead. As it turned out, during his second season on the coast in the summer of 1793 near Bella Coola, British Columbia, Vancouver barely missed Mackenzie, who had come overland in pursuit of the commercial and geographical vision inspired by Cook (inadvertently) and Pond.

Mackenzie's *Voyages from Montreal* was perhaps the most important account of North American exploration ever published. Its mercantilist vision, in which British fur-trading interests could encircle American fur-trading enterprises via a freshwater network extending from Lake Superior to the mouth of the Columbia River, alarmed Thomas Jefferson. He responded with what has come to be known as the Lewis and Clark Expedition. Theirs is an oft-told tale, at which I tried my hand with *River of Promise: Lewis and Clark on the Columbia* (2009). But in Chapter 8, I approach the American response to Mackenzie from the specific context that generated it – that is, as diplomatic history. Accordingly, Jefferson, not the explorers in the field, is placed at the centre of the story. This has never been attempted before. Lewis and Clark are typically treated as a frontier adventure story, and even only recently as a facet of Enlightenment science, though in fact Lewis approached the task as a proto-Romantic.

After several ill-fated attempts, including sponsorship of Ledyard from his ambassadorial post in pre-revolutionary France, when Jefferson became the president of the United States in 1801, he was finally in a position to direct his nascent country's engagement in scientific exploration of the American West. This included a search for the third-generation Northwest Passage. In addition to Mackenzie's account (after it was brought to his attention by a colleague from the American Philosophical Society in Philadelphia), Jefferson was also familiar with Cook's third-voyage narrative and Vancouver's report and cartography. After Mackenzie read Vancouver's account (published in 1798), he concluded in *Voyages,* published three years later, that the large southward-flowing river he encountered near present-day Prince George, British Columbia, was the same river whose mouth Gray and Vancouver had visited in 1792. Mackenzie was mistaken in this regard. In truth, before he shortened his trip in 1793 by heading directly to the Northwest Coast at Bella Coola, he had actually stumbled upon the upper reaches of what would later be discerned as the Fraser River, named after a fellow fur trader, Simon Fraser, who coursed the full length of the river to the Salish Sea in 1808.

Neither Jefferson nor his principal agent, Meriwether Lewis, could know that, however. They operated on the assumption that Mackenzie had found the headwaters of the Columbia. This "fact," within the paradigm of the doctrine of discovery guiding international law about sovereignty over previously unexplored land, attenuated the American claim (via Gray) to the waters drained by the Columbia. In response, by drawing on the archaic geographical principle of equipoise still regnant at that time, Jefferson's War Department commissioned a map for Lewis's use in the field. It posited, hypothetically, a major southern fork to the Columbia to counteract, cartographically, the challenge laid down by Mackenzie's image showing its northern headwaters. Drawn by Nicholas King, this chart depicted a southern fork exactly the same length as Mackenzie's northern stream, preserving the potential that it might be equal to or larger (hydrologically) than its northern twin.

Accordingly, when Lewis headed west with William Clark in 1803, he, like Jefferson, operated on the naive assumption that by following the Missouri River to its headwaters and crossing the Continental Divide, he would find in close juxtaposition the Missouri's western opposite (the Columbia). Drawing on the same equipoise principle that informed King's conjectural southern fork of the Columbia, the Rocky Mountains were presumed to be the same general height and breadth as the Appalachian Mountains, whose foothills Jefferson could see from Monticello. Lewis and Clark began to drop this illusion once they actually reached their zone of discovery: first, after discussing western geography with fur traders in St. Louis; next in dialogue with the Hidatsa people of present-day North Dakota, who often ventured to the Rockies; and finally by personal experiment. In the event, Clark (not Lewis as is commonly perceived) would discern the geographical complexity of the upper reaches of what has come to be known as the Snake River system, which is tributary to the Columbia. When Lewis and Clark reached the spot where these two major forks joined, in what is now the southeastern corner of the state of Washington, they were stunned to discover that their southern fork was distinctly smaller than Mackenzie's northern fork. This was such a shocking development to Lewis that he was unable to resume writing his journal for several months.

Nevertheless, Lewis gamely did his best to please his patron, writing from St. Louis upon the voyage's conclusion in 1806 that he had matched Mackenzie's interlocking river system with a practicable Northwest Passage that had the additional advantage of being further south. In reality, Lewis had done no such thing. The several ranges of the Rocky Mountains in western Montana proved as stout a threat to commercial traffic from the Atlantic Basin to the Pacific as the ice Cook struck off the northern coast of Alaska. Thus the third version of the Northwest Passage also came to naught.

In truth, a practicable Northwest Passage was never *discovered,* but one was eventually *built* in the industrial form of the transcontinental railroad. Each of the three preceding versions of the Northwest Passage had its prophets – Daines Barrington, Peter Pond, and Alexander Dalrymple in succession. The railroad had its own seers – Asa Whitney in the United States and Sandford Fleming in Canada. Their stories are outlined in Chapters 9 and 10, as is that of Isaac I. Stevens, the first explorer to marry mercantile philosophy with practical geography since Lieutenant James King. In Cook's third-voyage account, King literally wrote the business plan for what became the Northwest Coast maritime fur trade. Over the course of 1853–55, Stevens personally conducted or guided others in the reconnaissance of a potential railroad corridor – what became the Northern Pacific Railway – in addition to his obligations as the Washington Territory's first governor.

Of course, others had reconnoitred the northern west since Lewis and Clark. Between 1808 and 1812, Mackenzie's heirs, North West Company fur trade explorers Simon Fraser and David Thompson, discerned the juxtaposition of the upper reaches of the rivers named after them vis-à-vis the Columbia River. In the early 1840s, the combined efforts of US Navy commander Charles Wilkes and Army lieutenant John C. Frémont scoped the navigability of the Columbia and Puget Sound harbours, the fur trade and missionary districts in the Willamette Valley and Columbia Plateau, and the emigrant route to Oregon. Given this book's thematic arrangement and concomitant episodic nature, we leapfrog to Stevens's expedition because of his distinguishing characteristic as a Northwest Passage theorist and explorer. This previously untold facet of his career is rooted in the noteworthy contributions he made to the twelve-volume *Pacific Railroad Reports,* which appeared between 1855 and 1860.

The *Reports* were an amalgam of travel narratives by several US Army explorers looking for railroad routes across four distinct east-west corridors from the Mississippi Valley to the Pacific. These were supplemented by scientific studies conducted by civilian scientists covering such topics as the geology, flora, fauna, and climate of the American West. The cartography of all four surveys and allied explorations were conformed into a map drawn by Lieutenant G.K. Warren. In the same respect that the composite map tracking Cook's three circumnavigations, published in the atlas to his third-voyage account in 1784, was the first modern map of the world, Warren's 1858 chart was the first truly comprehensive map of the American West. In these respects and others, the *Reports* capped the era of scientific discovery first popularized by Cook and his teams of naturalists almost a century earlier.

In his geographical summary, Stevens frequently invoked the earlier work of Lewis and Clark in the western half of his designated corridor. More importantly, and uniquely among his Army explorer peer group, he was fully conversant in the mode of thought seen in the history of Northwest Passage theory and practice, and used that legacy as his route's principal selling point. His final report contained a prospectus for a triangular trade between North America and Europe with Asia, grounded in the linkage between the Atlantic and Pacific Basins that a northern transcontinental railroad could create. This kind of strategic thinking had not been seen in exploration literature since the time of Cook and Dalrymple.

Indeed, Stevens was such a forceful advocate of the trans-Pacific potential of his northern route that it was used against him by some of his detractors. The proponents of other lines further south argued that the point of a railroad to the Pacific was military defence against tribal depredations and internal colonial development, not gauzy visions of a passage to India. One critic, surveyor Frederick West Lander, was commissioned by Stevens's own territorial legislature to devise a route that was more conducive to emigrant colonial settlement from the lower Missouri Valley as an alternative to the governor's straight line west from St. Paul. Ironically, and perhaps unwittingly, Lander's railroad analogue to the Oregon Trail invoked some geographical concepts dating to Mackenzie.

As recounted in Chapter 10, a parallel set of commercial theorists and geographers in Canada seized upon the same prospect of extending industrial-scale transportation technology to the Pacific Slope found in the American *Pacific Railroad Reports.* The principal figure here was Sandford Fleming, best known as the leading international advocate for the standardization of time and the adoption of Greenwich, England, as home to the prime meridian. Fleming, however, was not merely a provocative theorist and advocate like Asa Whitney in the United States; he was also a practical geographer like Stevens.

The other key personage in the Canadian story was William Cornelius Van Horne, who also encompassed within himself a combination of skills, in his case that of geographer and construction engineer for the Canadian Pacific Railway. Van Horne said he would have avoided the Fraser River route for Sir John A. Macdonald's national dream if he had not inherited it from previous surveyors and contractors. Alexander Mackenzie (the explorer, not Macdonald's successor as prime minister) would have agreed. Indeed, in another echo from a previous generation's fascination with the concept of the Northwest Passage, Van Horne chose "Vancouver" as the name of his line's western terminus.

In the late 1880s, the Northwest Passage as a transcontinental "communication" was finally realized by two rail lines, one on each side of the border, the Northern Pacific Railway and the Canadian Pacific Railway. The Canadian press, in particular, consciously hailed the latter as the long wished for passage. To this day, cargoes to and from port cities in East Asia are carried across the muscular chest of North America via the Northern Pacific's successor lines and the Canadian Pacific. Not only was the latter Canada's first transcontinental railroad, but through modern mergers and acquisitions it became the first line to traverse all of North America, with routes through Mexico, the United States, and Canada.

But the story does not end there. Ironically, just as Stevens and Fleming were delineating the paths by which the iron horse could cross the mid-latitudes of the North American continent, the search for the classic, salt-water connection linking the Atlantic and the Pacific was reaching its culminating phase in the Canadian Arctic. This story is usually told with the ill-fated Sir John Franklin in the central role. In the 1820s, Franklin led two overland expeditions to the Arctic coastline via routes first blazed by Samuel Hearne and Alexander Mackenzie. As with Cook, Franklin's third expedition was his undoing. In 1845, he attempted to sail through the passage via Lancaster Sound opposite Greenland. Like all previous efforts at the passage from the Atlantic side since the time of Hudson, Frobisher, and Davis, Franklin failed. Worse, he and all hands were lost, though they live on in Arctic legend.

Instead of Franklin, Chapter 11 focuses on his fellow British navigator Robert McClure. After decades' of mixed results, including Franklin's last ill-starred endeavour, the Admiralty directed the expedition that McClure was a part of to approach the Northwest Passage from the Pacific side of the continent. Explicitly following Cook's 1778 track north from Hawai'i, in 1853 McClure claimed the disputed title of discoverer of the Northwest Passage. The corridor had been delineated at last, and with that goal in hand, the community of explorers turned its attention to reaching the North and South Poles and the summit of Mount Everest.

Britain's post-Napoleonic, multi-decade search for the ice-bound passage bore little resemblance to the Enlightenment-era quests of Cook and Vancouver. This was not exploration as science, but rather exploration as Romantic bravado. It resembled an athletic contest more than anything else, in part because, unlike the corridors that Cook, Vancouver, and Mackenzie sought, everyone knew there would be no productive utility to the one Franklin and McClure chased. They pursued it only because they presumed it was there.

And that's where the history of the Northwest Passage might have remained, but for changes to the global climate that became evident in the late twentieth century. This book concludes with an assessment of the environmental and diplomatic consequences of the recession of the Arctic ice pack. It could very well be that, in an ironic twist of fate, the ice-free North Pole of legend will be realized (at least seasonally) within a generation or two. If this happens, a new Mediterranean, like the one George Vancouver once mocked, will render the Panama Canal obsolete and vitiate the importance of the North American transcontinental railroads. The saltwater Northwest Passage that Cook sought may yet be revealed after all.

Key Sources

This study is based primarily upon the journals and post-expeditionary accounts of the explorers at the centre of the narrative. The full citations are found in the bibliography, but here I cite in short form the most important of them: the *Journals of Captain James Cook,* edited by J.C. Beaglehole (1955–67); *The Last Voyage of Captain Cook: The Collected Writings of John Ledyard,* edited by James Zug (2005); John Meares's *Voyages Made in the Years 1788 and 1789* (1790); George Vancouver's *A Voyage of Discovery to the North Pacific Ocean and Round the World, 1791–95,* annotated by W. Kaye Lamb in a Hakluyt Society edition published in 1984; and *The Journals and Letters of Sir Alexander Mackenzie,* edited W. Kaye Lamb (1970), including an annotation of the Scotsman's *Voyages from Montreal* (1801); Isaac Stevens's *Narrative and Final Report of Explorations* (1860); and George Grant's *Ocean to Ocean: Sandford Fleming's Expedition through Canada in 1872* (1879).

Interpretive treatments based on readings in primary materials have gone somewhat out of fashion in favour of "critical readings" of texts. There are risks associated with taking these aforementioned journals and accounts at face value, because explorers and their original patrons and publishers had clear narrative objectives they wished to convey, or hide. I previously addressed this aspect of the internal dynamics of travel literature, including the authorial conveniences derived from the retrospective casting of events, in my *River of Promise: Lewis and Clark on the Columbia* (2009) and *Captain Cook Rediscovered: Voyaging to the Icy Latitudes* (2020). I do so again in this book by intermittently applying a lens that reveals the underlying meanings of what various explorers had to say. However, to understand what the successors of certain explorers were to trying to do at sea or in the field, we have to assess what their predecessors conveyed in print about their geographical experiences, conclusions, and projections. This is the only way to determine what any particular explorer knew about his geographical task and when he knew it, whether in the act of physical discovery or in a library drafting his own account. Geographical comprehension was based upon an empirical building-block approach that was pursued until the reality of the globe's natural features was finally revealed. Besides which, and contrary to common misperception even among some scholars who should know better, there are still many new and interesting exploration stories to be gleaned from the reading of primary materials. As a genre of literature they are far from being "mined out."

The same dictum on the use of primary materials also applies to the hypotheses of speculative geographers who were not themselves explorers. The single most important theoretician encompassed by this study is Alexander Dalrymple, as exemplified in his *Plan for Promoting the Fur-Trade* (1789). This title was misleading because Dalrymple's pamphlet was actually a treatise attempting to reconceive the Northwest Passage. Herein I emphasize what he actually wrote, not how historians have encrusted it within an interpretive orthodoxy critical of Cook.

Nevertheless, a host of secondary interpretive accounts have been helpful in the framing of my arguments. Several are enumerated below, but it's important to note at the outset that what I have attempted in these pages – what might be called a cross-expeditionary analysis – is very rare in exploration literature. The historiography of travel literature is dominated by vertical treatments, that is, the story of one explorer or expedition from beginning to end. The work of two scholars has informed my revisionist approach: Richard Van Orman's *The Explorers: Nineteenth-Century Expeditions in Africa and the American West* (1984) and Felipe Fernández-Armesto's *Pathfinders: A Global History of Exploration* (2006). The single best cross-expeditionary treatment of the Northwest Passage saga is, in my opinion, Glyn Williams's *Voyages of Delusion* (2002).

Among the more particularized studies that have proved especially helpful are the series of books published by Barry Gough: *First across the Continent: Sir Alexander Mackenzie* (1997), *Juan de Fuca's Strait: Voyages in the Waterways of Forgotten Dreams* (2012), and *The Elusive Mr. Pond: The Fur Trader and Explorer Who Opened the Northwest* (2014). The last of those is nicely complemented by David Chapin's *Freshwater Passages: The Trade and Travels of Peter Pond* (2014). More than a generation ago, in his *Army Exploration of the American West* (1959), William Goetzmann pioneered the study of the *Pacific Railroad Reports.* Little has been done since to build upon his foundation until the chapter on that topic in this book. An intervening biography of Isaac Stevens failed to make a single mention of the appendix he attached to his report touting the trans-Pacific potential of his proposed northern railroad route. As might be expected, I have relied upon my *Captain Cook Rediscovered* and *River of Promise* for much of the background information reapplied here relative to Cook's third voyage and the Lewis and Clark Expedition, respectively.

For insight into Dalrymple's principal successors in Northwest Passage geographical theory and promotion, namely, Thomas Hart Benton and Asa Whitney in the United States and Joseph Howe in Canada, I found Henry Nash Smith's *Virgin Land* (1950) and Clark Blaise's *Sir Sandford Fleming* (2000) to be particularly helpful. These books have abundant insights about political and cultural conditions that informed the evolution of the Northwest Passage upon the onset of the Industrial Revolution and railroad technology.

As for the conclusive phase of post-Franklin Northwest Passage exploration in the Arctic, and contemporary circumstances in that basin, the most fruitful works are Pierre Berton's *Arctic Grail* (2000), Glyn Williams's *Arctic Labyrinth* (2009), and the essays by Lawson Brigham, Gudrun Bucher and Robin Inglis, and James Delgado found in *Arctic Ambitions* (2015), co-edited by James Barnett and myself. The definitive stand-alone study of the ecological and diplomatic dynamics in the new north is Shelagh Grant's *Polar Imperative* (2010). No study better evokes the history and aesthetics of the region than *Arctic Dreams* (2001) by Barry Lopez.

A Cautionary Note

The main title of this book is an obvious play on words. The Northwest Passage that speculators dreamed of and explorers sought was never discovered. Nevertheless, their serial quests provided modernity with its normative understanding of regional geography, a valuable contribution by any measure. Yet, even that wasn't true discovery, as the lands and waters of North America's Pacific Slope had been known for eons by Indigenous Peoples. As Fernández-Armesto trenchantly observed in *Pathfinders,* the greatest explorers the world has ever seen were the seekers of routes who, over centuries, guided our species from its homeland in Africa to every corner of the earth.

In a related irony, *Discovering Nothing* is a distant echo of *terra nullius,* the Latin term meaning "land belonging to no one." As Robert Miller explained in *Native America, Discovered and Conquered* (2006), *terra nullius* was one of the principal precepts of the doctrine of discovery. In the European world view, this concept implied that discovered lands, even if inhabited, were not being used in a productive fashion and, besides, were populated by people not considered a part of the family of nations. Their homelands, accordingly, were claimable as a matter of international law. None of the explorers discussed in this account made overt sovereignty claims, in part because they considered that their predecessors, dating in time back to Francis Drake in 1579, had already done so. Nonetheless, George Vancouver, Peter Pond, Alex Mackenzie, and the rest were part of a colonial legacy that is becoming another normative understanding.

Acknowledgments

CONDUCTING A BOOK FROM concept to publication can be a labour of love, though occasionally laborious nonetheless. In these latter instances, the guidance of counsellors and helpful assistance from friends, in the many forms it can take, are essential to completing a writing project.

First, I would like to express my gratitude to UBC Press acquisitions editor James MacNevin for showing interest in a book so soon after the publication of *Captain Cook Rediscovered*. I appreciate his confidence in me and the many suggestions made by the Press's anonymous peer reviewers and publications committee. I am also indebted to Meagan Dyer and the rest of the Press's editorial team. She has been wonderful to work with. More particularly, I am appreciative of George Kirkpatrick's imaginative cover design, Eric Leinberger's cartographic reconstructions, and a special thank you to copy editor Frank Chow whose work was superb.

As has been the case for nearly twenty years now, Clay Jenkinson has offered constant encouragement during my transition from a historical administrator to a history writer, including reading an occasional chapter and making other suggestions on how to navigate life as an author. I've also been continually cheered by other friends during the course of my daily rounds to the athletic club and local coffeehouses. In particular, I would like to recognize Dick Pust, Grey Brogdon, Phil Northcraft, and Alden Johnson for their interest in my writing and narrative subjects. During the darkest days of the COVID-19 pandemic, when most of this book was written, these gentlemen provided much of the only direct human interaction I could count on.

The staff of the Washington State Library were especially helpful in making research materials available to me. Sean Lanksbury, who has since moved on to other pursuits professionally, and Mary Schaff were very accommodating to my needs along these lines. Similarly, the Washington State Historical Society's reference archivist, Eileen Price, went out of her way to facilitate access to historical maps and to make them available for publication here.

Working as an independent historian without institutional support can be a challenge, but both Darlene Kemery and Beth Schatz Kaylor helped bridge that divide, for which I am especially grateful. I also want to thank Derek Hayes for sharing some hard-to-find historic cartography for use as illustrations in this book.

Lastly, I wish to acknowledge Robin Inglis, to whom this book is dedicated. In the 1990s, Robin introduced me to the deep cartographic record of European exploration of the Northwest Coast of North America. He was also the curator and publications manager for the *Arctic Ambitions: Captain Cook and the Northwest Passage* exhibition and anthology (2015), which I co-edited with Jim Barnett. During the course of that project, Robin tutored me in the literature of Cook and the great navigator's successors along the Northwest Coast. This book is a fruit of the inspiration and mentorship that Robin generously provided.

Discovering Nothing

Prologue

The Remarkable Tale of John Ledyard's Attempt to Copy Cook

An important part of James Cook's legacy is that he served as an inspirational model for upward of a generation or more. Only one direct successor – fellow Englishman George Vancouver – approached the depth of cartographic performance Cook executed on any one of his voyages, let alone three of them. But many, including explorers from other nations, sought to emulate the great navigator in one way or another. Usually this took the form of a nautical voyage, notably those of La Pérouse of France and Malaspina from Spain, but one terrestrial venture stands alone – that of John Ledyard. Even though Ledyard's extraordinary vision came to naught, his astonishing tale is worth recounting as a postscript to Cook because it sheds light on: (1) the allure of the Northwest Passage in the Euro-American imagination of the late eighteenth century; (2) Cook's role as a pioneering demographer and climatologist; and (3) the origins of the nascent American interest in the northwest corner of the continent.

Ledyard's most important duty (for which he volunteered) during Cook's third voyage was to make contact with the Russians at Unalaska in the Aleutian chain in the fall of 1778. More importantly, he was one of several members of that expedition to beat the Admiralty to print with an account of the voyage. Ledyard would have been a graduate of Dartmouth in the class of 1776 if he had stayed in school; instead, in March 1774, he sailed to Great Britain from New York. In Bristol, he avoided impressment into the Royal Navy (on a ship that was intended for the African coast) by enlisting in the army. When Ledyard's detachment was ordered to Boston to suppress the burgeoning revolt, he carried out a form of conscientious objection by volunteering to serve aboard Cook's *Resolution,* which was then docked at Plymouth. This duty, Ledyard wrote later, was "a favorable conjuncture" allowing him to avoid "coming to America as her enemy & prompted also by curiosity & disinterested enterprise embarked in [Cook's] expedition."[1] Cook mustered him aboard on July 9, 1776 – not coincidentally one of the more interesting weeks in American history.

When Lieutenant John Gore brought Cook's final expedition back to Great Britain in the fall of 1780, the war was reaching its conclusive phase (the battle of Yorktown took place the following summer). Ledyard was immediately promoted to sergeant of marines and posted with the division in Plymouth, England. Gore, also born in America (possibly Virginia), wrote a testimonial to the Admiralty for his fellow continental in June 1781 advocating further advancement: "Serjeant Ledyard is a Young active Man [and] hath done Good service on our late Circumnavigating Voyage, and ... is (I think) Properly Qualified and justly merits a higher Rank Than That which he holds at present. My wish is to recommend him to your Lordships notice In hopes That by your means a Good Officer may not be Lost by being continued the Serjeant." In the letter that Gore's correspondence enclosed as an attachment Ledyard played the Loyalist card, stipulating that "I am a native of North America, of a good Family and once had considerable connections there, untill the rupture with England in 1775 when I abandoned all and entered the Army here, where I flatter myself as I have been so I shall be encouraged." A promotion was not forthcoming by the time Ledyard's new ship came to America in October 1781. There, Ledyard later claimed, he waited for "an opportunity to renounce the service," eventually deserting the Royal Navy at Long Island in 1782. During the winter of 1782–83, he wrote his account of Cook's last voyage in Hartford, Connecticut. Historians who have matched texts consider his narrative to be heavily reliant on Lieutenant John Rickman's book, which was one of the first into print.[2]

Decades later, James Burney, who had been first lieutenant to Charles Clerke on Cook's third-voyage consort, *Discovery,* disclosed in his memoir that Ledyard had literary aims from early on. According to Burney, after Cook was killed, Ledyard proposed to Clerke that he become the official "historiographer" of the expedition. To help pitch the idea, Ledyard composed a sample narrative that Burney said was written with "a passion for lofty sentiment and description." Pursuant to the "palm-tree paradigm" that has dominated the study of Cook from the late eighteenth century to the present, Ledyard chose Tahiti as the venue for his textual "specimen." Nor was Ledyard the only one on board with such aspirations, Burney disclosed. "Literary ambition and disposition to authorship led us in each ship to set up a weekly paper," he declared. The crews exchanged editions in good weather, and Cook himself "was always glad to read our paper," Burney added, "but never favoured our editors with the contribution of a paragraph." Ledyard was one such contributor, though his writing was deemed "too sentimental, and his language too florid." In the end, Clerke rejected Ledyard's preposterous idea of becoming the expedition's historian, a role that fell to the erudite Lieutenant James King.[3]

Ledyard was not content with being a mere writer. His book seems to have been a means to another end. Even before it appeared in print, Ledyard canvassed business circles in Philadelphia, New York, and New London, Connecticut, trying to organize a fur-trading expedition to the Northwest Coast. His plan was a variation of the triangular trade outlined by James King in the Admiralty account of Cook's final voyage, substituting an American home port for one in Great Britain. It was concurrent with schemes being arranged by a host of English privateers from East Asia and a joint stock concern headquartered in London. The most prominent in the first category was John Meares (a major figure in the evolution of the Northwest Passage as a cartographic image), and in the latter group it was Richard Cadman Etches and Company, doing business as the King George's Sound Company. That was Cook's name for Nootka Sound, on the west coast of Vancouver Island, which he surveyed in 1778 on his way to the Arctic. The Etches firm employed a supercargo, William Beresford, to represent the investors' commercial interests, while the practical navigation to the Pacific Northwest was conducted by competent seamen. These were *Discovery* alumni George Dixon, formerly an armourer, and Nathaniel Portlock, who served as master's mate under Clerke. All of these figures and others are covered in more detail in Chapter 1.

Ledyard was not a navigator so he proposed to serve as the supercargo for the advancement of the concern on behalf of his prospective investors. Oddly for a Connecticut Yankee, Ledyard had his most promising discussions in Philadelphia. There, in November 1783, he met with Robert Morris, famous in American history for his work in financing the revolt against Great Britain. Ledyard also met merchants in New York. He regaled them with the story, as related in in his soon-to-be-published book, that at Nootka

Sound upward of 1,500 fine furs were acquired, but no one had "thoughts at that time of using them to other advantage than converting them to the purposes of cloathing, but it afterwards happened that skins which did not cost the purchasers six pence sterling sold in China for 100 dollars." Ledyard added that he and his mates could have easily secured four times as many pelts to sell in China at "an astonishing profit."[4]

Though these businessmen were intrigued with Ledyard's proposition, he was never able to secure conclusive support for a commercial venture, or at least one with him as supercargo. Later that winter, Morris told John Jay, secretary for foreign affairs during the era of the Articles of Confederation, that he intended to send a ship to China "in order to encourage others in the adventurous pursuits of commerce." In February 1784, his *Empress of China* sailed from New York, reaching Canton the following August, the first American ship to do so. It seems American entrepreneurs were ready to engage in the China trade but Ledyard did not inspire confidence as a leader. A New London captain turned him down, "not daring to trust" Ledyard's "sanguine representations." It didn't help that Ledyard embellished his scheme by proposing to "settle a colony on the N.W. Coast." Nevertheless, in his own way, Ledyard was a pioneer. Daniel Parker, a ship captain operating out of New York, recalled in 1822 that Ledyard "was the first person who proposed that trade to the merchants of the U. States," half a decade before the American innovators in the Northwest maritime fur trade, Robert Gray and John Kendrick, secured their capital from New England investors.[5]

Ledyard was a man not easily discouraged from achieving his dreams of greatness, and his vision for a way to become a noble successor to Cook began to evolve. After his book was published, Ledyard met with Ezra Stiles, the president of Yale University in New Haven, Connecticut. (Stiles was also a confidant of terrestrial fur trader Peter Pond, discussed more fully later in this book, who likewise figures prominently in the evolution of the Northwest Passage after Cook.) From Stiles, we discern the first glimmer's of Ledyard's adaptive strategy. In his diary for February 3, 1784, Stiles recorded that Ledyard's imagination was struck by "the narrow passage of 14 Leagues between Asia & America." And even this small gap, Ledyard told Stiles, was "frozen across over the streight from America to Asia." The cartographic definition of the physical particulars of the Bering Strait (see Figure P.1) is one of Cook's great accomplishments, but neither he nor Clerke (nor Ledyard) actually saw ice from one side of the narrows to the other. They learned of its seasonal appearance from Russian fur traders they met

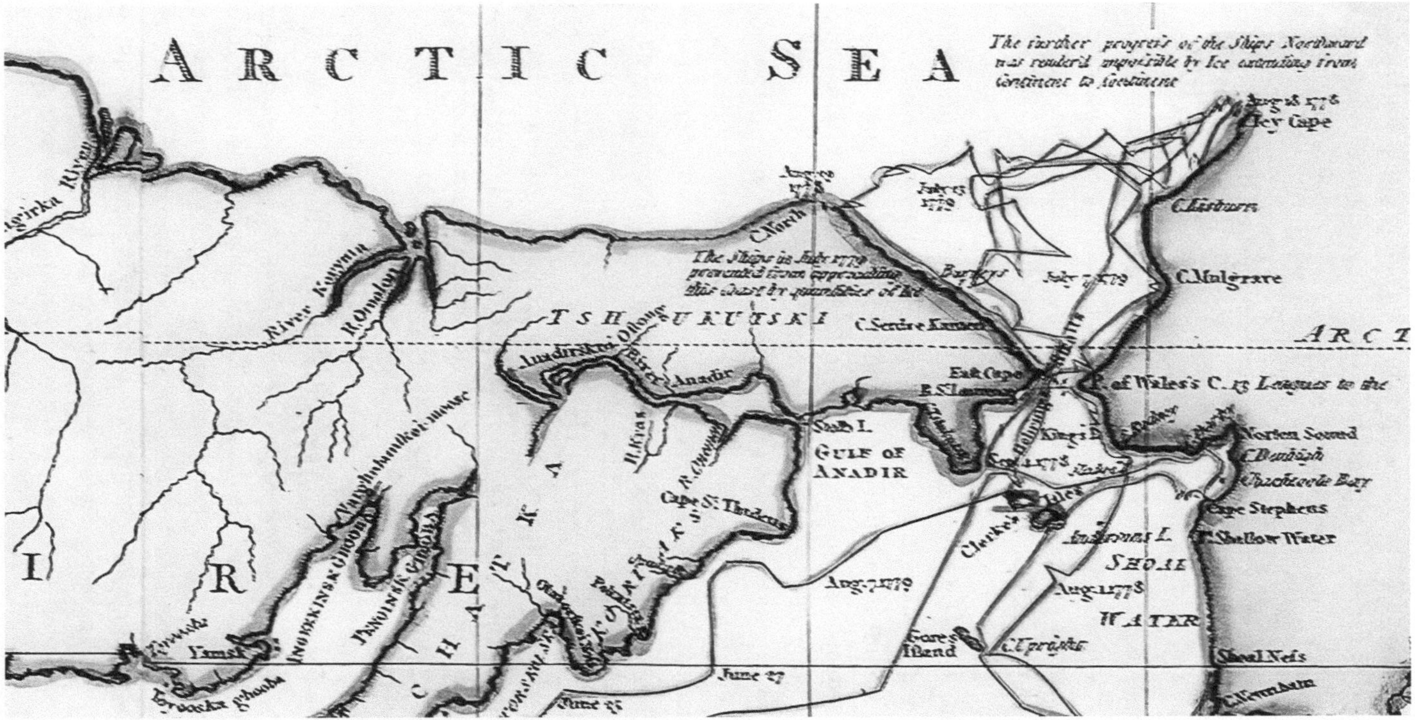

Figure P.1 Henry Roberts, *A General Chart Exhibiting the Discoveries made by Captn James Cook*, 1784 (detail). This inset is from the larger work completed after Cook's final voyage. Among its many noteworthy features, the map accurately depicted the close juxtaposition of Asia and North America for the first time. The relative proximity of the two continental headlands spurred Ledyard's imagination. | Washington State Historical Society, 2011.0.60.3.3.1

in the Aleutians and later in Kamchatka. In conversation with Stiles, Ledyard moved beyond the idea of a commercial fur-trading venture, having struck upon a fallback plan: the grandiose idea – one of Cook-like proportions – to circumnavigate the world by himself.[6]

We cannot be sure whether Stiles encouraged the next steps, but in June 1784 Ledyard went to the Spanish port city of Cadiz and then north to the bustling French port of Lorient, on the south coast of Brittany. At Lorient he tried one last time to entice financiers and maritime figures to his original fur trade scheme. France had an interest in the Pacific, but its orientation ran toward a new scientific voyage matching Cook – eventually taking the form of the La Pérouse expedition – not a commercial venture. In the spring and summer of 1785, Ledyard was pitching his ideas in Paris to John Paul Jones, the famous American naval captain, who would have been the ideal figure to command the navigational side of a trading voyage to the Northwest. Jones toyed with the idea of sailing under French sponsorship, but the project fell apart for several reasons. First, La Pérouse's expedition, which left Brest in July 1785, eliminated any possibility of governmental support. Nor was Jones able to procure a French merchantman. But perhaps the most important factor impeding a Jones/Ledyard joint venture was the news that Portlock and Dixon had sailed to the Northwest Coast under the auspices of the Etches concern in August. Simply put, Jones was not enamoured of the idea of competing with the British in those waters. For Ledyard, though, Paris held other rewards. He soon met a fellow visionary, Thomas Jefferson, who had succeeded Benjamin Franklin as the American ambassador to the French court. Ledyard's chronicler said of the former Cook crewman's conversations with Jefferson that the future American president's "hospitality was unfailing; his eagerness for new information on many subjects was contagious."[7]

It is easy to see how Jefferson would find Ledyard a compelling figure in at least two respects. First, Ledyard had served in the Royal Navy but was an American patriot nonetheless because he refused to fight against his native countrymen. Second, he was an explorer, having sailed with the great Cook, no less. Indeed, it is commonly believed by historians that Ledyard helped to seed the third president's interest in transcontinental exploration. This came to fruition in the form of the Lewis and Clark Expedition of 1803–06, yet another quest for the Northwest Passage – not the one that Cook had searched for, but in its evolved form of interconnecting rivers. This will be explicated in the ensuing chapters. Jefferson took great interest in Ledyard's ambitions. Like Stiles, he encouraged the Connecticut Yankee to take a non-commercial direction, and over the winter of 1785–86, Ledyard abandoned his fur-trading plan, substituting in its place a blueprint for travel that would take him from his base in Western Europe across Siberia to Kamchatka. There he would link up with Russian traders whose commercial links, he knew from personal experience, crossed to the Aleutian chain and possibly by that point in time now even further eastward toward the continental mainland at Cook Inlet or Prince William Sound. From there, Ledyard imagined travelling with a maritime trader of Russian or British origin to the more temperate latitudes of California. After reaching that point, his plan was to walk to the Atlantic seaboard by way of Kentucky, at that time the western frontier, intending to reach New York or, given Jefferson's encouragement, the Virginia tidewater.

In his unpublished autobiography, written late in his life, Jefferson claimed credit for Ledyard's projected itinerary. "I suggested to him the enterprise of exploring the Western part of our continent, by passing thro St. Petersburg to Kamchatka, and procuring a passage thence in some of the Russian vessels to Nootka Sound, whence he might make his way across the continent to America." Jefferson was so enthusiastic about the idea of global circumambulation that he introduced Ledyard to the Marquis de Lafayette in February 1786, telling the marquis of Ledyard's prospective "journey through the Northern parts of Asia and America." Jefferson lent his personal copy of Ledyard's book to Madame de Lafayette later that summer. The new plan, of course, countenanced the idea of a companion narrative. What was the point of such a trip unless it could be related in literary form? Jefferson told Lafayette of Ledyard's personal qualities, which in some measure anticipated the skills he sought in the person of Meriwether Lewis, to wit: "a spirit of enterprize," a "genius" with "an education better than the common, and a talent for useful and interesting observation." Ledyard was "a man of truth," in Jefferson's opinion, in addition to which he possessed "much singularity of character, and of that particular kind too, as was necessary to make him undertake the journey he proposes." Should Ledyard "get safe through," Jefferson told Lafayette, "he will give an interesting account of what he shall have seen."[8]

Jefferson was sufficiently intrigued by the idea of this one-man expedition that in the interest of Enlightenment science he tutored Ledyard, in what might be termed a correspondence course, on the importance of observing magnetic variation. Ledyard's proposed jaunt involved little by way of logistical requirements, and may have had celebrity backing from Jefferson, but it still required some capital and Russian permission. Jefferson gave him 476 francs and Lafayette provided an additional 600 francs to defray future expenses, but the passport was slow in coming from St. Petersburg. Ledyard was prepared to travel across Russia without a visa, but Jefferson and Lafayette counselled against it. Both interceded with Russian emissaries in Paris and the French ambassador to the Russian court. On August 15, 1786, while at the weekly meeting of the diplomatic corps at the palace at Versailles, Jefferson received word that Ledyard's application had been rejected.

As conveyed to Jefferson, Empress Catherine's opposition lay, in part, in her perception that Ledyard was tardily attempting to join the Arctic exploring expedition already commissioned under the leadership of Joseph Billings, yet another Cook alumnus. (Billings had signed on *Discovery* as an able seaman and later joined the Russian navy as a midshipman. He departed for points east long before the date of Ledyard's rejection.) But that may have been a cover story to hide a more deep-seated wariness, because the official communique from St. Petersburg transmitted the opinion that she had concluded that Ledyard's plan was "a chimerical dream." She advised that Ledyard "would do well to take another route than that of Kamchatka." Jefferson counselled Ledyard, in writing, to the same effect, recommending instead a trip exploring America from east to west, the earliest intimation of his grand idea that eventually reached fruition under Lewis and Clark. By the time Ledyard received Jefferson's note, however, he was already in London pursuing different avenues.[9]

Ledyard seems to have tired of waiting for permission in Paris and abruptly departed for England in August 1786. In a tactical retrogression, he revisited the idea of voyaging directly to the Northwest Coast on a British merchantman. Ledyard first tried to latch on to the trading voyage of James Colnett aboard the *Prince of Wales* since a letter from him mentions a voyage about to depart that was to be commanded by "An Officer of Capt Cooks." (Colnett was a midshipman on Cook's second voyage, and his trading venture, part of the Etches concern, was only the fourth to sail from England to the Northwest Coast. His exploits at Nootka Sound, which in large measure led to the Vancouver expedition, are recounted later in this book.) Ledyard wanted to join Colnett's enterprise as a supernumerary, accompanied by "two great Dogs," taking as his principal possessions only "an Indian pipe and a hatchet." From London, he promised Jefferson: "If I find in my

Travels a mountain as much above the Mountains as he is above ordinary men I will name it La Fayette." Such patriotic sentiments aside, Ledyard's plan was patently naive and came to naught, if for no other reason than the fact that Great Britain considered him a deserter. The request was presumptuous in the extreme, besides. Colnett sailed without him. Another ship Ledyard hoped to join, the *Harriet,* was seized by the British customs office and sold for back taxes. No less a figure than the great speculative geographer and Cook provocateur Alexander Dalrymple played a hand in this denouement. Colnett's trading expedition, and the three ships from other companies that had sailed from England to the Northwest beforehand, had all been issued licences to trade in China by the East India Company, the firm that held the royally chartered monopoly in that part of the world. Dalrymple, an East India Company man going back to the 1760s, inadvertently provided the *Harriet* with charts and other information in his possession under the mistaken impression that she was similarly licensed. When he learned that it was not, he called in the authorities.[10]

After these further disappointments, Ledyard wrote Jefferson – addressing him as "My friend, my brother, my Father" – on November 25, 1786. In this letter, he reverted to the idea of a land tour across North America by way of Siberia. It turned out, however, that the trip to England was not a complete loss. Ledyard made the acquaintance of Joseph Banks in London, and Banks too was taken by the same qualities that impressed Jefferson. Banks volunteered some of his own financial resources and raised money from other parties, including William Smith, the son-in-law of John Adams, the American ambassador. Now with sufficient funds to reach Hamburg and St. Petersburg, Ledyard advised Jefferson of his intention to go "across Siberia as I had once before intended from Paris." He thought his resources were adequate "to reach those dominions where I shall not want money." By this he presumably meant the fur trade districts of the Russian Far East and North America, where he would survive on his own wits. This, he told a relative, was his "road to fame."[11]

Ledyard reached St. Petersburg in March 1787, and wrote Jefferson again, thanking him for his friendship and reporting on his travels and recent developments. The most important of these was the exploratory venture in the Russian Far East led by Billings, with whom, somewhat remarkably, he would later engage. Portentously, Ledyard also told Jefferson he was waiting anew for his "*royal passport* without which I cannot stir." For this he was still relying on the French ambassador to Empress Catherine's court. Two months later, he wrote William Smith in London advising that "I am likely to obtain a passport ... if so I set out immediately." On June 1, 1787, after a three-month delay, Ledyard departed Moscow for points east. Neither his journal nor surviving correspondence makes specific reference to his securing a passport. It is possible, if not likely, that he did, but given diplomatic practice then and now, such permission would have been considered revocable, which is what eventually happened. The first of Ledyard's two planned cross-continental treks was conducted by way a series of carriage rides along the Great Siberian Post Road running between St. Petersburg and Irkutsk, which, despite its grand moniker, was barely more than a rutted track. Writing from Barnaul in southwestern Siberia that summer, Ledyard sent Jefferson a purplish passage about the remaining distance to "that Ocean which I hope will bear me on its boosom to the coast of America." In this same letter, he invoked one of the great anthropological lessons that Cook's third-voyage account did much to popularize, now enhanced by his own travels, to wit: "Tartars resemble the aborigines of America." Indeed, Ledyard announced that "they are the same people" and that had it not been for "a small sea [that] divided them, they would all have still been known by the *same name*." In his journal, he asserted that "America was peopled after Asia."[12]

At Omsk, halfway across Siberia, Ledyard's journal for August 3, 1787, offers detail on another – indeed the primary – late-career scientific interest of Cook the pioneering climatologist: ice. Crossing several Siberian rivers on his travels east, Ledyard tried measuring their depth with a view to estimating the volume of water they carried to the Arctic. Here, though, he proved that he was not in Cook's class as an observer. He specified that the "vast quantity of water" he saw flowing north in these streams "must be the cause of the great quantity of Ice about the Pole," an idea Cook abandoned during his second voyage and confirmed on the third. Proving that he had not shared in shipboard conversation on this subject with Cook, King, or any of the more educated sailors on the *Resolution,* Ledyard said his "calculation must be allowed to be as vague" as any other theory. His objective was only to "enable the reader to conceive of a very great flux to the Ocean, and the ingenious *observer* to form his opinion on the above hypothesis."[13]

On reaching Irkutsk on Lake Baikal, the Siberian capital and fur trade centre for the Russian Far East and Alaska, Ledyard made an extended stopover that gave him a chance to catch up on his journal and correspondence. There, in the middle of north central Asia, he found himself re-immersed in the world of Cook the global explorer. He saw "some Sandwich Island

Cloth" that had been obtained from Cook's ships in Kamchatka after the expedition's second season in the Arctic. He also learned from the Russian merchant who showed it to him that there were now upward of 2,000 Russian fur traders "on different parts of the Coast of America," meaning the subcontinent of Alaska. Promisingly, this same contact had a "Vessel of his own at Okotsk which leaves that Country for America next Summer, and offers me a passage on her." On August 20, he penned a letter to Adams's son-in-law William Smith in London. Like all effective fundraisers and publicists in need of patronage or support, Ledyard knew the value of flattery, as we saw earlier in his correspondence with Jefferson. He told Smith, from his vantage point two-thirds of the way across Siberia, that "in Russia I am treated as an American with politeness & respect and on my account the healths of Dr Franklin and General Washington have been drunk at the tables of two Governors: and at Irkutsk the name of Adams has found its Way." (The absence of Jefferson's name in this litany is noteworthy.) Ledyard did not wish to elevate himself to the pantheon of American greatness, but with barely a hint of modesty he professed to Smith that "among the middling Class of People, I am kind of Phenomenon, among the peasantry a right down wizard." Thus feted, his country toasted, he said that a passage from Okhotsk to North America had already been promised. At that point, it seemed to Ledyard that he would actually be able to fully execute the Siberian phase of his amazing journey, enabling him to move next to its North American segment. To Smith, he admitted having some anxiety that with the approach of fall and winter soon after, the travel season might soon close. He noted that "no post goes after Snow falls – at least very seldom: I suppose either on a Sledge drawn by dogs or reindeer I am told so." "As for going on foot," Ledyard continued, "it is ridiculous in this Country."[14]

Beyond Irkutsk, travel conditions were even more primitive than along the Great Siberian Post Road. After a "very fatiguing voyage" of twenty-two days, Ledyard reached Yakutsk, Siberia's second-largest town and its coldest, on September 18, 1787. This was the same month the new American constitution was signed and La Pérouse reached Kamchatka in his Pacific follow-up to Cook. Over the course of two and a half months covering 5,500 miles, a series of carriage rides had taken Ledyard 90 percent of the way across Russia from St. Petersburg to his intended destination at Okhotsk. But he never moved another mile toward his destination on the Pacific. Initially, Ledyard thought his transit of the remaining 700 miles to the Pacific seaboard was impeded by the prospect of the forthcoming Siberian winter or just bad luck, but it was neither. That fall, Ivan Iakobi, the governor of Irkutsk – probably one of the figures who toasted Washington and Franklin – had become convinced that Ledyard was a spy for British fur-trading interests, and he asked the commandant in Yakutsk, Grigori Marklovski, to quietly detain this curious traveller until further guidance could be secured from St. Petersburg. Notwithstanding Ledyard's avowals that he was an American patriot, the suspicion against him was probably grounded in his association with Cook, which he was never shy about disclosing, or the immediate circumstance – setting off from London with a subvention from such notables as Joseph Banks.[15]

While his monitors waited for word from Empress Catherine, Ledyard was (benignly, he thought) prevented from proceeding eastward before eventually being escorted back to Irkutsk in December. Ironically, his principal minder was Joseph Billings, who was still in the preparatory stage of the maritime survey that was intended to provide clearer definition to the extreme northeastern parts of Siberia. Billings returned to central Siberia in mid-November 1787 for overwintering when he received the order to keep an eye on Ledyard. Initially, Ledyard believed the retrograde movement was a temporary setback, but when Billings turned him over to Russian authorities in Irkutsk, it finally occurred to him that the Russians had had him under suspicion.

During his detention, Ledyard made a virtue of adversity. Before leaving Yakutsk, he filled his journal with a host of entries that echoed common concerns from Cook's third voyage, principally the attributes of a frigid climatic zone and ethnography, or more particularly, the demographic dispersion that had led to the peopling of the Americas. Billings lent him his Arctic journal as a reading supplement, but the baseline was always Cook. "Here at Yakutsk the Atmosphere is constantly charged with Snow," he recorded, adding that the "air is much like that we experienced with Capt Cook in *Mare Glaciale.*" Ledyard's interest in the fur trade on the Northwest Coast was his original motive and never far from his mind. He wrote an extended entry that alternately argued against or affirmed aspects of Cook's thinking. "No Sea Otter Skins have ever been seen in either Hudsons or Baffins Bay, or any of the Eastern Coasts of America," he began. To Ledyard this was proof that the Native Peoples of the Northwest Coast had never had "Commercial Intercourse" with either with the Hudson's Bay Company "or with the eastern Indians of America," even if this was "supposed by Captain Cook." Ledyard then asked: "Does not this Circumstance tend to invalidate the Authenticity of several pretended Voyages & Discoveries on the N.W. Coast of America?" Ledyard was here referring to the legends of Juan de Fuca and Admiral de Fonte, whose dubiety Cook had proved, by hammering home a rhetorical query of his own: "Why should Captain Cook & the Russian Voyagers be the only ones who found" sea otters, while "others [i.e., mythical voyagers] have not even made mention of them, & yet they are not only very plenty with the Natives, but we with Captain Cook also saw them frequently in the Waters of the Coast alive."[16] Ledyard's last observation was astute but his dismissal of Cook's theory on Native trade routes running across the width of North America was simply a failure of imagination.

In a letter to Joseph Banks, Ledyard derided the goal of the Billings Arctic expedition, of which he now had ample opportunity to learn. Given the experience of Cook's last voyage, he considered it "ridiculous to attempt." Nevertheless, Ledyard was very glad to see an old acquaintance, especially since Billings had taken him under his wing, while "having time on my hands at Yakutsk." Ledyard's letter regaled Banks with stories of the impact of cold temperatures on brandy and scientific instruments. The mercury in his own thermometer, he said, could be cut like lead metal. As for his imagined future, Ledyard told Banks he expected to return to Yakutsk with Billings in the spring of 1788, and then proceed with him to Okhotsk, "where we shall do well to arrive before another winter." From that spot, he would ride on the ship promised previously by the merchant in Irkutsk across the Sea of Okhotsk to Kamchatka, and thence to the Aleutians. If he was not able to make that voyage in the short term, he mused about spending the ensuing summer visiting either Korea or the "Tschutokoi Indians" he had met on the Cook voyage. Concluding his letter to Banks, Ledyard boldly promised that by the summer of 1789, "I shall most certainly be at Nootka sound."[17]

Ledyard's remarkable ambitions crashed on February 1, 1788, when a team of imperial guards send by Empress Catherine arrested him in Irkutsk and escorted him to the Polish border, which they reached on March 18. The trip took a mere six weeks covering 6,000 miles. Summing up the entire experience in a passage found at the end of his Siberian journal, he damned his nemeses: "Let no European put entire confidence in a Russian of whatever Condition and none at all in the lower & middle ranks of People." After reaching the Baltic, he booked passage back to London, arriving in May. Once there he wrote a letter to his cousin ascribing this stunning turn of events to "jealous envy & malice."[18]

Historian James Gibson asserts that the empress's decision was rooted in the dawning awareness that her Far Eastern frontier was poorly defended. Why allow Ledyard, already a somewhat accomplished literary man, to further advertise that fact? Gibson reasons further: "The less that foreign powers knew of this weakness, the better, otherwise they would take advantage of it."[19] Nevertheless, the well-read Cook third-voyage account, especially James King's supplement about how to capitalize on the sea otter trade of the North Pacific, more than hinted at this vulnerability, so one more report was probably not the dispositive factor per se.

Here some productive speculation may be in order because there are two more curious twists to consider within a saga already replete with them. The first relates to Ledyard's suspicion that he was done in by personal jealousy. As it turned out, one of Catherine's principal nautical advisers at this time was James Trevenen, formerly a midshipman aboard the *Resolution* on Cook's third voyage. Trevenen's career, like that of Billings, had also been in the doldrums at the end of the American War for Independence and, doubting his chances of securing the *Bounty* command that eventually went to William Bligh, he offered his nautical skills to the Russian ambassador to Great Britain in February 1787. The empress told her envoy in April that she was "very willing to employ this officer [Trevenen] in my service." Seizing on James King's fur trade prospectus in the Admiralty account, Trevenen developed a strategic plan calling for a shift in the fur trade of the Russian Far East. His scheme called for bypassing the monopolistic Chinese merchants at Mai-mai-cheng (opposite the Russian inland port of Kyakhta in present-day Mongolia) by sailing directly from Kamchatka to the more competitive Chinese markets at Canton and Macao, while opening trade with Japan besides. For Russia to become an adept maritime power capable of conducting a triangular sail from St. Petersburg to Kamchatka with stops in Japan and China, Trevenen's blueprint called for "a nursery of Russian seamen."[20]

Trevenen believed, on the basis of the Cook model of which he was an example, that the best teacher is experience, and the voyage of discovery Billings had been commissioned to lead was a start. Thus, one of the two known occurrences between the disclosure of Ledyard's well-advertised intention in St. Petersburg to travel to the Russian Far East and his eventual arrest was the empress's hiring of Trevenen. Trevenen was a staunch British patriot, and an adversary of the American John Gore, Ledyard's advocate during Cook's last voyage and the man who succeeded to command of the third voyage after Cook and his principal adjutant, Charles Clerke, died. (It was Gore who brought the expedition's ships home.) Trevenen was also a vitriolic public critic of John Paul Jones's depredations against Great Britain during the recent war. Thus, it's possible that Trevenen nurtured the idea in Catherine's mind that Ledyard was not a romantic ambulatory traveller but was instead off scouting fur trade prospects in her Far East domains, if not for the British then perhaps on behalf of some of his countrymen. In such a scenario, Ledyard's ambitions might have rivalled Trevenen's own scheme for a Russian maritime fur trade that would obviate long overland treks across Siberia. Even though he may have had a hand in sabotaging Ledyard's venture, Trevenen never had a chance to implement his vision because war between Russia and Sweden diverted his career from exploration and the fur trade in the Russian Far East to nautical combat in the Baltic, where he died in action.

Despite his suspicion that he was done in by some jealous conspiracy, there can be no doubt that the voluble Ledyard was his own worst enemy. He must have raised eyebrows when he arrived in Irkutsk and began laying plans for his trek to Okhotsk to be followed by a voyage to North America. Aleksandr Baranov, the future governor of the Russian-American Company, arrived in Irkutsk in 1781, and by 1786 he had joined the concern of Grigorii Shelikhov, one of the major fur trade brokers operating in the Aleutians. Shelikhov returned to Irkutsk from Alaska in 1786 on his way to St. Petersburg to confer with the empress about crown support for a full-fledged North American colony. In the hypercompetitive environment of the sea otter fur trade, Shelikhov's enterprise, which was already in a contest of wills with Billings over what the Englishman considered the inhumane treatment of Indigenous Peoples in the northern commerce, may have suggested to the Siberian governor that the American interloper be detained and expelled. From the Russian point of view, the fact that Ledyard was a veteran of the Cook expedition, one of whose chroniclers (James King) had publicly touted its friendly relations with the Chukchi, one of the most intractable Indigenous populations in the Far East, would have only accentuated this opinion.[21]

Though disappointing in its conclusion, Ledyard's truncated voyage was not fruitless in terms of environmental history. In his Siberian journal, which he continued to keep during his transit out of the country, Ledyard returned to Cook's legacy in the *Mère glaciale.* Crossing a number of rivers travelling west, he revisited the question of the Arctic ice pack that had stymied Cook a decade earlier. He observed that, other than the Volga, every river in Russia emptied "immediately or collaterally into the Northern Sea." A number of large lakes also paid "their Tribute in the same line to the Icy Sea," foremost among them the "princely Baikal." Ledyard thought the combined "Column of Water" that Asia conveyed northward was theoretically sufficient to provide, in the appropriate season, a volume of ice to cover the surface of the shallow ocean basin "that at present barely covers the Earth beyond the Latitude of 70 North." Here we see Ledyard drawing on a key insight from Cook's final voyage – the shallow depth of the Arctic Ocean – but a larger question was raised by the carrying capacity of these rivers that merited "the attention of the curious since it naturally leads to a reflection on the quantity of Ice there, of its cause, duration, formation, diminution, etc." He was now entering a discourse joined previously by Cook and his principal second-voyage naturalist, Johann Reinhold Forster. In his journal, Ledyard averred that "Men have supposed [the Arctic] Ocean to be supplied with Ice from those Rivers."[22] This is an important point because it shows how Cook's second and third voyages and early observations about polar climatology energized and informed Euro-American thought about these issues.

Ledyard began his long reflection on the subject by expressing admiration for the "few observations" Cook made on ice formation theory, which, like much of his former captain's work, showed his "Abilities & good Sense." According to Ledyard, Cook asserted "that the Waters of that Ocean froze & that it did not receive its Ice from Rivers." In truth, Cook never addressed the first half of that proposition quite so concretely. The latter part, however, Ledyard got right, for he agreed that Cook had "a very excellent reason" for coming to that conclusion, namely, "that Ice formed in Rivers floating in that Ocean must necessarily bring along with them Vegetable Substances of some kind." Since Cook never saw such evidence, Ledyard believed that the great navigator made the proper deduction. If Cook's travels had taken him to the mouths of Siberia's rivers, Ledyard stated, "his Observations would have been perfect." This backhanded compliment was based upon Ledyard's reading of Billings's journal describing the Kolyma River, no doubt supplemented by conversations with his former shipmate over the winter of 1787–88, including the ride from Yakutsk to Irkutsk. Ledyard related "that at the Mouths of every great River in Siberia emptying themselves into [the Arctic] Ocean there were Islands, Spits, & Bars of Sand across them that rendered a Communication of Ice to or from the Sea impracticable." Ledyard hadn't been to these confluences either, but Billings had, and there was yet one more salient point to make. Kolyma River ice at its mouth was only four or five feet thick, perhaps a third of the depth seen in the Arctic ice pack. Ledyard supplemented his reading of the Billings journal by conducting an "Enquiry" among Russians and Siberian Natives as he travelled toward the Polish border. No informant reported seeing river ice thicker than the Kolyma's. This circumstance, Ledyard reasoned, "obliged me to change my Opinion," impliedly toward the view first propounded by Cook.[23]

Ledyard admitted to having never "reflected on the circumstance of Congelation generally & particularly of the thickness of Ice," but now it struck him "forcibly." Indeed, the concluding sections of his journal are dominated by the same topic of sea-ice formation that had fascinated Cook late in his career. Reflecting on his experience growing up in New England and "having also travelled to the NW among our cold Lakes & Rivers in the Winter" (seemingly a reference to the Champlain Valley), Ledyard could not remember seeing ice of greater thickness than what he saw or learned about in Russia. He recalled transiting the Strait of Belle Isle separating Newfoundland and Labrador with its "high Sea Ice" (of unspecified depth). In the Baltic, he never saw anything "so thick" as the ice that was reported to be at the mouth of the Kolyma in Siberia where it emptied into the Arctic Ocean at latitude 70°N, citing "the best observations made by Captain Billings there during a Winters Residence." Nevertheless, the Arctic "Ocean Ice" he witnessed with Cook consistently exceeded what his own experiences and trusted sources instructed him about the thickness of freshwater ice emanations. From this Ledyard inferred, sensibly, that the further north one went, the thicker freshwater ice would become in proportion. His more complex theory was that the upper limit of any ice, fresh or salt, "formed by Congelation" was fixed at four to six feet. Any greater thickness was "adventitious accumulation not owing to Frost only, but to Snow Hail, Sleet & far to the Northward heavy mists which fall on it & add rather a coagulated than a frozen conjunctive Mass." In this roundabout way, Ledyard circled back to Cook's preferred way of thinking of saltwater ice packs: the rivers of the Northern Hemisphere do not issue ice in sufficient quantity to form the polar ice sheet, and therefore sea ice accretes mostly from atmospheric sources, such as snowfall. Ledyard concluded, in a fashion Cook often did when he came to the verge of propounding scientific theory, by hoping that his observations would be studied in "comparative Detail of a number of facts to ascertain their Merit or Demerit."[24]

Ledyard's ill-fated journey of American exploration to the North Pacific seemed to leave that zone in the control of Russian and British fur traders. Nevertheless, during his Siberian sojourn, two fellow New Englanders – Robert Gray and John Kendrick – had stirred and were joining the soft gold rush to the Northwest Coast. When Ledyard surfaced in London in May 1788, he announced, in Cook-like fashion, that he was done with the North Pacific but not with exploration. He soon attached himself to the newest discovery fad by joining the Association for Promoting the Discovery of the Interior Parts of Africa. Cook was dead, and though the era of long distance ocean voyaging was not over (as witness the La Pérouse and Malaspina expeditions conducted by France and Spain, respectively, trying to emulate Cook), in Great Britain the new frontier for future government-sponsored discovery was not yet perceived to be the Northwest Coast of America; rather, it was Africa. Indeed, as detailed in Chapter 5, what became known as the Vancouver expedition was originally intended to foster African discovery. In a sense, then, Cook's circumnavigations largely completed the work of discovering continental coastlines and major island groups, at least in the tropical and temperate zones, and so geographical interest began to shift toward continental interiors. The expeditions of Alexander Mackenzie (1789 and 1793) and Lewis and Clark (1803–06) in North America are emblematic of this shift. Since rivers provided the principal means of access to remote hinterlands, source hunting became the operative dynamic, as indicated, for example, by Jefferson's instructions to Meriwether Lewis.

Ledyard was caught up in the first phase of this source-hunting excitement, which lasted for much of the nineteenth century. He wrote a family member in June 1788: "Some few weeks only before my arival now at London a number of Gentlemen had been talking somewhat in earnest about exploring the Continent of Africa. My arival has made it a reality." Among them was the seemingly ubiquitous Joseph Banks who, Ledyard reported, had been told by "the king" that "no expence should be spared" on the effort. Having proved his bona fides on the Siberian venture, Ledyard secured underwriting for another solo expedition – a search for the source of the Niger. He planned on reaching the African interior by way of the Nile (whose headwaters were another mystery to Europeans). Once inland, he intended to take a Nubian trade caravan westward toward the upper reaches of the Niger, near exotic Timbuktu. After floating downstream to colonial outposts on the west coast of Africa, he would find passage to England. In what biographer James Zug calls a "stunning volte-face," in a few months Ledyard had transformed himself from a Russian vagabond to a lionized celebrity sitting for farewell dinners and a portrait. On the way to Cairo he visited his old friend and sponsor in Paris, Jefferson. He knew of the ambassador's annual July 4 party but, "ill with a cold & fever," sent a note expressing his regrets. Jefferson, who was probably surprised that Ledyard was in the city, and not, as Zug put it, "striding across the hills of California," arranged for a separate dinner that same night. Ledyard outlined his plan to explore Africa under the auspices of the British association, which we know separately upset the future president. Perhaps as a concession to American sensibilities, Ledyard promised, as Jefferson later wrote James Madison, that if he survived the African venture "he will go to Kentuckey and endeavour to penetrate Westwardly from thence to the South sea."[25] Ledyard would not live long enough to keep that pledge, but here we see a prefiguration of Meriwether Lewis's assignment.

In another attempt to re-ingratiate himself with Jefferson, the next day an obeisant Ledyard provided the ambassador with a wide-ranging reflection upon his Russian travels that touched on many of the main currents of late-Enlightenment thought. It concentrated on one of that era's favourite topics, the racial origins of Native Americans. This subject was of great interest to Jefferson, and the two men's interlocutions, orally or in writing, could only have fed each other's fascination. Ledyard told Jefferson, "I am certain that all people you call red people on the continent of America & on the continents of Europe & Asia as far south as the southern parts of China are all one people by whatever names distinguished & that the best general one would be *Tartar*." This was a term that Ledyard had been introduced to along the Great Siberian Post Road. Broadening his scope, he added: "I am satisfied myself that America was peopled from Asia & had some if not all its animals from thence."[26] With that, Ledyard shifted to another topic that absorbed Jefferson's imagination and became central to his famous debate with the Comte de Buffon, the great French naturalist. Buffon had wounded Jefferson's New World pride by suggesting that the animal species in the Western Hemisphere were puny or otherwise immature compared with their European counterparts.

Ledyard was in Alexandria a month later, and from Cairo he dispatched his last correspondence with friends and sponsors, including Jefferson. It was at this point that he penned what was both a stout defence of his conduct and an effusive apology to Jefferson: "I shall never be entirely without a subject when it is to you I write: I shall never think my letter a bad one when it contains the declarations of the esteem I have for you: of my gratitude &

of my affection for you & this notwithstanding you thought hard of me being employed by an english association which hurt me very much while I was at paris." Yet, he asked Jefferson to forgive him, desperate as he was "not to loose [sic] the little regard you have in times past been pleased to honour me with. You are not obliged to esteem me, but I am obliged to esteem you, or to take leave of my senses & confront the opinions of the greatest & best characters I know of."[27]

Ledyard's last letter, fittingly it seems, was addressed to Jefferson from Cairo on November 15, 1788, the eve of his expected departure. In it he spoke ominously and somewhat prophetically about "my fate" and prospects for survival. He ended: "I shall not forget you. Indeed it would be a consolation to think of you in my last moments."[28] Unfortunately, his departure from Cairo for the interior was delayed, a circumstance that facilitated an illness, probably dysentery, from which he died on January 10, 1789, as the most famous explorer who never really discovered anything.

1

Reinventing the Northwest Passage

IT MIGHT HAVE BEEN expected, once the intelligence gleaned from two summers' worth of exploration (1778–79) in the Arctic under James Cook and Charles Clerke settled in, that the Northwest Passage would have disappeared as a geographical projection. After all, Cook's third expedition failed to find a saltwater route from the Pacific Ocean to Baffin Bay. The Admiralty's complementary attempt, led by Richard Pickersgill and Walter Young in successive command of the *Lyon,* to support Cook when he was expected to emerge on the Atlantic side of the supposed passage, was even less productive. *Lyon* fell many degrees short of previous northern sailing records in Greenland's waters. Thus, Cook's northern venture seemed to have destroyed yet another cartographic image, just as he had *Terra Australis Incognita* during his second voyage. In the immediate aftermath of his final expedition's return in 1780, there was no movement in Britain to launch another high-latitude investigation of a passage in the north.

This perception was reflected in the considered thinking of Johann Reinhold Forster, who, in 1783, published the original German-language edition of *Voyages and Discoveries Made in the North* (translated into English in 1786). No Cook apologist, Forster referenced the great navigator's most recent effort, asserting that "in our days new discoveries have been made, which have brought us better than ever acquainted with the North, and have left us little more to discover with respect to these regions." Cook, complemented by recent Spanish and Russian exploration, had "rendered it pretty probable that no passage is to be expected there." Like Cook, Forster deemed the legends of Admiral de Fonte and Juan de Fuca, which both posited the existence of mid-latitude access off the Pacific coast deep into the North American interior, "absurd and incongruous" and "rather fabulous." Hudson Bay had been "so often explored, that we need not seek any more for a passage in those parts." All that remained of the Northwest Passage concept, Forster added, "can only exist in the weak brains of idle visionaries." Howard T. Fry, Alexander Dalrymple's biographer, charged that a "cautious scholar would have waited for more evidence," but before the decade was over, Forster's castigation was vindicated.[1]

The passage may have been a moribund concept in Forster's mind, but it is nevertheless true that when Jean-François de Galaup, comte de La Pérouse left Brest in August 1785 at the head of France's national response to Cook's efforts in the Pacific Basin, he was tasked with looking for it on the Northwest Coast of America. French cartographic projections had long posited theoretical eastward projections off the Pacific, but, like Cook before him, La Pérouse was dismissive of such geographical tales. In an ironic twist and narrative convergence, in 1781, during the course of the war in which France was allied with the rebellious American colonies, La Pérouse captured Fort Churchill on Hudson Bay from none other than Samuel Hearne. He thereby had direct access to the same pivotally conclusive geographical findings that had worked their way from Hearne to the Admiralty's instructions for Cook the previous decade.

In July 1786, La Pérouse, hugging the Northwest Coast, found an indent near latitude 59°N – Lituya Bay. For a moment, the pulses of those on his two ships quickened when it seemed they had entered a large body of water. Inevitably, La Pérouse hit a wall of solid glacial ice emanating from high, snow-covered peaks, providing a dead-end moment reminiscent of Cook's entry into Prince William Sound in 1778. The French explorer joked that his ships "completed in a few hours our voyage into the interior of America." The Northwest Passage, he concluded (then, if not half a decade earlier in Hearne's company), "is a dream." Sailing away from North America for the balance of what would prove an ill-fated voyage, La Pérouse framed a reflection in his journal expressing frustration with speculative geographers. Channelling Cook's earlier sentiments, he railed against such obvious inventions as the de Fonte story, which he considered an English invention. In this he was not mistaken. But as we shall see below, just when the pre-eminent French explorer categorically rejected his nation's cartographic tradition for depicting the west coast of America, the de Fonte legend was about to return to life in Great Britain.[2]

From 1780, when John Gore returned to Great Britain with Cook's ships, until 1787, Cook's third voyage, like the second, appeared dispositive relative to its central mission. Intervening developments soon changed that perception. There was no short-term commercial consequence from Cook's Antarctic voyage, which banished the idea of *Terra Australis Incognita*. But, as a by-product of the maritime fur trade seeded in China and in print during and after the return of his last expedition, the Northwest Passage hypothesis was resuscitated. It is routinely posited in Cook historiography that this revival is evidence of the great navigator's deviation from the standard of accomplishment seen in his first two voyages, necessitating the follow-up voyage of George Vancouver to finish the work that Cook had left incomplete. But, like many aspects of Cook third-voyage interpretive orthodoxy, this conclusion is mistaken. The features of the Northwest Coast that Cook supposedly "missed" occurred because he was guided by what might be termed the Hearne dictum. To wit, that North America was a wider continent than previously believed and, concomitantly, its broad northwestern quadrant west of the 110th meridian was not bisected by salt water south of latitude 70°N. Pursuant to this understanding Cook was directed to begin his quest near latitude 65°N. For this reason, he sensibly avoided pursuing the legendary straits of Juan de Fuca and Bartholomew de Fonte in more southerly latitudes.

Before I assailed the standard interpretation of Cook's supposedly laggard third-voyage conduct in *Captain Cook Rediscovered,* Barry Gough was one of the few historians who had insisted that Cook should not be faulted for having "dismissed the possibility of Juan de Fuca's strait" because "Cook had his orders, and he followed them to the letter." In his estimation, Cook's great modern biographer J.C. "Beaglehole, and other historians, have been unduly and carelessly guilty of attacking Cook for missing the entrance to the strait."[3]

Compounding matters, bad weather happened to force Cook far off shore after he left Nootka Sound for the icy northern latitudes. Thus, Cook's charts depicted the Northwest Coast from latitudes 49°N to 58°N as a simple and slightly wavy line trending to the northwest, until it was broken into detail when he reached the discovery zone and investigated Prince William Sound and Cook Inlet. But in the last half of the 1780s, British maritime fur traders detected many inlets and a complex array of islands dotting a

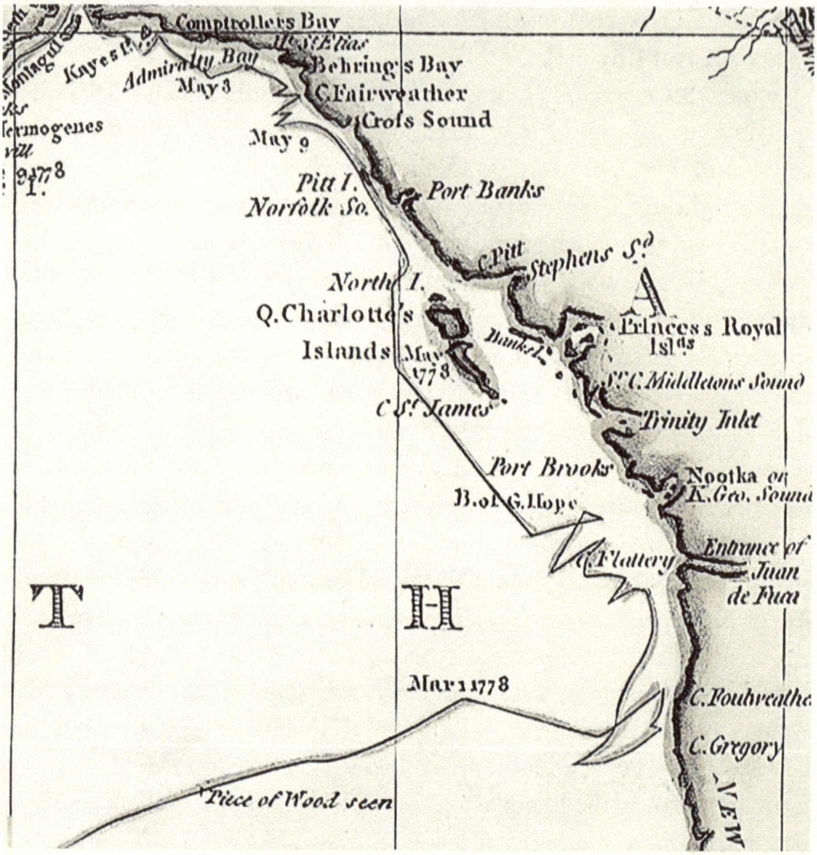

Figure 1.1 Henry Roberts, *A General Chart*, 1784 (detail). This inset is from the master map detailing the track of Cook's three voyages. Among this chart's many attributes is that it accurately recorded the northwesterly trend of North America's Pacific coastline for the first time. However, the lack of complexity in this depiction of the coast was later seized upon by subsequent maritime fur traders to suggest that Cook had missed one or more openings into the continental interior that could form a temperate latitude Northwest Passage. | Washington State Historical Society. 2011.0.60.3.3.1

coastline that Cook saw only intermittently and, by Admiralty order, surveyed at a distance and in a fleeting manner. The newly revealed intricacy of the Alaska littoral and the coast of present-day British Columbia became, to less thoughtful and capricious observers, an indictment of Cook's thoroughness. Most historians, failing to discern the actual nature of Cook's mission as it relates to the Northwest Coast, have also neglected to make allowance for the idiosyncrasies of Cook's contemporary fault-finders and the nuanced nature of their impeachment.

Cook had proved, at least until the twenty-first century, when the effects of global warming became apparent, that the classic version of a Northwest Passage that he was sent to find – a saltwater conduit from ocean to ocean over the top of North America – did not exist. But as the historian and geographer John Wright has noted, the history of exploration is replete with *idées fixes,* which he defines as concepts resistant "to change in the light of factual evidence or rational thought." The Northwest Passage is an example of this phenomenon, arguably one that endures to this day. Though some of Cook's contemporaries seized upon the gaps in the great navigator's coverage to further their commercial prospects or for self-aggrandizement, the underlying reality of Hearne's dictum had *not* been obviated by any subsequent exploratory disclosure. But as Richard Van Orman observes, with sufficient incentive there was a tendency in matters of geography "to push aside unpopular facts that did not fit into the established image."[4]

As a function of this dynamic, the notion of a navigable Northwest Passage was able to survive Cook, though with an important distinction. Historians, in their enthusiasm to abide with standard criticism of Cook's third voyage, have missed how the passage subtly evolved as a cartographic image between 1784 and 1801. These bookend dates are a function of two landmark publications: the Admiralty account of Cook's third voyage and terrestrial fur trade explorer Alexander Mackenzie's *Voyages from Montreal,* respectively. In less than two decades the Northwest Passage was transformed from (1) the imagined high-latitude saltwater throughway Cook was sent to find, initially succeeded by (2) the concept of inland seas thought to "communicate" with each other mid-continent in some undefined manner, and finally to (3) a set of interconnecting rivers.

The key figures in this conversion process were the propagandist fur trader John Meares, theoretical geographer Alexander Dalrymple, and practical investigators Peter Pond, George Vancouver, and Mackenzie. Though these men quibbled with the findings of Cook's last voyage, none of them, contrary to the convention of Cook historiography, not even Cook's supposed archnemesis Dalrymple, took issue with the fundamental truth about the impracticability of what Cook had been sent to find on his third voyage. Even Vancouver did not correct Cook in any truly substantial fashion. Indeed, the Northwest Passage as an *idée fixe* endured long after Vancouver and Mackenzie, to attain, in the words of Felipe Fernández-Armesto, the status of a "pathological quest, an irrational obsession" in the middle of the nineteenth century, when explorer after explorer traversed the Arctic "as if blinded by the snow and maddened by the ice."[5] This ancient dream was not truly realized until the completion of the Canadian Pacific Railway in 1885, and, on the American side of the border, the Northern Pacific and Great Northern lines.

The Quest for Soft Gold, by Sea

The history of the Northwest Coast fur trade, and the Vancouver expedition it spawned, is both complicated and best told in detail elsewhere. A few key developments stand out for presentation here, starting with the observation that by 1784 several published narratives from Cook third- voyage veterans highlighting the promise of an extensive sea otter trade in the North Pacific were already in circulation. Of these, James King's contribution to the official Admiralty account, issued in June of that year, was particularly noteworthy. It served, in the words of Barry Gough, as "a blueprint for commerce." Merely one year later, Gough adds, "almost identical projects had been launched simultaneously from London, Bombay, Calcutta, and Canton" in response to what Cook's last voyage had revealed. It is impossible to exaggerate the importance of this phenomenon for the subsequent history of North America. In the estimation of Peter Stark, "Cook's surviving officers and crew ... made a discovery that would determine the future of the Pacific Rim."[6]

The first to set sail was James Hanna from Macao, the Portuguese colony at the mouth of the Pearl River downstream from Canton, not surprisingly since that is where the enchanting news of this promising trade had first arrived via Cook's ships. Captaining, fittingly, the *Sea Otter,* Hanna left port on April 15, 1785, and reached Nootka Sound on August 18. His ship was flying Portuguese colours, a ruse to avoid violation of the monopoly charter of the East India Company (EIC). In December of that year, James Strange, an EIC merchant who had read the official account of Cook's last voyage, sailed from Bombay as supercargo for a trading expedition to the Pacific Northwest. It consisted of the equally apt *Captain Cook* (piloted by Henry Laurie) and the correspondingly resonant *Experiment* (led by Henry Guise). These ships were another private venture but indirectly assisted by the EIC as a test voyage. They reached Nootka Sound on June 25, 1786.

Back in the mother country, Nathaniel Portlock (master's mate aboard *Discovery* on Cook's third voyage) and George Dixon (armourer on the same vessel), respectively captained the *King George* and *Queen Charlotte* (after the reigning monarchs) for the King George's Sound Company. This new British joint stock association was named after Cook's transitory name for Nootka Sound, and was formed by Richard Cadman Etches with support from the influential Joseph Banks. These ships left Great Britain about the same time that Hanna left Macao, but, with a greater distance to travel, they did not arrive in the North Pacific until July 1786. They called serially at Cook Inlet, Prince William Sound, and Nootka Sound, the precise itinerary Lieutenant James King had mapped out in the Cook third-voyage account.

William Beresford, the company's supercargo managing the business side of the Etches venture, in a post-voyage narrative of the trading expedition published in 1789, explained their route with a testimonial to Cook's thoroughness: "To say that these different harbours are laid down with accuracy and precision, would be unnecessary; it is sufficient to observe that they were surveyed by *Captain Cook.*" Beresford added that Nootka Sound had been "pointed out by the late Captain Cook, as the most eligible spot where a valuable cargo of furs might be collected," but this was a misreading of history because, as noted earlier, it was King's narrative, not Cook's, which seeded the fur trade. Elsewhere Beresford volunteered that Cook's discovery of Hawai'i, where the ships overwintered, was "sufficient alone to render his name immortal." Portlock and Dixon, having studied under the master, plied the coast of Hawai'i for an extended period rather than "coming to anchor." Beresford considered this a "somewhat extraordinary" circumstance until he learned that the company's ships "easily procured" the hogs and greens "we chiefly wanted ... by plying off and on."[7] In yet another unintended vindication of Cook, who was criticized by his crews and later by historians for laying off the island for what was perceived as an inordinate amount of time, Beresford recorded the great difficulty Portlock and Dixon had trouble approaching the islands because of the same troublesome winds that had bedevilled Cook and Clerke in 1779.

Captain Hanna reached Nootka Sound for the second time in August 1786, in what could be called the fourth sea otter trading voyage to the Northwest Coast. There he met Strange and his two East Indiamen – the first competitive meet-up. That summer, John Meares arrived in the North Pacific, having sailed from Calcutta to Prince William Sound in command of the *Nootka,* accompanied by the *Sea Otter* captained by William Tipping. The Meares venture represented the fifth trading expedition and the seventh and eighth ships to respond to the promise outlined by King. Captain

Strange and his East Indiamen encountered the Meares vessels at Cook Inlet, representing Strange's second competitive meet-up. During the winter of 1786–87, Meares made the fateful decision to overwinter at Prince William Sound rather than retreat to Hawai'i like Portlock and Dixon, or to his home port in Asia. Dixon found Meares's expedition in May 1787 under extreme duress, with many men dead from starvation or scurvy, having spent the winter ice-bound. Dixon was a rival but Meares welcomed his providential appearance "as a guardian angel with tears of joy." Having learned his lesson, Meares limped to Hawai'i and back to China with his first cargo of furs. Captain Tipping and all hands on the *Sea Otter* were lost at sea. Meares returned to the Northwest Coast in both 1788 and 1789 in a new partnership arrangement. As a consequence, he became a pivotal figure in a sequence of events that almost resulted in war between Great Britain and Spain and led directly to the commissioning of the Vancouver expedition.[8]

In June 1787, the *Imperial Eagle,* a British ship but outfitted at the Flemish port of Ostend with partial financial support from former East India supercargoes and also flying a false flag, captained by William Barkley, became the ninth ship to reach Nootka Sound. There Barkley met Captain Charles Duncan of the *Princess Royal* and James Colnett in command of *Prince of Wales* (the tenth and eleventh ships). These craft had been readied by the same King George's Sound Company that sponsored Portlock and Dixon, who themselves re-attained Nootka Sound in August after rescuing Meares. (Colnett was another Cook veteran from the Antarctic voyage; it was his venture that John Ledyard had tried to join.) Barry Gough's comprehensive catalogue of trading voyages gives the impression that by the summer of 1787 the Northwest Coast, particularly Nootka Sound, was getting crowded because "too many vessels pursued too few sea otters."[9]

Rivals hoping to cash in on the lucrative trade were coming from all directions, seemingly criss-crossing the Pacific and each other. Traders left Asian or British ports with goods of value to Indigenous hunters who would acquire pelts, which after one or two seasons of coastal commerce were transported to Macao and Canton and exchanged for silk and tea that would be subsequently conveyed back to Europe. Some of these ventures were more profitable than others, as the sea otter pelts could not always be procured from local hunters with ease or at an advantageous price. And what the traders could not acquire through fair exchange they did through ever more violent means, including the destruction of entire villages. But it is not the commercial dynamic or tyrannical deportment of the traders that most interests us here, but rather the geographical information they secured, as well as its contemporary interpretation and propagation. In this regard, the voyages of Dixon, Duncan and Colnett, Barkley, and especially John Meares are probative.

During the 1787 trading season, Dixon discovered the Queen Charlotte Islands, named after his vessel but now known as Haida Gwaii after its Indigenous People. The strait to the north of this archipelago is named after Dixon. In his account, Beresford noted that this was the zone of de Fonte's supposed gateway eastward across North America, a purely mythical concept that had been resurrected by French cartographers in the 1750s. Cook thought he had banished this idea from serious consideration for all time and, with the authoritative publication of the Admiralty account in 1784, most sensible observers, including Johann Reinhold Forster, agreed. Beresford, like Cook, termed the de Fonte legend "almost incredible." Nevertheless, when Dixon brought the *Queen Charlotte* closer to the continent, Beresford wrote that "it appears very probable, that there are deep inlets into the country." Noting the many islands and inlets that faced out toward the Pacific, he "doubted whether we have yet seen the main land." That proposition, Beresford mused, was "to be determined by future navigators."[10]

Concurrently with Portlock and Dixon, Colnett and Duncan also discerned that the mainland coast between latitudes 53°N and 58°N was strewn with islands (including two large ones that they named after their ships, the *Princess Royal* and the *Prince of Wales*). This was the latitudinal range Cook "missed" on his way north from Nootka Sound (at approximately 50°N) before re-establishing contact with the coast at Mount Edgecumbe (neighbouring present-day Sitka) near 57°N. More astonishingly, Captain Barkley believed he had stumbled into the legendary Strait of Juan de Fuca, which, as Barry Gough put it, "Cook had emphatically declared could not exist."[11] Since Cook had not explored this coast, who really knew what this labyrinth of islands in the mid-50s concealed, or where the strait sighted by Barkley might lead?

Traders avidly exchanged geographical information in ports like Nootka, and so after Duncan learned of Barkley's find in August 1787, he sailed the *Princess Royal* into de Fuca's strait the following year to see for himself. Entering, he spied a rock pillar, a geological feature that Fuca had described to Michael Lok in 1596. Duncan then proceeded to the southeast as far as an imaginary line running from present-day Victoria, British Columbia, to Port Angeles, Washington. When Duncan returned to England in 1789, he showed his chart of this voyage to Alexander Dalrymple, who found it of such interest that he had it engraved the following year. At this point, de Fuca's strait was literally back on the map (and the subject of much public discussion, to which we will return). Correspondingly, Duncan's chart also provided the first cartographic hint of Vancouver Island's separation from the mainland. The wide saltwater gap between Cook's "Cape Flattery" and Neah Bay on the south side of the opening ran inland for nearly 100 miles along what would later be discerned as the Olympic Peninsula. The south coast of this inlet was now seen to lie opposite terrain that had heretofore – that is, since Cook – been presumed to be part of the continent. If the apparent breadth and depth of this gap was not propitious enough, Duncan gleaned intelligence from Makah villagers near Neah Bay that at its eastern limit this passage led to a "Great Sea" that thence ran north and south. This was an intimation, of course, of Puget Sound to the south and the northerly Strait of Georgia, both later mapped by Vancouver.[12]

The Quest for Soft Gold, by Land

Here we must shift the orientation of discovery away from the maritime sphere to consider an explorer with a terrestrial orientation who also exerted great influence on the evolution of the Northwest Passage – Peter Pond. A Connecticut Yankee like Ledyard, Pond fought on the side of his colonial militia against France during the Seven Years' War. After that conflict, he entered the fur trade. From his base in New York City, he worked his way west into the Great Lakes region before migrating to Quebec, a move that, prior to the American Revolution, did not have patriotic implications. Given the cellular proliferation of the fur business as traders moved from one lake or river to the next, Pond ultimately trapped his way to the northwest of present-day Alberta and into the Northwest Territories. As an independent trader, he conducted his business through a series of intermittent and ephemeral partnerships. However, he was so energetic and effective that in time he became a principal in the North West Company, headquartered in Montreal. This firm became the great rival of the Hudson's Bay Company (HBC) once it was formally organized over the winter of 1783–84. Pond's discoveries, once they were relayed by British colonial officials in Canada to Great Britain, and to major figures in geographical circles there such as Joseph Banks and Alexander Dalrymple, added even more yeast to the ferment in London over what might have been missed when the Admiralty sent Cook so far north.

The Pond connection to the Northwest Passage begins on July 1, 1776, while Cook was preparing for his third voyage at Plymouth and the Second Continental Congress was framing the Declaration of Independence in Philadelphia. That day, a fur trader from Quebec, Alexander Henry (the elder, to distinguish him from a nephew by the same name), recorded in his journal that he had met Natives from a "Lake Arabuthcow" (Athabasca) who had travelled south and east of there to barter with him. Writing from what was then the most distant post in the continental trade, Lac Île-à-la-Crosse in the northwestern corner of present-day Saskatchewan, Henry claimed: "They informed us, that there was, at the further end of that lake, a river, called Peace River, which descended from the Stony or Rocky Mountains, and from which mountains the distance of the salt lake, meaning the Pacific Ocean, was not great." This same Lake Athabasca, Henry was informed, "emptied itself by a river, which ran to the northward, which they called ... Slave River; and which flows into another lake, called by the same name [i.e., Slave Lake]; but whether this lake was or was not the sea, or whether it emptied itself into the sea, they were unable to say." Furthermore, Henry's Indigenous informants told him that "they also made war with the Indians who live at the bottom of the river; where the water is salt," implying an outlet into the Arctic. All this geographical information was a reasonably accurate description of the complex Mackenzie River Basin, including the relationship between the eastward-flowing Peace River, with its headwaters on the Continental Divide, and the great inland lakes (Athabasca and Great Slave). These lakes are in turn linked by the Slave River, whose waters do in fact drain to the Arctic via the Mackenzie River. Finally, Henry's informants from Lake Athabasca told him that "they also made war on the people beyond the mountains, toward the Pacific Ocean, to which their warriors had frequently been near enough to see it."[13] Henry, it seemed, found a people who knew pathways to the ends of the continent and salt water in two different directions.

These key insights inspired one of Henry's associates, Peter Pond. Two years later, as part of the constant quest for new fur country and the geographical information that would enhance such endeavours, Pond was the first Euro-American to cross the Methye Portage into the Athabasca district. His successor in the field, Alexander Mackenzie, later described this traverse's vista opening into the Athabasca country below it as "a most extensive, romantic, and ravishing prospect." Historian David Chapin, defining the significance of Pond's discovery, writes: "Like Cumberland Gap in the Appalachians or South Pass in the Rockies, the Methye Portage is a geographical gateway that provides the easiest access between separate regions of the continent." Barry Gough, explaining the commercial significance of Pond's extension into the Athabasca country, states: "He had opened to his partners an entire new realm of opportunity. He had outflanked the Hudson's Bay Company, stolen from them precious Native trading partners, and increased the value of the trading shares of all the partners."[14]

Pond built a fort, "Pond House," on the Athabasca River, about thirty-six miles above where it entered Lake Athabasca. This was the first fur trading post in the Arctic watershed (not counting the Hudson Bay), and its development accelerated the diversion of furs from the interior away from the HBC's main post on the bay at Churchill to Montreal. The Chipewyan hunters were happy with Pond's incursion because they no longer needed to travel hundreds of miles downstream to Churchill to unload their pelts. Even if they got less value in trade with Pond and his early associates, Benjamin and Joseph Frobisher, than they could have negotiated with the HBC men, it was far less trouble for them. Pond's outpost was the only one in the Athabasca country until 1785, and its monopoly standing made it fabulously successful. Pond's success in the Canadian Northwest laid the foundation for the incorporation of his enterprise under the rubric of the North West Company in 1783.

Pond's path-breaking Methye Portage gateway was first ascertained in 1778, the same year that Cook was off the Northwest Coast. Pond was inland twelve hundred miles at the same latitude as Cook's Mount Edgecumbe. His discovery precipitated a rush for the soft gold of beaver pelts every bit as lucrative as the maritime trade in sea otters that followed in Cook's wake. When Pond returned to the great fur mart at Grand Portage near Lake Superior in July 1779, to drop off packs of fur and to procure supplies for another season of trading, he disclosed his findings. Other traders joined his enterprise and followed him across the divide into the Athabasca country. At the same time, word of his exciting discovery of the Methye Portage and verification of Henry's earlier intelligence spread eastward and into the intellectual currents of the Atlantic seaboard and eventually all the way to London. Henry, who had retired from the trade and was then travelling in England, wrote Joseph Banks in 1781 proposing that a formal scientific expedition be sent west from Hudson Bay to authenticate the Indigenous geographical understanding he had first gleaned, and which Pond seemed to have confirmed. He suggested that it was by this network of northwestern rivers, lakes, and adjacent land routes that objects of European manufacture that Cook had discerned at Nootka had made their way west. In the event, a government-sponsored investigation proved unnecessary because of further explorations by both Pond and especially his successor, Alexander Mackenzie.

As with the maritime fur traders introduced above, our intention with Pond is focused on his geographical insights and their distribution to the outside world in the immediate aftermath of Cook's third voyage. Not that Pond thought of himself principally as an explorer; the commercial requirements of the fur trade always took precedence over discovery. Nevertheless, reconnaissance was an essential part of a trader's job. Pond was unusually good at it because he was always attentive to Indigenous insights about topography, which he combined with his own study of the land. Pond overwintered several seasons in the Athabasca country, during which he refined his understanding of how the rivers and lakes of the northwest interior might flow alternatively to the Arctic or the Pacific.

Pond himself never travelled to the extremities of these rivers. During the winter of 1783–84, he queried his Chipewyan neighbours about waters beyond the immediate horizon of Lake Athabasca. He received thereby his first real intimation of Great Slave Lake and its contribution to what would later be termed the Mackenzie River Basin. He also learned that this system's principal western tributary, the Peace River, had its headwaters in mountains not far from what his informants described as a salt lake – the Pacific Ocean. When he finally returned to the Atlantic seaboard in late 1784 or early 1785, Pond began circulating the first of several maps that provided the first cartographic insight on the question of the Northwest Passage, from a terrestrial vantage, since Samuel Hearne's expedition to the mouth of the Coppermine River northwest of Hudson Bay in the summer of 1771.

Pond's first visual representation is known among historians as the "Congress Map" because in March 1785, after returning to Connecticut, he presented it to that body in New York City (then the capital of the nascent United States). Scholars have pondered Pond's motivation for this maneuver since he was working for an enterprise headquartered in a British colony. It is conceivable that he submitted his map to Congress merely out of a native son's pride – if not the usual patriotic variety then a corollary – as his own modest contribution to Enlightenment-era enthusiasm for discovery. His often contentious relationship with partners in the fur trade (he had been

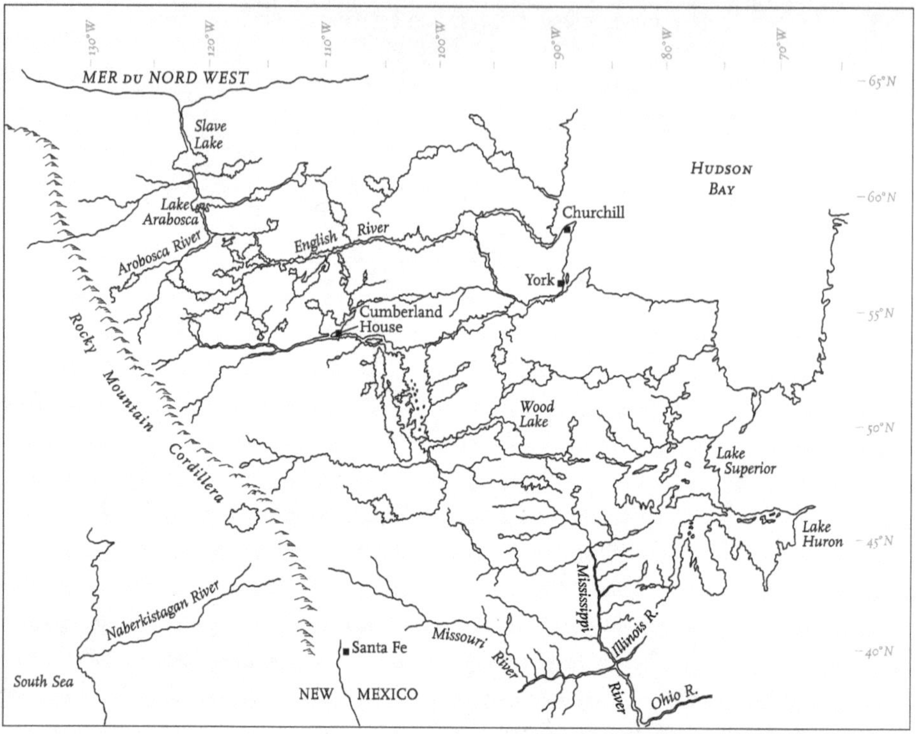

Figure 1.2 Peter Pond's "Congress Map," 1784. Pond's original manuscript maps did not survive, though some, like the one he submitted to Congress, were copied. Because of their size and often faded nature, plus stark divergence from normative geographical understanding, their use in a small book format is limited. Pond's cartographic presentation to Congress depicted a vast network of lakes and rivers to the northwest of the Great Lakes. | Based on David Chapin, *Freshwater Passages*, 208; redrawn by Eric Leinberger

seriously implicated in a frontier murder in 1782) and the subsequent reaction in Quebec to his sojourn in the United States suggest that Pond was trying to lay the groundwork for a combined American discovery/trading venture of an overland variety. This would have been similar to what John Ledyard was pursuing from a maritime angle at almost the same time, and possibly with some of the same individuals. Pond's manuscript map was probably drawn over the winter of 1783–84 at his post on the Athabasca River. The original tendered to Congress was lost, probably due to the immature record-keeping capabilities of the new nation. We know what it depicted because Hector St. John de Crèvecouer (otherwise famous for his *Letters from an American Farmer*), the French consul to the Mid-Atlantic States, made a copy of it. (See Figure 1.2.)

Advertised, according to Crèvecouer's inscription, as representing the findings of an "Extraordinary Man" who had "resided 17 Years in those Countries" of the far Northwest, Pond's map represented several geographical concepts, some of them previously mentioned by Henry. It showed Lake Athabasca and Great Slave Lake feeding a river flowing north where it emptied into the "Mer du Nord West." At first glance, this was hint of the Arctic Ocean, but its positioning in juxtaposition with the Great Lakes put it at a latitude too far south of the pole to represent that body of water. The French nomenclature is probative because it evoked that nation's long cartographic association with fanciful notions about the Northwest Passage, including transitional inland seas from which one could navigate to and from the Pacific. Another river, the "Naberkistigan," from a narrow range of mountains that ran transversely from the Arctic to New Mexico, emptied into the "South Sea," i.e., the Pacific, near Cape Mendocino and the 42nd parallel. This was a long-resonant depiction, lasting until Thomas Jefferson's time as president and the origin of the Lewis and Clark Expedition, showing the headwaters of the Great River of the West theoretically accessible from the continental interior after crossing over a single chain of mountains.[15]

Pond's strategic purpose with the Congress Map seems to have been the captivation of an audience in New York. As a cartographic image, it projects a line running diagonally across the continent via lakes and rivers from (using modern geographical terminology) Duluth on Lake Superior to Tuktoyaktuk, Northwest Territories, straight through his rich fur-bearing district of the Athabasca. The visual effect was unmistakable: Pond's Athabascan system was the means by which the young United States could hem in the British, leaving them bound to the increasingly trapped-out basin of Hudson Bay. The far Northwestern interior, all the way to the Arctic, could be dominated by interests in New York through intermediating entrepôts in Detroit and Michilimackinac.

Pond's depiction of the Pacific coast gives no hint of any awareness of Cook's last voyage, a function of its preparation in the wilderness. Though his image was insightful, in another sense it was already archaic because of Cook's charts. While Pond was in the Athabasca country, the unauthorized accounts of Cook's last voyage by William Ellis (1782) and John Ledyard (1783) had been published, copies of which may even have been in circulation at Grand Portage when Pond emerged from the backcountry. At the time of his presentation to Congress, it was almost a year since the Admiralty published the official account. At home in Connecticut or in New York, Pond probably would have now had a chance to read James King's seismic prospectus on the maritime trade, and time to study Cook's maps. So informed, as David Chapin observes, Pond began "to modify his own theories in response."[16] Cook's third voyage was central to Pond's subsequent cartographic projections of North America.

Pond's incipient proposal to Congress was ahead of its time. The new country's business and political leaders were overwhelmed by their fresh responsibilities, and its most visionary figure, Thomas Jefferson, was on assignment in Paris (fashioning John Ledyard's quest). So, with no reciprocal interest in his concept, Pond turned to Quebec. A month after leaving New York, he was back in association with some of his former partners. Their new gambit was an attempt to fend off their mutual nemesis, the Hudson's Bay Company, with a proposal to the colonial government for a ten-year trade monopoly in the rich Athabasca district that Nor' Westers like Henry and Pond had opened up over a decade earlier. In April 1785, as a part of a larger North West Company delegation, Pond met with the lieutenant governor of Canada, Henry Hamilton, the second-highest authority in the colony, and presented a new chart. (See Figure 1.3) Chapin asserts that "Pond and his map were brought in as evidence of the importance of the company to the well-being of the Empire."[17]

The revised map that Pond showed Hamilton was one clearly informed by Cook because it shows King George's Sound (Nootka Sound), Prince William Sound, and "Cook's River" on the Pacific coastline. But there was another important evolution from the Congress Map that Cook also influenced. The river draining Lake Athabasca and Great Slave Lake now emptied into a supposed "Ice Sea" close to the Arctic Circle, this in lieu of the classic French idea of the *Mer de l'Ouest,* adapted by Pond to Mer du Nord West in the chart presented to Congress. Curiously, its mouth was seven degrees of latitude south of where Hearne placed the outfall of the Coppermine River. The "Naberkistagan River" flowing into the North Pacific, for which Cook had seen no evidence, disappeared altogether. But the "River of Peace" that empties into Great Slave Lake more clearly points to its source in the western mountains, now depicted as several ranges deep. However, south of the 52nd parallel, it retains a single ridge line, an enduring concept that in time proved so attractive to Thomas Jefferson's imagination. It seems, then, that Cook had the effect of convincing Pond that the continent's coastline was difficult to penetrate. At the same time, however, Pond bent several

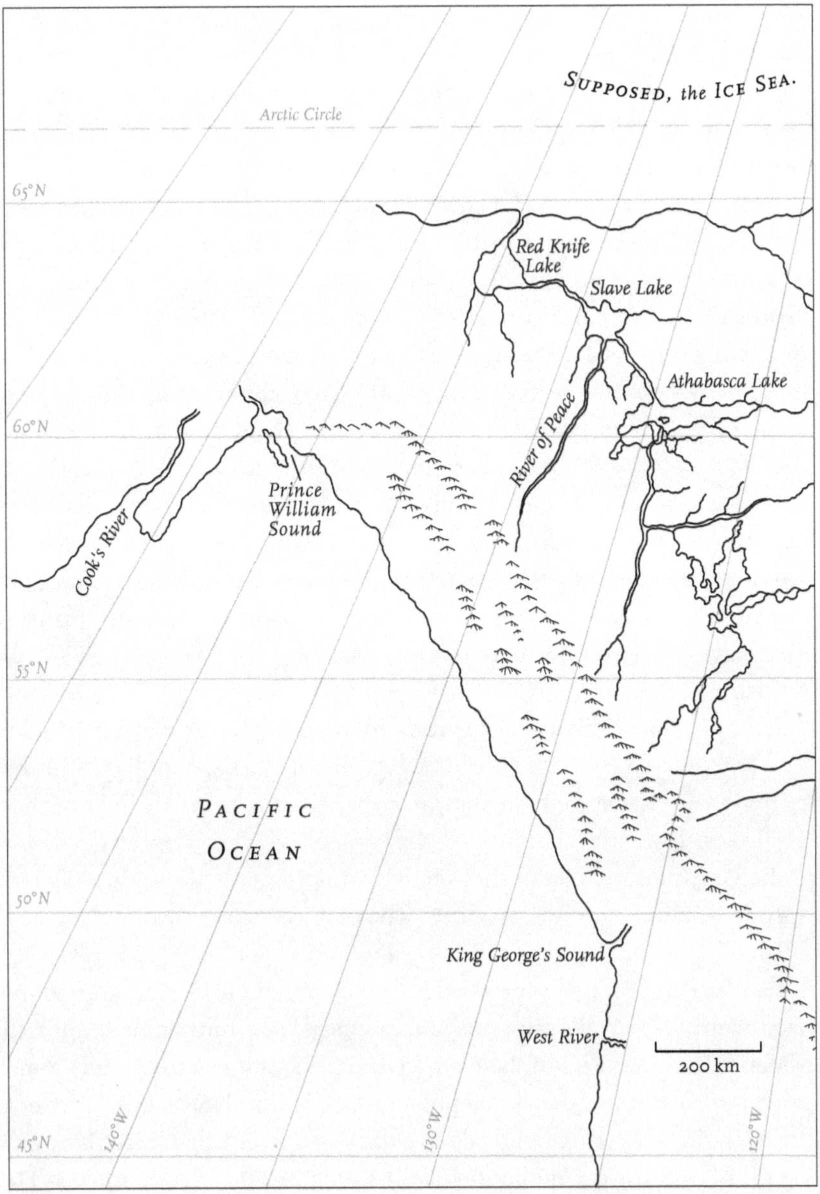

Figure 1.3 Peter Pond's "Hamilton Map," 1785. As redrawn here, Pond's revised conception of the continent west of Hudson Bay drew upon his cognizance of Cook's account and charts, showing both "King George's Sound" (Nootka Sound), Prince William Sound, and "Cook's River." Waters from Lake Athabasca and Great Slave Lake now reach the "Ice Sea" that Cook encountered instead of the legendary "Mer du Nord West" shown on the Congress Map. | Based on Derek Hayes, *First Crossing*, 56–57; redrawn by Eric Leinberger

elements of his Athabascan river system closer to Cook's newly defined Pacific coastline. On Pond's second map, "Cook's River" was aimed north into the interior of the continent far deeper than Cook showed it, while at the same time, per the insight of his mentor, Alexander Henry, the headwaters of the Peace River were depicted tantalizingly close to the Pacific.

Like the Congress Map before it, the "Hamilton Map" had its own strategic vision to convey, to wit: there may be discoverable connections to Cook's three main coastal harbours by way of the Athabascan system Pond had discovered. Again, Barry Gough explains: Pond "had unlocked an essential secret of western waters and drainage basins. He had revealed a vast interior space, one with a door to the Arctic in the north, the Coppermine River in the northeast, the Churchill River system in the east-northeast, and possibly some form of communication west." Thereby, Pond also provided the first glimmer of an eventual successor to the classic Northwest Passage that Cook sought: a system of interlocking rivers and lakes that could carry commerce from the Atlantic to the Pacific and vice versa. This cartographic image would be popularized by his protege Alexander Mackenzie and pursued in turn by Meriwether Lewis and William Clark.[18]

Nothing spurs activity more promptly than the threat that others may seize an opportunity before it can be capitalized upon. Accordingly, in the written memorial accompanying the chart, Pond apprised Hamilton that he "had positive information from the Natives, who had been to the Coast of the North Pacific Ocean that there is a trading post already established by the Russians." This was not news: Cook's third-voyage account had long since communicated that fact. But the implication behind the word "Coast" is that the Russians had moved east of the point where Cook had encountered them. In fact, the Russians had extended their fur trading empire to Kodiak Island in 1784, but there is a legitimate question as to whether Pond could have learned of this development so soon. But the Russians were not the main threat because Pond added what he hoped might clinch the argument in favour of the monopoly grant to the North West Company. He asserted that there was credible information "that Ships are now setting out from the United States of America, under the Command of Experienced Sea-Men who accompanied Captain Cook in his last Voyage in order to establish a furs trade upon the North West Coast of North America, at or near to Prince William's Sound."[19] Though the language is overly expansive in the pluralized "Sea-men," this must have been a garbled reference to the scheme John Ledyard was trying to put together, news of which Pond would have picked up during his visit home in Connecticut and around Congress in New York. Robert Gray, the first American to look for furs on the Northwest Coast, was not a Cook man and did not leave New England until two years later.

The Nor'Westers did not succeed in gaining their charter but Pond emerged from the meeting with a burnished reputation as a practical geographer. Hamilton wrote Home Secretary Lord Sydney in the wake of the meeting. After allowing that Pond was "an American by birth," he thought the results of his "active mind" might "prove of infinite utility to this country, consequently of great importance to the parent state," Great Britain. This was important because the British, he felt, needed a presence on the Pacific shore to fend off the "encroaching spirit of our neighbors," i.e., the Americans, echoing the hint that Pond had laid before him about the scheme Ledyard was propagating. Given Pond's dalliance with Congress, Hamilton was eager to keep him attached "to British interest, a want of it may drive him to court employment elsewhere."[20]

As it turned out, Pond's American flirtation was short-lived. He returned to the far northwest of Canada in the summer of 1785 for three years' worth of trading and exploration, near the end of which sojourn he made the acquaintance of a new trader. As a consequence of the North West Company reorganization in 1787, Alexander Mackenzie, a recent immigrant from Scotland, was sent to Lake Athabasca to become Pond's deputy. Pond tutored Mackenzie in the intricacies of northern geography during the course of their one year together in-country before they parted company. Mackenzie would in many respects realize the promise that Alexander Henry the elder and Pond had envisioned, though he was stinting in his recognition of their contributions to his accomplishments and attendant fame. But that lies ahead. Our immediate interest is focused on the third map that Pond prepared, which, like the first one, was also completed in the field, most likely in July 1787 (see Figure 1.4).

That summer, Pond and his *engagés* left the fort on the Athabasca River and after reaching the western end of Lake Athabasca they entered the Slave River. On their downstream course to Great Slave Lake, they crossed the mouth of the Peace River where it flows into the Slave, eventually reaching Great Slave Lake itself, a body of water larger than either Lake Erie or Lake Ontario but little known today because of its remoteness. Pond may have travelled as far east as Yellowknife Bay.

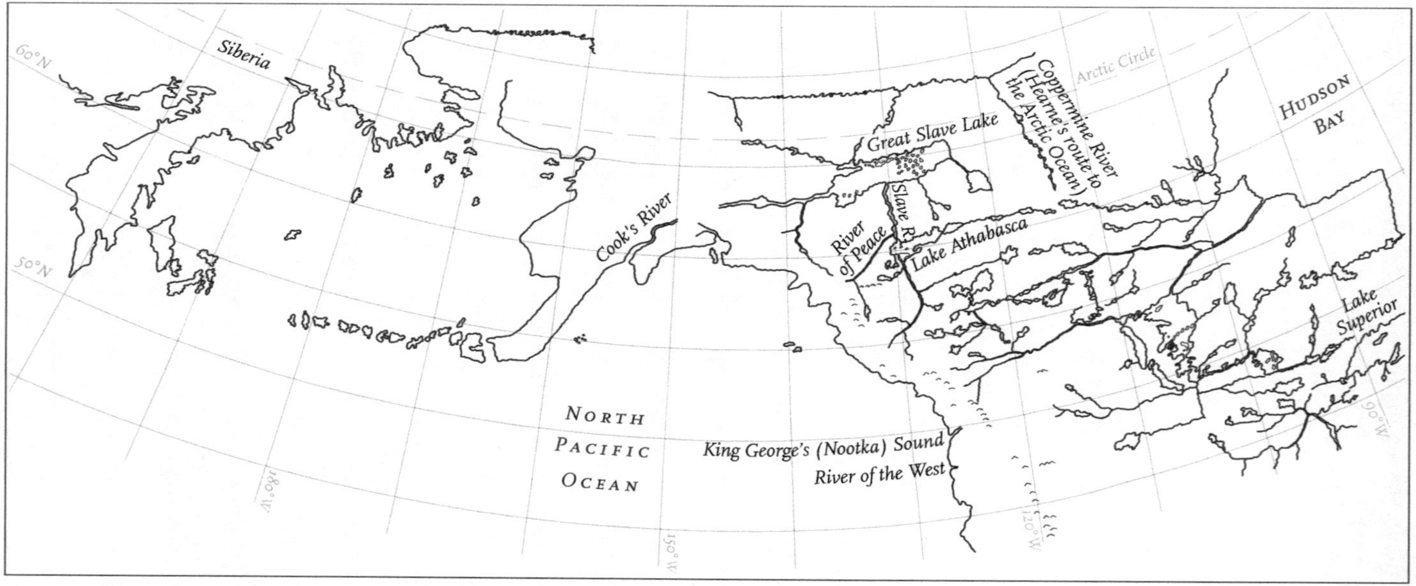

Figure 1.4 Peter Pond's "Empress Catherine Map," 1787, so called because he gave a version to Alexander Mackenzie to present to the ruler of Russia if his fellow trader got that far. This image, redrawn from a copy of the original chart that Pond shared with Mackenzie, depicts the continued evolution of his theory for an inland Northwest Passage. It shows the continued relevance of Cook's third-voyage cartography, including a new addition not found in his previous charts – an outlet from Great Slave Lake heading toward, but not connected to, "Cook's River." | Based on Chapin, *Freshwater Passages*, 280, and Hayes, *First Crossing*, 64; redrawn by Eric Leinberger

On the basis of this and his previous journeys, plus the continuous procurement of Indigenous intelligence about the hydrology of their homeland, Pond refined his cartographic projection of the region. The most significant evolution depicted in the 1787 version was the unusual idea, evoking the spirit of the Admiral de Fonte fable, that Great Slave Lake had three outlets. Two of them, reflecting his original presentation to Congress, flowed respectively to the northwest and northeast and emptied into "water that ebbs and flows" upon which the "Eskimaux" lived. This clearly was an allusion to the Arctic. The main outlet, however, flowed west in the unmistakable direction and approximate latitude of "Cook's River." David Chapin reckons that "no one looking at the map could fail to conclude that they were the same" concept. Notwithstanding "a great Waterfall" (also in the de Fonte tradition), Pond's insinuation about the interior origins of "Cook's River" conveniently passed to the north of the chain of mountains extending north from Mexico, terminating at latitude 62°30′N. Pond was hedging his topographic bets. His third projection clearly conveyed the promise of a tie-in to one of Cook's major Pacific coast inlets while still, ever so deftly, preserving the potential of one or more outlets for the Athabascan system of lakes and rivers into the Arctic.[21]

Though Pond's third map utilizes place names derived from the great navigator, his definition of the coastline for the Alaskan subcontinent echoed archaic Russian projections more than it did Cook's, though this may have been a function of Pond's rudimentary map-making skills. Cartographic deficiency is also evident in the compression of the longitudinal spacing between the actual location of Great Slave Lake and Cook Inlet, amounting to seventeen degrees of longitude or nearly 700 miles. Here another explanation applies, one that resonated throughout the long history of the Northwest Passage both before and after Cook, though excluding the great navigator himself – the geography of hope, or, to be less kind, wishful thinking. On the map presented to Hamilton in the spring of 1785, Pond was content to use Cook's west coast harbours as markers, as a mere intimation of possibility. The 1787 version now tilted in favour of probability. Pond believed he was merely six days of travel from his Athabasca post to the outfall of the "river" Cook spied in an inlet that others later named for him.

By the spring of 1788, Pond had spent almost three full years at his post in the Athabasca country building up the trade – a long time for such duty. He was now forty-eight years old and his continuously problematic relations with partners in the trade, including allegations that he was responsible for the murder of a colleague, and possibly a second one, complicated his stay in the interior. And so he withdrew, turning over matters to a younger generation, whose principal figure was Alexander Mackenzie. Before leaving, Pond gave Mackenzie one of the four copies of his third map known to exist. But that was not the most remarkable element of this transition, for Pond also imparted to Mackenzie a Promethean vision that was remarkable in its coincidental imitation of John Ledyard's conversation with Jefferson. Since setting out from Quebec in the spring of 1785, Pond had cultivated the dream that it was theoretically possible to travel across Canada to the Athabasca country, thence to the mouth of "Cook's River," and then, once linkage was made with merchants from Russia, sail to Kamchatka. Going in the opposite direction to what Ledyard was envisaging at exactly the same time, he merely had to cross Siberia to reach St. Petersburg.

Pond never set his reverie in motion, but he fired Mackenzie's imagination and largely formed the Scotsman's subsequent plan of action. To others, Mackenzie may have mocked Pond's idea of laying his chart "before the Empress of Russia" as an "incomprehensibly extravagant notion," but in his practical field operations he followed Pond's formula. This included taking a copy of Pond's third map and a letter of introduction to any Russian fur traders he might meet so that at least these two documents, if not the circumambulator in person, could wend their way to Petropavlovsk if not St. Petersburg. In the event, Mackenzie's 1789 expedition, borne along by the river that now carries his name, took him from Lake Athabasca to the Arctic, not Cook Inlet. Even though his first voyage carried him to the wrong ocean, four years later, on the eve of his departure for the Pacific via the Peace River that Pond had first charted, the Scotsman told his cousin Roderic that he took some money so as to "traffick with the Russians." In 1793, Mackenzie didn't make it to Cook Inlet or Unalaska either, coming out further to the south, near Bella Coola on British Columbia's central coast. But the larger point of Pond's plan, and Mackenzie's attempt to implement it, is that Cook had so stirred the imagination of his generation that, if the great navigator had done all that was possible at sea, others wanted to boldly carry the banner as far as they could by land. Another fur man, Sir George Simpson, near the end of his tenure as governor of North American operations for the Hudson's Bay Company, would end up being the first person to circumnavigate the world overland, in 1841–42.[22]

Pond left his fort on the Athabasca River on May 15, 1788, hoping to renegotiate the terms of his partnership in the North West Company. It was his intention to return to Athabasca country but he never did. Pond permanently severed his relationship with the firm in 1790. His whereabouts during the interim are largely unknown. Pond may have spent time in the Great Lakes region, or possibly he returned to Connecticut. We do know that he was in Montreal in late October or early November 1789, and there discussed his most recent three years of northwestern travel with two colonial figures, John Mervin Nooth and Isaac Ogden. The former, a physician and a member of the Royal Society then serving in Canada as an army doctor, was a frequent correspondent with Joseph Banks on matters botanical, zoological, geological, and geographical. Ogden was a judge of the Admiralty court in Quebec (and was also the father of a famous explorer for the HBC's Columbia River district, Peter Skene Ogden). Pond's conversations with these men were important because their gist was conveyed to the most prominent geographical circle in Great Britain, landing there at precisely the same time the wave of new information derived from the Northwest Coast maritime fur trade did.

Isaac Ogden was a guest at Montreal's famous Beaver Club, where he encountered Pond and saw the latest version of the fur trader's map. On November 7, 1789, he wrote to his father, David Ogden, also a judge but living in London. He related what he had learned from his conversation with Pond. This information, he wrote, was "exclusive of the Map itself, which I could not get a Copy of." As conveyed to the elder Ogden, Pond asserted that out of the Great Slave Lake "runs a very large River, which runs almost South West, and has the largest falls on it in the known World, it is as least two miles wide where the Falls are, and an amazing Body of Water." This was pure fantasy of the de Fonte variety (as Mackenzie would prove). The more important point, which Ogden related with the confidence Pond had instilled, was that "there can be little doubt, but that the Source of Cook's River is now fully discovered and known." Several supposed proofs were offered, the most salient being Pond's claim that in 1787 he had seen two Natives on the shore of Great Slave Lake "who came (as they said) up a River from the Northern Pacific Ocean" and presented him a blanket "which they received from Vessels which were at the Mouth of the River; they said that the River he was in is large to the place of Discharge and Navigable." There is no reason to doubt that Pond had actually met Indigenous People in possession of trade items, but such artifacts may have come inland from any number of places, such as Hudson Bay, or from the Pacific Ocean by way of the Stikine River. This stream's headwaters are opposite the upper reaches of the Liard River, which itself joins the Mackenzie just downstream from where it flows out of Great Slave Lake.[23]

Ogden's narrative explains why Pond's third map depicts a prodigious western flow from Great Slave Lake versus the trickle of the two outfalls into the Arctic. According to Ogden, Pond never asserted reaching to the far "north ... where the Waters ebb and flow," but Indians he knew did in 1787, where "they met with & killed a number of Eskimaux." This was perhaps of little consequence in itself except, Ogden added, that those killed were the same people "to be found only on the Banks or Boundary of the [Arctic] Ocean from the Labrador Coast Northward, and they are found on the whole Extent of that Coast as far North as we have any Knowledge." Ogden concluded by noting that via this inland network of lakes and rivers, "an Advantageous Commerce may be carried" to "the mouth of Cook's River to be thence carried to China." In furtherance of that objective, Pond had left a man in the Canadian interior "with orders to go down the River and from thence to Unalaska and so to Kamskatska, and then to England through Russia." Ogden advised his father to expect to see this man, who went "by the Name of McKenzie," "next year," presuming "he meets with no Accident." This news was too exciting to keep private, so on January 23, 1790, the senior Ogden forwarded its contents to Alexander Dalrymple, Evan Nepean, the minister in charge of colonial affairs, and the redoubtable Joseph Banks, president of the Royal Society, who kept an abstract of it in his files.[24]

Nooth's letter, written the same week as Odgen's, was sent directly to Banks. His was a shorter missive, and the Northwest Passage was not the main focus. Nooth referred to Pond as "a singular person" who had been "some years in the western parts of America" and had recently presented a map (a copy of his third one, dated December 6, 1787, drawn while he was still in the wilderness) to Hamilton's colonial successor, Lord Dorchester. Nooth averred that Pond "positively asserts that he had discover'd an immense Lake nearly equal to Great Britain that communicates in all probability with Cooks River" or Prince William Sound. To cement Pond's notion, the Nooth letter centred on the same trade blanket story that captured Isaac Ogden's attention: "In the River which was form'd by the Water that was discharg'd from this Lake [Great Slave Lake] he met with Indians that had undoubtedly seen Cooks Ships and who had with them a variety of European Articles evidently of English manufacture." Nooth also viewed the latest version of Pond's map, "in which the Latitude is well laid down." Pond disclosed, or Nooth intuited, that the fur trader "was very capable of ascertaining that circumstance" but "the longitude seems to be guess-work and not in any respect accurate enough to be depended on."[25]

Unlike Nooth (or the Odgens), Banks had heard of Pond earlier. When he received Ogden's letter, it prompted a memory that he recorded on his copy of the correspondence, to wit:

> A Mr. Pond set out from Quebeck in the Spring of 1785 Alone with the intention of Crossing America Westward & returning [to Europe] by Siberia, he had before traveld till he met the Tide in a river which ran to the west & supposes he then was within three days Journey of the Sea in the Neighbourhood of Jesuits Harbor.

This report, Banks noted, came to him from a "Capt. Bentincks ... when at Revesby 1788." (Revesby was Banks's country estate in Lincolnshire, and he divided his time between it and his home in London.) Captain (later Admiral) William Bentinck (1764–1813) had served in the American War for Independence and later in Canada. The most relevant issue is not Bentinck's respect for Pond's ambition but the relative importance he placed on the story of the European goods he saw in Indigenous hands at Great Slave Lake. We know that Cook and his men did not use blankets in trade, and it is unlikely that the first wave of maritime fur traders in those waters did. (Indigenous trading patterns, we now know, were truly transcontinental in nature, and far more extensive than Pond's generation believed possible. For example, Lewis and Clark were surprised to see how far and how quickly manufactured trade goods could cross the continent.) The more important point is how this intelligence was interpreted in England. Chapin writes: "Here was an appealing notion that seemed to tie together two disparate strands of British exploration: Native people meeting Cook on the shores of the Pacific, and a few years later meeting Pond at Great Slave Lake."[26]

If Pond's supposed link to Cook on the Pacific were not already provocative enough, he propagated yet another beguiling idea. It was self-evident to those who spoke to Pond, or read of his theories through intermediaries, that he was well travelled in the same latitudes that Cook had been sent to explore from a seaborne vantage. Yellowknife, Northwest Territories, on the north shore of Great Slave Lake is just above the 62nd parallel, roughly equivalent to the latitude of Norton Sound off the Alaskan subcontinent. Pond's map, and what we glean indirectly through Judge Isaac Ogden's correspondence, indicates that Pond thought the Arctic coast lay in navigable latitudes. This, of course, is equivalent to what Hearne had found nearly two decades earlier, which was in itself sufficient to guide Cook's exploratory parameters. But in his letter to his father in Great Britain, the younger Ogden added: "My Informant [Pond] tells me, that if he [Cook] had been a month later, he would have met with no obstruction from Ice in that Quarter." Modern understanding of polar climatology confirms what Pond's Dene guides, on the basis of their war parties against the Arctic Inuit, said: the maximum recession of the northern ice pack is in September. The upshot, Ogden related, is that if Cook had simply been able to get past Alaska's Icy Cape (named by the great navigator), he "might have passed the Northern Coast of America, and have returned to Europe by that Rout." According to Ogden, Pond believed that "on the North Western Coast of America there is a large or long Point of Land that extends to the Lat. of 71 or further, and then the Coast trends South Easterly" to such an extent that the Arctic eventually "washes the Land in Lat. of 68 1/2."[27]

Since the parties to this discourse presumed no European had yet visited the western portion of the Arctic coast, this was merely interesting speculation. But unbeknownst to Pond, Ogden, Banks, or anyone living in Quebec or Great Britain in the winter of 1789–90, Alexander Mackenzie had reached the Arctic on July 14, 1789, news of which took time to reach the Atlantic seaboard of North America or Europe. Nevertheless, Pond's prefiguration of the "long Point" and the southeasterly trend of the Arctic coastline east of its northernmost point, and the Indigenous reckonings it was based upon, proved true. During the nineteenth-century wave of maritime Arctic exploration, Alaska's northernmost projection, Point Barrow, was discerned at latitude 71°23′N, as forecast by Pond. Hearne had reached the northern sea where the Coppermine River falls in, far to the east of the Mackenzie River at a latitude he mistakenly calculated to be just shy of 72°N, later determined to be four degrees of latitude too high. Thus Pond correctly comprehended from his own experience and Indigenous intermediaries (or simply guessed correctly) that in places the Arctic coast actually ran lower than 68°N and that the principal impediment to reaching Hearne's coast from the Pacific was the northern shore of Alaska where Cook hit the Arctic ice pack.

Of course, the Boothia and Melville peninsular extensions north of Hudson Bay, to say nothing of the extensive complication of the northern Canadian Arctic Archipelago, would have remained obstacles for Cook, just as they were for another hundred years and more until Amundsen made it through early in the twentieth century. However, the important point to consider within the Cook context is that although Pond seemed to have reopened the practicability of an ocean-to-ocean Northwest Passage, this prospect – the original vision of the Northwest Passage – was not an idea that captivated the minds of geographers in the late 1780s. The notion that Pond had found a freshwater route to the Pacific was far more appealing. Barry Gough asserts that Pond's discoveries in the Canadian interior "again raised new hopes of the Northwest Passage."[28] True, but this was not the passage Cook was sent to find, demonstrating how thoroughly Cook had demolished the classic formulation of the Northwest Passage. As an *idée fixe,* the fabled route from the Atlantic to the Pacific was evolving.

Thus, with the convergence of two streams of geographical information, we see that Pond's continental reports trickled back to attentive audiences only a year or two after those from Dixon, Barkley, Duncan, and other seafarers. Fertile minds in Great Britain were well stimulated by the end of 1789. For example, Richard Cadman Etches, the managing partner of the King George's Sound Company, which sponsored the Portlock and Dixon trading voyage, along with the peripatetic Joseph Banks, floated a scheme in London to emulate the penal colony plan that had already been launched toward Australia's Botany Bay. In their vision, a critical mass of felons intermixed with some soldiers would sail in a Royal Naval vessel to the Northwest Coast and after the human cargo was landed at Nootka Sound, a more detailed survey might commence from there to Cook Inlet. Captain Dixon endorsed this idea in a July 1789 letter sent to Evan Nepean, the British undersecretary handling colonial affairs, arguing for an expedition to the northwest that would promote both trade and settlement. It was in this fashion, Glyn Williams notes, that "casual observations along the coast raised the first doubts about Cook's conclusions."[29] Both from the sea and from Pond's inland vantage, the Northwest Passage was taking new forms.

2

Dalrymple's Revenge?

The seminal characters in the culmination of what became known as the Vancouver expedition were geographer Alexander Dalrymple and maritime fur trader John Meares. The former is better known for his role in the Captain Cook story. He had been proposed by the Royal Society to lead the voyage to the South Pacific to observe the transit of Venus, an assignment that went to Cook. Not only did Cook supersede Dalrymple as expeditionary leader but on that voyage and the one that followed, he routinely demolished Dalrymple's speculative geographical theories, principally the existence of *Terra Australis Incognita.* For that reason, Dalrymple is conventionally interpreted in Cook historiography as a caricature – a proponent of half-baked geographical nostrums who had no real reason to presume he was capable of commanding the *Endeavour* voyage – and Cook's adversary. As I demonstrated in *Captain Cook Rediscovered,* Dalrymple's candidacy for leadership of the *Endeavour* was not as preposterous as is commonly supposed. Even if he was not as skillful as Cook proved to be, Dalrymple knew his way around ships, distant seas, and exotic harbours. And he had an inquisitive mind. When James King returned to London at the conclusion of the third voyage, one of the first to debrief him was Dalrymple – a conversation that no doubt focused on points of both geography and commerce. The fur trade that Cook's crew had stumbled onto and which King later wrote about at length was of vital concern to Dalrymple's employer, the East India Company (EIC).

Yes, Dalrymple was the principal promoter of the lost southern continent in the English language, but it is equally true that Cook was an equal-opportunity destroyer of many geographical myths. Historians routinely postulate that when the fur traders began to poke holes in Cook's simplistic depiction of the mid-latitudes of the Northwest Coast, Dalrymple seized an opportunity to gain a measure of revenge against his supposed rival. Dalrymple would call for a reopening of the geography in question, but in describing this turn of events, some Cook scholars now shift their interpretive allegiance to Dalrymple as a function of their commitment to the underlying paradigm touting the great navigator's (supposedly) lacklustre performance during his third and final voyage. Having joined in the joke at Dalrymple's expense about *Terra Australis Incognita,* historians have tended to switch sides when it comes to Cook's purported failure to find a Northwest Passage, rallying to Dalrymple's cause.

Dalrymple's argument about a revived Northwest Passage was far more nuanced in its criticism of Cook than is usually understood to be the case. Fundamentally, it is not critical of Cook at all. It is true that in February 1790, in the heat of the renewed debate about the Northwest Passage in the run-up to the Admiralty's commissioning of a new voyage of discovery to the Pacific Northwest, Dalrymple famously averred in *private* correspondence with Evan Nepean, in reference to Cook, that "I cannot admit of a Pope in Geography or Navigation."[1] But in his far more widely read pamphlet published in the spring of 1789, when and where he began his public advocacy for a follow-up investigation, Dalrymple *never* suggested that what Cook was sent to find – the classic formulation of the Northwest Passage as a saline corridor from the Atlantic to the Pacific – existed in the latitudes where the maritime fur traders found numerous inlets.

Nonetheless, as historian Howard Fry relates, these reports made it imperative, to Dalrymple's way of thinking, "that a new close survey of the northwest coast should be undertaken to discover whether any germ of truth might be found to underlie the legends of de Fuca and de Fonte, a matter not clarified by Cook's last voyage."[2] Now late in his career, with his reputation still intact, holding the title of Examiner of Sea Journals for the EIC, Dalrymple was in a position to act on the matter. He had access to the map collections and reports of both the Admiralty and the Hudson's Bay Company (HBC), including the data provided by their fur trade explorer Samuel Hearne in the early 1770s.

Dalrymple was a voluble pamphleteer on matters of geography and that discipline's intersection with commerce and international affairs, publishing three such items over the course of 1789–90. His thinking on the Northwest Passage and Cook is not to be found in snide asides in private correspondence, but in his comprehensive assessment, which was titled *Plan for Promoting the Fur-Trade, and Securing It to This Country, by Uniting the Operations of the East-India and Hudson's Bay Companys* [sic]. Dalrymple's essay (which he started writing in January 1789 and published that March) stands in stark contrast to the breezy optimism of Daines Barrington's Northwest Passage manifesto, *The Probability of Reaching the North Pole, Discussed,* published in 1775 on the cusp of Cook's last voyage. Nevertheless, Dalrymple's title is misleading. Though he had a long association with the EIC, and was profuse in his thanks to the HBC "for their very liberal communication of the many *Surveys* and Observations that have been made at their expence," the tenor and extent of his remarks show a far greater interest in geographical problem solving than in commercial concerns.[3]

The interests of the royally chartered companies and their trade monopolies were under siege from independent adventurers coming to the Northwest Coast from ports in China, or heading there overland from Montreal. Dalrymple, Fry notes, was troubled by "the influential new economic school of laissez-faire free trade" propagated by yet another major Enlightenment figure, Adam Smith. Dalrymple, Fry asserts, "was invariably a doughty champion of the East India Company's monopoly," but he sensed that the HBC in particular might be damaged by the growth of free trade. But commercial considerations were mere pretext for what truly interested him: geography. Of his *Plan*'s thirty-four pages of text, the Northwest Passage puzzle dominates the first twenty-eight. Only the last six deal with the dynamics of the terrestrial and maritime fur trades and his fear that Spanish and Russian expansion, coming from opposite directions, could close the Northwest Coast to English vessels. A further complication was the fear that the recently independent Americans might jump in too.[4]

Dalrymple believed the best way to avoid the consequences of this maritime pincer movement was for the HBC and EIC to combine forces. A joint venture between the royally chartered companies would also help fend off the growing number of independent British merchants whose ventures were testing the limits of their monopolies. Nominally, the South Sea Company was to control commerce on the west coast of North America, but as a practical matter it was the HBC that had an established trade in furs. The principal market for these pelts was in China, part of the EIC's zone of control. Enterprising independents who wanted to sell their furs in Macao or Canton had to secure an EIC licence, which is what the Etches concern did. Most, however, chose to engage in the subterfuge of flying foreign colours.

Dalrymple paid close attention to the interesting news about the Northwest Coast trickling back to London during the late 1780s from the likes of Nathaniel Portlock and George Dixon, and from what his company friends in Canton could relay. He folded these reports into his intimate knowledge of Cook's third voyage, the account of which he helped prepare for publication. As Fry notes, it became evident to Dalrymple "that the coast of America differed materially from that described by Captain Cook." Even before the *Plan* was issued, he published charts that displayed the new geographical insights secured from fur trade sources, the most important of which was Charles Duncan's survey of Juan de Fuca's strait after he learned about it from William Barkley. Indeed, it was Barkley's discovery of a strait reasonably close to where long-distant chroniclers had placed de Fuca's supposed opening into the continent that prompted Dalrymple to shift from merely collating reports into taking up his pen as an advocate for a new round of exploration in the Pacific Northwest.

Cook had been so utterly dismissive of this prospect that the actual hint of such an opening, coming in the following decade, made him an easy target. This is where Dalrymple's private dismissal of Cook as a "Pope" establishing cartographic doctrine, a snarky aside privileged by scholars in Cook historiography, applies. A close reading of Dalrymple's public text in his *Plan,* on the other hand, reveals a range of geographical thinking that has been overlooked by historians to the detriment of Dalrymple's reputation, and, ironically, Cook's as well. Cook may have rejected his hypotheses about *Terra Australis Incognita* on the basis of practical experiment at sea but, as Barbara Belyea observes, the penchant of modern historians to resort to "*ad hominem* arguments" to discredit Dalrymple's speculative hypotheses obscures his "essential contribution to late eighteenth-century exploration."[5]

Dalrymple began his treatise with a key line, almost entirely overlooked by scholars, that correlates directly with the main theme of this book. He stipulated that the very idea of a Northwest Passage "has been much misconceived." Elaborating on this point, he asserted that the meaning of the term had changed over the centuries, including his own thinking on the matter "in modern Times." Dalrymple allowed that the early Renaissance idea of the Strait of Anián, facilitating a passage from the Atlantic to the Pacific, was a hazy prefiguration of what was later discerned as the Bering Strait. This "*alledged Passage*" was thought to exist "on the *North* of AMERICA, by what may be called the *Hyperborean Sea*" (i.e., the far northern, or Arctic Ocean). This was the classic version of the Northwest Passage that Cook sought. However, in Dalrymple's new post-Cook idiom, he envisioned a "NW Passage, *through* AMERICA." This is an important point worth dwelling on for two reasons. First, it shows how the idea of the passage had long been – and even into the post-Cook era was still – a malleable concept. Second, because Dalrymple's pamphlet was the seedbed of Vancouver's forthcoming instructions, it is vital to understand its actual argumentative thrust and geographical framework. Contrary to conventional understanding, it was *not* an indictment of Cook's third-voyage effort. Indeed, Dalrymple exculpated Cook, observing that even though "Capt. *Cook* and his Successor [Charles Clerke], were obstructed by Ice," they were merely the victims of "*unlucky* seasons." Here Dalrymple adheres to an old theory that if an explorer found the right opening in the right year, a successful transit of the "Hyperborean" route was not only possible but likely.[6]

To further his argument, Dalrymple invoked another hoary geographical tradition, the equipoise principle. This theory had also served as the foundation for the premise of the great southern continent; indeed, Dalrymple himself popularized the idea in his writing before Cook's *Endeavour* voyage. Even though Cook had demolished that prospect during his first two expeditions, Dalrymple resurrected the equipoise premise for his new Northwest Passage. On the basis of HBC reports and charts, Dalrymple determined that the presumed coastline of "the NW part of *Hudson's Bay*" was actually a line of "*broken Islands.*" The lands bordering the west side of Foxe Basin north of Hudson Bay are indeed a complex geography that was not yet fully discerned in Dalrymple's time. From this murk, Dalrymple conveniently posited an analogue to what the maritime fur traders following James King to the Northwest Coast were beginning to discern upon closer inspection of a shore that Cook only glanced at and assumed to be a solid continental coastline. In the absence of an exhaustive survey of the Foxe Basin and adjoining inlets, which the HBC was only intermittently committed to for competitive reasons, Dalrymple suggested there were some still "*undiscovered Channels*" leading westward.[7]

Dalrymple's principal interest in the *Plan* was the commissioning of a comprehensive expedition to the Pacific Northwest, but his discussion of Hudson Bay geography was sufficient to force the HBC to eventually underwrite the voyage to those waters by Charles Duncan, late of the King George's Sound Company and commander of the *Princess Royal.* Duncan was supposed to leave England in 1790 but did not depart until May 1791, reaching Fort Churchill on the west coast of Hudson Bay aboard the *Beaver* in July. His cursory exploration along the west coast of the bay and briefly into Chesterfield Inlet failed to find one of Dalrymple's "channels."

Dalrymple's larger argument, however, was actually more complex than the historiography of Cook and the Northwest Passage normally allows. A forthright statement, universally overlooked by historians who prefer to use Dalrymple's privately expressed criticism of the (presumed) laxity of Cook, ensues: "What I have here said, must not be misunderstood to imply, that I *believe,* or *even suppose,* there is a *Sea-Communication* from *Hudson's-Bay* to the *Pacifick Ocean.*"[8]

Here it must be noted that in addition to exaggerating Dalrymple's criticism of Cook, historians have also overlooked the speculative geographer's assessments of Samuel Hearne. Gordon Speck writes that Hearne's position on the practicability of a Northwest Passage was not dispositive because he was "only a commercial employee and his irrefutable proofs were not good enough for the deeply entrenched and conservative" British naval officialdom.[9] It is nonetheless true that after studying all the materials at hand, including the work of the HBC's crack surveyor Philip Turnor, Dalrymple correctly deduced that Hearne had placed the mouth of the Coppermine River too far north. Another factor Dalrymple used in this assessment was Hearne's reported sighting of grass further above a latitude (65°N) than it grew along Atlantic shores. Even though Dalrymple critiqued Hearne constructively, the HBC man's key insight was not the latitude of the Arctic coast (which Hearne had admitted was as estimate) but rather the width of North America. This more significant deduction, based on his own observations of the landscape, and, perhaps more importantly, by what his Indigenous informants (who had travelled much further west into the depths of the continent) told him, clearly indicated that there was no mid-latitude saltwater shortcut from the Pacific to Hudson or Baffin Bay. The Admiralty and Cook planned accordingly in 1776, determining not to begin a search for the Northwest Passage until his ships reached the high latitudes of the North Pacific.

Hearne was still alive, living in England, when Dalrymple published his *Plan.* He appears to have been inspired to publish his journal of the Coppermine River expedition in response to the criticism. Hearne was aided by William Wales, Cook's second-voyage astronomer, who had met Hearne at Fort Churchill during his observation of the transit of Venus in 1769, a few years before he trekked to the Arctic shore. Wales took exception to Dalrymple's dismissal of Hearne's skills and spruced up the text, especially the preface, which directly responded to Dalrymple's published criticism in the *Plan.* He also served as his agent in securing a publisher. Indeed, Wales was Hearne's literary executor as well. The latter died in November 1792 before his book was published.

Thus, the first two words in Hearne's book are "Mr. Dalrymple," whom he found "very particular in his observations of my Journey." Hearne/Wales defended the Coppermine River venture, noting that the expedition was not a purely scientific venture but a more practical exploration initiated by "my employers," who sought results "satisfactory to them" alone. Hearne never imagined that his trip "would ever come under the inspection of so ingenious and indefatigable a geographer as Mr. Dalrymple." Hearne/Wales took exception to Dalrymple's extrapolation from the grass line in Greenland, but the published journal's strongest point was the assertion that the Coppermine River expedition "put a final end to all disputes concerning a North West Passage through Hudson's Bay."[10]

Dalrymple was within his rights to correct Hearne's latitudinal calculation for the outfall of the Coppermine River into the Arctic, but in his *Plan,* he affirmed that there was "no circumstantial report" that could undo the HBC man's principal finding. Indeed, Dalrymple asserted that the Hearne dictum, which had guided Cook's course, still held. The fur trader's voyage to the Arctic was "*Proof… positive* … that there can be no *Sea-Communication* under 72° North Latitude." Consequently, Dalrymple began to deftly shift the focus of his Northwest Passage thesis from a saline "communication" to the very intriguing region of large freshwater lakes and rivers that laced the Canadian Northwest. He referred to the map of the "*Canadian Traders*" showing a drainage system with water flowing from "*Arathapescow Lake,*" through another river into Great Slave Lake, and ultimately into the "*Hyperborean Sea.*" This image is perfectly conformable to the map Peter Pond presented in Quebec in 1785 on behalf of the North West Company, a "Canadian" firm in contradistinction to the royally chartered Hudson's Bay Company.[11]

Dalrymple's application of the malleable linguistic term "communication" facilitated the evolution of the Northwest Passage as a literary and cartographic image. It allowed the ancient concept to survive the otherwise determinative conclusions derived from Hearne and Cook – the impracticability of a direct saltwater channel passage from one ocean to another. In his *Plan,* Dalrymple accelerated the process of shifting the passage from a navigable corridor to a membranous construct within which the eastern (Atlantic) and western (Pacific) segments "communicated" with each other in the interior of the continent.

This evolution was *anything but* a refutation of Cook's third voyage. Indeed, the opposite was true. Dalrymple relied upon the accuracy of Cook's cartography to establish the projected utility of his emerging scheme, the practicability of which came down to a measurement of the gap between saltwater segments as bridged by inland freshwater lakes, rivers, or portages. Dalrymple could rely on the latitude and longitude Wales established for Fort Churchill on the west coast of Hudson Bay when he was there to observe the transit. On the Pacific coast, Dalrymple had a choice between Cook's known stations at Nootka Sound and Cross Sound, the opening just south of Mount Fairweather in Southeast Alaska. He logically chose the latter as the analogue for Fort Churchill since both were approximately at 58°N. Starting with the meridians for these two places, fixed by the two most reliable authorities of that age, Dalrymple drew a line between them and calculated that they were separated by 1,300 miles. This "extent of the Land," Dalrymple added, "from *one Sea* to the *other,* is therefore now as well known, as any equivalent Space in Europe, or Asia."[12]

This was the gap that Dalrymple's *Plan* for a new "NW Passage, *through* AMERICA" had to span. Cook had already established the northwesterly trajectory of the Pacific coast as one proceeded from south to north. So, Dalrymple posited that by moving south from Cross Sound by five to seven degrees of latitude, the distance to Hudson Bay could be shortened by 100 miles. To narrow the gap further, Dalrymple relied on extrapolations from the elusive, ephemeral, sometimes fugitive, never entirely scientific discoveries, or mere notions, propagated by the fur traders who followed James King's call to the Northwest Coast. The report of George Dixon was especially appealing because, as Dalrymple phrased it, he was of the opinion "that all the *Lands,* yet discovered, on the *West Side* [of North America], at least from *Nootka* inclusively Northward, are *Islands,* and *not* the *Continent.*" Pushing his theory forward, Dalrymple argued that if openings such as

Dixon's in the vicinity of the Queen Charlotte Islands (Haida Gwai) were to prove extensive, "then, the distance between the *two Seas* will be still more reduced." A deep bay reaching eastward would be the most salutary outcome, but absent that, Dixon and James Hanna had detected "indications of a *large River*" that might serve as an effective substitute for access to the continental interior. Dalrymple reminded his readers that it was precisely in these latitudes (51°–55°N) that just such a feature had been placed by de Fonte. "This gives some countenance to that too hastily exploded Narration," he added plaintively, coming as close to an overt criticism of Cook (or La Pérouse, for that matter) as one can find in his treatise.[13]

No navigator (real or imagined), ever came in and out and back in favour with such rapidity as de Fonte, or "De Fonta," which is how Dalrymple preferred to spell his name. (Vancouver would later find the multiplicity of Fonte, Fonta, and even Des Fuentes an opportunity for derision.[14]) In Dalrymple's *Plan,* the mythical explorer enjoyed one last moment of currency before Vancouver banished him for all time. Dalrymple purported to document that there was a bona fide Portuguese navigator by that name who, on behalf of Spain, had explored the tip of South America. In itself, Dalrymple admitted, this was not proof of claims that de Fonte was supposed to have made on the west coast of North America in 1640, but it was a ledge, thin or otherwise, and he chose stand on it. The basis for Dalrymple's leap of faith was the large number of inlets that the fur traders found on their inspection of the shore, plus the profusion of driftwood in waters such as Fitzhugh Sound. This was taken as evidence of the river upon which de Fonte supposedly coursed eastward into the depths of the continent (see Figure 2.1). In Dalrymple's mind, the constellation of islands, inlets, bays, and possibly a river "strongly indicates the reality of *De Fonta's* account," which the *Plan* replicated again in all its bizarre detail. Explaining how de Fonte could have the time in one summer to sail around coastal islands through "*crooked channels*" up a river to a chain of freshwater lakes, separated by rapids but influenced by tides, through a strait where he met a ship sailing from New England commanded by a Captain "*Shapley,*" and back to his port of entry on the Pacific, all within twenty-nine days, Dalrymple announced glibly that "the length of the days at that season would leave but a short night."[15]

That truly would have been a fantastic journey, via features defying the laws of nature or common sense. Dalrymple conceded that some thought de Fonte's saga was a "*Forgery,*" not the least of them, by implication, Cook, but he chose "to add a few words" in defence of its resurrection. Dalrymple admitted that the de Fonte story originated in 1708 as a second-hand account decades after the purported voyage, and that it contained "many mistakes and inconsistencies." There being no original source document, he argued that many of the difficulties in the text were problems in translation from Spanish into English, or errors in copy-editing. Therefore, an "injudicious, or careless, Editor" was to blame for "such an *incoherent Piece.*" Dalrymple's contention that there really was a figure named De Fuentes in Chile at the time of the supposed voyage was supposedly sustained by information that came, in part, from Benjamin Franklin, his friend and associate in the Royal Society. Franklin scoured records in Boston, where he was able to document that a "Capt. *Nicolas Shapley*" had once sailed to Hudson Bay from New England. Dalrymple testified that he had seen a manuscript map "at the *British Museum*" showing "part of the Coast of *New-England*" drawn by a "*Nic Shapley.*" Thus armed, so he thought, Dalrymple ended his apologia with this coda: "The affectation of being wiser than our *Forefathers,* is the *great Characteristick Distinction* of *Those Men* who assume the *authority* of *Modern Oracles:* and it is so much easier, to *treat* with *derision,* than to *investigate.*" It seems safe to conclude that Cook is thus discreetly posited as an oracular figure, a prefiguration of Dalrymple's subsequent biting comment about the "Pope" of Northwest geography. Still, Dalrymple's target in his *Plan* was not Cook but those resisting the idea of further exploration for the new version of the Northwest Passage that he was advocating.[16]

Maximizing his theoretical options, Dalrymple turned to Barkley's discovery, or, as it was thought, rediscovery of de Fuca's strait near latitude 47°N. As historian Barry Gough has recently pointed out in his diligent recapitulation of that story, the de Fuca voyage is both better documented and more credible than the de Fonte legend, if only because it had fewer bizarre elements undercutting it. Indeed, Gough maintains that a de Fuca voyage north from Mexico in 1592 was "eminently possible."[17]

Cook had also summarily dismissed the de Fuca narrative as preposterous, but Captains Barkley and Duncan discerned a wide strait trending to the southeast from its opening on the Pacific at almost exactly the same latitude highlighted in what Cook presumed to be the de Fuca fable. This lent far more credibility than Dixon had provided for the de Fonte myth. De Fuca's strait, however, had a structural problem that the de Fonte story did not. Unlike the latter, with its putative links to bodies of water that

Figure 2.1 Leonhard Euler, *Tabula geographica partis Septentrionalis Maris Pacifici,* 1760 (detail). This inset is from a larger map that recapitulated the work of Philippe Buache, an influential French cartographer. It incorporated both the long-standing French cartographic notion of the *Mer de l'Ouest,* an analogue to Hudson Bay, but also, further north, the network of inlets, rivers, and lakes that framed de Fonte's imaginary Northwest Passage. | Washington State Historical Society, 2003.16.19

Detroit d'Hudson
LABRADOR
ou Pays des Esquimaux
BAYE D'HUDSON
NOUV
CANADA
FRANCE
NOUV ANGLETERRE
MER DU
Boston
Acadie
Baye Françoise
Chesapeack
C. Hatteras
Caroline
Virginie
Georgie
Charlestown
Londres
Bahame
Floride
Apalache
Embouchure du Mississippi
C. de la Floride
LOUISIANE
GOLFE DU MEXIQUE
NOUVEAU MEXIQUE
Mississippi
Missouris
Ozages
Akansas
Chicachas
Illinois
Chicagou
Renards
Mascoutins
Padoucas
Quivira
Maha
L. des Tintons
Sioux de l'Ouest
Moqui
Taos
Apaches
Cadadoquios
Natchitoches
Pte. Coupée
N. Orleans
S. Bernard
Mobile
Pensacola
I. Dauphine
Calos
325
320
315
310
305
300

Peter Pond had discovered on his westward migrations into the Canadian interior, the more southerly location of de Fuca's strait stretched the bounds of plausibility for any "communication" with the Atlantic seaboard. The Missouri-Mississippi basin clearly required an extensive land mass in the mid-latitudes of the continental interior to propagate its prodigious hydrological capacity. Besides which, as Gordon Speck, an early Hearne biographer has written, French and British explorers had traversed "every foot of North America from Churchill south in a straight line to the Gulf [of Mexico] ... without finding any passage between the oceans."[18] Thus, since de Fuca's strait could not credibly result in a direct "communication" with the Atlantic (though even that presumably settled fact would come back into play), the best that Dalrymple could hope for was that the channel sighted by Barkley and mapped by Duncan led to an even larger inlet, which body of water would narrow his mileage gap to Hudson Bay.

As with de Fonte, Dalrymple's *Plan* recapitulated the de Fuca saga, albeit focusing on those aspects that seemed to mirror the reality encountered by Barkley and Duncan, as opposed to its more far-fetched elements. This literary maneuver carried the additional burden of the contemporary international situation, specifically Britain's contestation of Spanish sovereignty in the North Pacific. At some level, de Fuca's story had utility in validating Spanish hegemony in those waters, an issue Dalrymple took extreme exception to and would eventually pamphleteer against in the wake of John Meares's forthcoming book. De Fuca reputedly found his strait in 1592, merely a century after Columbus, and over a course of twenty-seven days emerged in "the vicinity of *Hudson's Bay*," Dalrymple reported. The *Plan*'s clever textual allusion to the "vicinity" of Hudson Bay preserved the potential for a mid-latitude inland sea without explicitly proposing a saltwater passage that Hearne and Cook had proved impossible. This image of a western analogue to Hudson Bay, as discussed earlier, was a long-favoured idea in French cartographic circles, operating under the name *Mer de l'Ouest.* Dalrymple did not define "vicinity." He wanted only to tantalize his readers and policy-makers in the cabinet and the Admiralty with that dreamy prospect. Every inlet, from north to south along Cook's former route, was worthy of exploration because any one of them could prove to be the Pacific portal to the next version of the Northwest Passage. Dalrymple was careful to point out in a footnote: "De Fuca's information only proves an inland Sea, and does not imply that he sailed beyond America, altho he himself drew that inference."[19]

Like the de Fonte legend, Dalrymple was not aware of any reliable primary document to sustain the reality of de Fuca's supposed voyage. What served as the larger foundation was told to, or embellished, by one Michael Lok, an English merchant traveller who reportedly learned of de Fuca's exploits in another maritime roadstead, Venice. Thereafter, Lok related the story to chronicler Samuel Purchas, who in 1625 published the first account of de Fuca in English or any language. The one element in de Fuca's story that preserved the potential for verification was Barkley's report (actually written by his wife, Frances) that he had found "an *exceeding High Pinnacle,* or *spired Rock,* like a *Pillar*" just barely north of where legend had placed de Fuca's discovery of a similar monolith. In Dalrymple's telling, Barkley came to the logical conclusion that the wide entrance he found while foraging for sea otter furs was *"De Fuca's Strait."* Barkley's discovery, mirroring that of Dixon (or even new Spanish claims thought to sustain the verity of the de Fuca story, as related below) was interpreted by Dalrymple to imply that the Pacific coast from 47°N to 58°N "is only broken *Lands* and *Islands*." Combined with a "multiplicity of Reports" of long-standing duration "concerning an *Inland Sea*, on this part of AMERICA," the fur traders' explorations yielded the inevitable conclusion – "the existence of a *Mediterranean Sea*, on this side of AMERICA," and the need for a new authoritative voyage of discovery that could find it. It was this particular line of text regarding a western "Mediterranean" that Vancouver, the man ultimately tasked with that mission, derided with great regularity, certainly more pointedly than any criticism of his mentor Cook.[20]

At this point in his argument, on page 25 of a 32-page pamphlet, Dalrymple finally turned to the ostensible subject of his treatise, to wit: the "Union of the Operations of The *East-India* and *Hudson's Bay Companys.*" He stipulated, with little passion, that the preceding "long discussion" on northern geography was a prerequisite to his commercial argument, but in retrospect it is clear that the fate of the chartered companies was secondary to what he wanted to write about: an evolved cartographic image of the Northwest Passage. Simply put, the maritime fur traders had created a variation on the original vision of the passage in which a "*Sea Communication*" of some undefined dimension extended eastward far into the interior of North America. Once discerned, this connection would provide "greater facility" for conducting "joint Operations" between the companies. Still, the gap between Hudson Bay and its Pacific analogue (presuming it existed) had to be bridged if it was to serve as new version of the passage. Citing

Philip Turnor, whom Peter Pond had once encountered in the hinterlands, Dalrymple asserted that the terrain west from Hudson Bay had been scientifically surveyed for 530 miles to a point near longitude 106°W. This, he computed, left only an 800-mile gap to be transected that would link Cook's fixed points on the Pacific with Hudson House, the HBC post furthest west from the bay's salt water. This establishment was on the North Fork of the Saskatchewan River upstream from the present-day city of Prince Albert. Per Indigenous reports that Turnor conveyed, it was thought "navigable as far *above* Hudson's House as *below* it." If true, this would have shaved hundreds of miles more off Dalrymple's gap. But the Saskatchewan had the singular demerit of flowing into Lake Winnipeg, albeit a part of the Hudson Bay basin but dangerously close to Lake Superior and the other Great Lakes, which the rival North West Company dominated from its headquarters in Montreal.[21]

The commercial competition with the Nor'Westers was reason enough to abandon that model, besides which Dalrymple was more firmly wedded to his idea of a second-generation Northwest Passage consisting of more northerly inland seas. He considered it "probable that the Communication between *Hudson's-Bay* and the *West Coast* of AMERICA would, with more facility, be made in a higher Latitude." The more northerly projection of this new version of the passage melded the insights of Hearne and Pond with the de Fonte fable, or the northernmost extent of the supposed inland sea that de Fuca's strait accessed. Finally turning to the mercantile dynamic implied in his title, Dalrymple catalogued all the ships that were known to have visited the Northwest Coast in the years following Cook up to 1787. He noted that the sea otter furs from Prince William Sound were "held in the most esteem by the Chinese, being the largest Skins, and the finest Fur." Informed by practical experiences not available at the time of James King's prospectus, Dalrymple also offered some advice on the best time to arrive in the hunting waters, but more importantly where to go. The vicinity of Prince of Wales Island, near present-day Ketchikan and north to Mount Edgecumbe near present-day Sitka, was the prime sea otter zone because that coastline had an abundance of "excellent *Harbours.*" But even in this discussion, Dalrymple was inevitably drawn back to his principal interest, for Indigenous reports gleaned by Nathaniel Portlock suggested that these anchorages facing the Pacific had "*Water* behind them," which made these lands insular and their adjoining seas "*Sounds.*"[22]

Reaching a narrative crescendo, Dalrymple asserted that the promise hinted at by the combined efforts of Pond, Portlock, Dixon, Barkley, and others, including the possibility of an inland sea accessed from the Pacific that mirrored Hudson Bay off the Atlantic, could be fully appraised only by "a more exact knowledge, not only of the *Coast,* but of the interiour *Lakes* and *Rivers,* than we at present have. But the Election should be made of that Port, where the communication can be most conveniently kept up, by Posts or Factories, at regular and expedient distances from Hudson's-Bay." This reference to a set of interior stations is in itself a clear indication that Dalrymple's pamphlet was projecting a far different passage from the one Cook was sent to find. Vancouver's voyage, which logically ensued from Dalrymple's treatise and the Meares affair (described below), was *not* an attempt to compensate for any shortcoming on Cook's part. It was instead merely an evolutionary step, among several, intended to preserve the utility of an *idèe fixe* in European thought – a shortcut to Asia. Dalrymple's conceit was that his new passage would not only facilitate mercantile "intercourse with the East, but ... open a *new branch of Commerce,* in the Countries thro' which the Passage was expected to lead the *adventurous Navigator.*" The key British exports were woolen goods and iron products, which he considered a line of merchandise superior to the wares carried by the "Canadian Traders" from Montreal. He asserted that because of the limitations of their canoes, they could move only easily portable items, like ammunition and alcohol. That trade, Dalrymple added sensibly, was so deleterious it risked "the destruction of the Indians." He also viewed a lower-latitude river and land route susceptible, "in the course of Time," to the influx of "Emigrants from the United States of America." Though often astoundingly wrong in his geographical projections, this demographic insight, later also seized upon by Alexander Mackenzie (thereby both annoying and spurring Thomas Jefferson into action), presciently anticipated the developments of the 1840s, when Great Britain was pressured to conclude a sovereignty agreement with the United States over the division of what was then known as the Oregon Country.[23]

3

Mackenzie Solves Pond's Puzzle

Alexander Dalrymple's pamphlet was a crystallization of the maritime geographical context prior to the arrival of the latest reports of Peter Pond's discoveries during the winter of 1789–90, via the correspondence of John Mervin Nooth and Isaac Ogden from Quebec. Their narrative constructions were combined into a graphic image that became the fourth, last, and most important version of Pond's several geographical projections. This was the only one to be published while Pond was alive, in the March 1790 edition of London's *Gentleman's Magazine* (Figure 3.1). This map was the most explicit visualization of Pond's new Northwest Passage, an arrangement premised on a system of interlocking rivers and lakes. It deviated in some respects from the preceding version, which he hoped might be presented to Empress Catherine by himself or his successor in the field, Alexander Mackenzie. A new element, the Queen Charlotte Islands, previously charted by the maritime fur traders, was added. The Great Slave Lake had shrunk in size. More importantly, the map showed the lake rather improbably as having two fully delineated outlets to salt water. One went to the "North Sea" (a placeholder for the Arctic Ocean), where "the water ebbs and flows." The other ran west into "Cooks or Slave R.," which itself had two channels implausibly flowing into the Pacific at Prince William Sound and Cook Inlet. Pond's preceding map had merely hinted at a possible connection between Great Slave Lake and "Cook's River," but by virtue of Ogden's exclamatory text cited previously – "the source of Cook's River is now fully

Figure 3.1 Peter Pond's *Gentleman's Magazine* map, 1790. Based on Pond's conversations with colonial officials in Quebec after he retired from the field, this map harmonized the full evolution of Pond's disparate geographical theories into a single coherent image. | Courtesy Derek Hayes

discovered and known" – the mapmaker for *Gentleman's Magazine* made the linkage explicit.[1]

The *Gentleman's Magazine* map was the high-water mark of Pond's influence over the evolution of the Northwest Passage. Historian Barry Gough has elaborated on its importance, arguing that it "disclosed the general geography of northwestern America" for the first time. As a by-product, "it advertised Pond's achievements" and "gave him a place in the history of discoveries." After his conversations with Ogden and Nooth in Quebec in November 1789, Pond retired to Connecticut, where he died in 1807.

Dalrymple, notwithstanding his suspicions about the loyalty of Pond as a native New Englander, trumpeted the Pond/Nooth-Ogden information to help spur the government into pursuit of another voyage of the discovery to the Northwest Coast. Ogden's note to the editor juxtaposed with the printed map was also the occasion when Mackenzie first became known to the wider public. He was the figure Pond had left behind in the Athabasca country, ostensibly "with orders to go down the River, and from thence to Unalaska," if not to Kamchatka and across Russia to England. As noted earlier, Pond left the Northwest for good after spending just the one winter of 1787–88 with Mackenzie. He was somewhat suspicious of Mackenzie, which seems to have been warranted because of the way the Scotsman later treated his predecessor's place in the unfolding saga of northwestern discovery. Be that as it may, their one year together was sufficiently inspiring to instill in Mackenzie his own "distant intentions."[2]

Mackenzie became Pond's full successor in long-distance terrestrial exploration through two river voyages mentioned earlier, one in 1789 to the Arctic on the river that bears his name, and the other to tidewater on the Pacific in 1793. For our purposes, it was the first of these that was most important, though the latter looms large in American history because it prompted Thomas Jefferson's vision for what became known as the Lewis and Clark Expedition. As with the maritime fur traders who followed in Cook's wake, we have space here to account only for that aspect of Mackenzie's travels that shed light on the reinvention of the Northwest Passage as a cartographic image. The significant point to bear in mind regarding Mackenzie's exploits is that his classic exploration narrative, *Voyages from Montreal,* was not published until 1801. Thus, parts of it were informed by a great deal of geographical information that came to light in the meantime, notably from the Vancouver expedition, the account of which was published by the Admiralty in 1798.

When Pond permanently left his headquarters near Lake Athabasca in May 1788, Mackenzie began to execute his mentor's plan for finding a route to the Pacific. As a first step, Mackenzie's cousin, Roderic Mackenzie, was directed to build a new post on Lake Athabasca during the ensuing summer. Eventually named Fort Chipewyan, it would serve as the jumping-off point for exploration toward the Russian fur trade district. In a letter dated February 15, 1789, written from Pond's old fort on the Athabasca River, Alexander Mackenzie told his North West Company partners assembled that winter in their mid-continent headquarters at Grand Portage: "My intention is to have the effects here transported there in the Spring, and to make that the principal Post for the Department" in the Athabasca country.[3] This subtle maneuver, it seems clear in retrospect, was part of a larger strategy to push Pond and his pioneering efforts and vision into the background, if not out of the story altogether.

In his published text, Mackenzie wrote of his first voyage, which took him to the shores of the Arctic Ocean in July 1789:

> I went ... in hopes of getting into Cooks River; [and] tho' I was disappointed in this [expedition,] it proved without a doubt that there is not a north west passage below ... [69°15′N, where he reached the Arctic] and I believe it will generally be allowed that no passage is practicable in a Higher Latitude the Sea being eternally cover'd with Ice.

Here Mackenzie is referring to the classic formulation of the passage as a saltwater corridor from the Atlantic to the Pacific that Cook was sent to find, not the evolved image that Dalrymple had begun to popularize in the late 1780s. Accentuating his own role in a debate that was essentially settled three years before the publication of his narrative, Mackenzie crowed that his first voyage "has settled the dubious point of a practicable North-West passage; and I trust, that it has set that long agitated question at rest, and extinguished the disputes respecting it for ever."[4]

Of course, Samuel Hearne, whose geographical memoir was published in 1795, thought that his 1771 expedition to the mouth of the Coppermine River had proved the same point Mackenzie was now making; likewise George Vancouver, whose journal was published in 1798. Here we see how the retrospective dynamic inherent in travel literature can create profound illusions. In his book, Mackenzie deleted several lines of text that he had inscribed in his field journal. Thinking initially that he was sailing down Pond's version of "Cook's River," he found that the voyage took an unexpected turn northward and away from a bearing toward the Gulf of Alaska. Mackenzie wrote: "I am at a loss here how to act being certain that my going further in this Direction will not answer the Purpose of which the Voyage was intended, as it is evident these Waters must empty themselves into the Northern Ocean." This was confirmed by his Indigenous conductors, who said they had never been to the "White Man's Lake [i.e., the Pacific Ocean] only the Eskmeaux lake, which is not far off." Though his first was a magnificent voyage, Mackenzie's greatest accomplishment was to prove a negative: he had unintentionally dismissed Pond's Athabascan basin from the Northwest Passage discussion.[5]

Returning from the Arctic Ocean on the river that had just taken him north, Mackenzie learned from Indigenous visitors to his camp "that there is another large River on the other Side of the Mountains to the S.W. which falls into the ... White-man's Lake in comparison to which this is but a small River." The white men there, Mackenzie was told, were in search of "a large Beaver the Skin of which is almost Red" – that is, sea otter pelts. The Scotsman inquired further and concluded that this new river, southwest of the eponymous Mackenzie River, did not extend to the "Eskmeaux Lake." That was good. The problem lay in the fact that there was "no Communication by Water to the above River, those of the Natives who saw it went over the Mountains on Foot."[6] This passage was inscribed on July 26, 1789, just south of the Arctic Circle, near present-day Fort Good Hope, Northwest Territories. South and west of this station across the Mackenzie Mountains lay the upper tributaries of the Yukon River, including the Teslin, Big Salmon, and Macmillan, and the legendary Klondike forks. This system is a significant one but it does not exceed the hydrological capacity of the Mackenzie River, so we can see that something was amiss in the transfer or interpretation of geographical information from Mackenzie's visitors. Besides which, since there were no fur traders at the mouth of the Yukon River at this point in time, it is probable that either the Stikine or Fraser River was meant.

The next day, Mackenzie wrote: "I endeavoured to get some further Intelligence respecting the River I heard of yesterday." He asked his visitors to sketch a map. This image showed a river running "into the Great Lake," at the mouth of which was a "White Mens Fort." Based upon his reading of Cook's last voyage, Mackenzie declared: "This I take to be *Unalaschka Fort* & of course, the River to the West to be Cooks River." This led the Scotsman to conclude that the stream he had just taken to its northern extremity emptied at "Norta Sound." This "body of dead water" was his characterization of the shoaly inlet Cook had discovered and named Norton Sound, which Mackenzie knew from third-voyage accounts was north of Unalaska. Even as published in 1801, with intervening time to reflect on Vancouver's journal, which clarified the true nature of Cook Inlet, Mackenzie let stand his surmise from 1789 that this mysterious stream to his southwest was "Cook's River."[7]

Little did Mackenzie know at this time, if at any point, that the new river across the mountains, the Yukon, itself flowed into Norton Sound. But one should not be too critical of an explorer at work. After all, he was getting information third- and fourth-hand. By the same token, one can readily understand Hudson's Bay Company (HBC) surveyor Philip Turnor's conclusion after meeting Mackenzie later that summer that the Nor'Wester did not really know where he had been.

After his return to Fort Chipewyan on Lake Athabasca late in the summer of 1789, word of Mackenzie's voyage trickled back to Canada and from there to England. It landed with dramatic impact. While Mackenzie contemplated, and then attempted to pursue, the "Cook's River"/Kamchatka chimera, concurrent with Dalrymple's agitation for a new maritime coastal voyage of exploration, a still older idea for a government-sponsored overland expedition to the northwest was being authorized. This scheme dated to Pond's presentation before colonial officials in Quebec earlier in the decade, subsequently reinvigorated by his conversations with Ogden and Nooth. Home Secretary Lord Grenville approved British army captain John Frederick Holland's plan in February 1790. Holland intended to follow the hints outlined by Pond's most recent map by conducting a scientific terrestrial survey from Hudson Bay to the Pacific. More specifically, he intended to examine the region between Lake Athabasca "and the Line of Coast discovered by Capt. Cook." After "Exploring the Interior parts of the Northern and Western Quarter of America," Holland expected to emerge (as did Mackenzie) at "Prince William Sound or Cooks River," preferably the latter. His design was to then "stretch along the coast" until he encountered a fur-trading vessel and thence proceed home to Great Britain.[8] This plan, in part, evoked John Ledyard's scheme for travelling around the world.

With government support sufficient to underwrite the expense of a sixteen-man party plus an array of scientific equipment (chronometer, telescope, thermometer, barometer, theodolite, quadrant), Holland left England for Quebec. He was establishing the logistical support for this expedition when news that Mackenzie had reached the Arctic in July 1789 arrived on November 10, 1790, and in London on December 18. (The source was not the secretive North West Company but rather the HBC's Turnor, who met Mackenzie on his way back to the company's field headquarters at Grand Portage.) Holland's expedition was immediately cancelled because Mackenzie's course proved without a doubt that there was no connection between

Figure 3.2 Alexander Mackenzie, *A Map of America between the Latitudes 40 and 70 [N], and Longitudes 45 and 180 West: Exhibiting Mackenzie's Track from Montreal to Fort Chipewyan, & from thence to the North Sea in 1789, & to the West Pacific Ocean [sic] in 1793*, 1801 (detail). This chart reflected the geographical insights gleaned during Mackenzie's two expeditions combined with George Vancouver's survey of the Northwest Coast. This image was the sum total of geographical knowledge about the northwest quadrant of North America on the eve of the Lewis and Clark Expedition. | Washington State Historical Society, 1918.15.8.1

the Athabasca country and Cook Inlet. "From that moment," Gough notes, "Pond's star faded rapidly and that of Mackenzie glowed ever brighter." Nevertheless, on its own initiative, the HBC sent two of its best surveyors, Turnor and Peter Fidler, into the Athabasca region in 1790 to fix with astronomical precision many of the places that Pond had introduced to drawing rooms and council chambers in Montreal and London over the course of the preceding decade. Summarizing matters, Gough adds that although Pond's "cartography was flawed," he correctly outlined "the general features of the Mackenzie River system."[9]

Mackenzie's first voyage had effectively determined that water flowing out of Lake Athabasca and Great Slave Lake and thence downstream to the Arctic Ocean was in a sense a commercial dead end. Certainly it did nothing to further the prospects of an adaptive Northwest Passage, a dream so durable that in 1793 Mackenzie essentially reversed course and from Fort Chipewyan determined to travel upstream to the south and west toward the headwaters of the Peace River, where he could cross the Continental Divide. (This episode is discussed in greater detail in Chapter 5.) There Mackenzie found a river that he thought would take him to the Pacific. (In his published narrative, he postulated that it was the Columbia River, but actually it was a stream later named after a fellow North West Company man, Simon Fraser). Fearsome talk of rapids on this southward-flowing river gleaned from Natives persuaded Mackenzie to shift his track westward, and that summer he reached Pacific tidewater near present-day Bella Coola, British Columbia, ostensibly the first man to cross North America north of Mexico (see Figure 3.2). The larger import of this voyage, or more particularly Mackenzie's narrative account of it once published, was the notion of a system of interlocking rivers that could connect Atlantic seaports with the Pacific coast. With this network British commercial interests could dominate the continental fur trade and circumscribe their American competitors. This was an idea that borrowed heavily from Alexander Henry the elder and Pond, but more importantly it seeded in print a new vision for the old idea of a Northwest Passage, the third in an evolving series of projections.

By all accounts, Mackenzie's strategic plan in the last few pages of *Voyages from Montreal* prompted Thomas Jefferson, who became president of the United States the same year the book was published, to pursue an American cross-continental expedition contesting British imperial goals. After Vancouver's expedition precluded the possibility of Dalrymple's concept of a Pacific analogue to Hudson Bay, the only remnant of the Northwest Passage as the nineteenth century dawned was Mackenzie's scheme, or the competing notion to the south that the freshwater concourses of the Missouri and Columbia Rivers might, with a practicable portage near their respective headwaters, allow the transit of goods from one side of the continent to the other. Meriwether Lewis and William Clark were to become only the next set of explorers to be disappointed in their quest.

4
John Meares and the Nootka Affair

The last figure to consider in the transition from Cook to Vancouver is the maritime fur trader John Meares. His activities were mentioned earlier, but here we need to consider the dispositive element – his role, and those of his agents, in the Nootka affair that almost led to war between Spain and Great Britain. Officials in Madrid had viewed Cook's third-voyage incursion into the North Pacific with alarm (orders were given to arrest Cook if he anchored in a Spanish port). Now, compounding matters, his countrymen were engaged in commercial trading in waters the Spanish considered a part of their domain. In this regard, the Englishmen joined the already worrisome Russians, whose advances across Bering Strait the Spanish first learned about from their diplomats in St. Petersburg. The initial accounts about the 1741 Bering and Chirkov expedition were followed by hints of an active fur trade in subsequent decades, but the information was garbled. The situation became all the more ominous in 1786, when the French explorer La Pérouse, on his ill-fated attempt at emulating Cook with a scientific voyage circumnavigating the world, visited Spanish California and spoke of sailing into the increasingly contested waters of the North Pacific.

In response to these disquieting developments, the viceroy for New Spain, Manuel Antonio Flores, sent Estéban José Martínez north in 1788. Martínez had been in the vicinity of (but not in) Nootka Sound before Cook, having accompanied Juan Pérez to the North Pacific on a voyage of exploration in 1774. (Indistinct accounts of this expedition, and a follow-up in 1775 by Bruno de Hezeta and Juan Francisco Bodega y Quadra aiming for but falling shy of latitude 60°N looking for de Fonte's imaginary strait, reached London in May 1776. Accordingly, Cook was vaguely aware that the Spanish had been active along this coast prior to his departure from England.) Martínez's new instructions emphasized an investigation of whether the Russians were building fur-trading establishments south of 60°N in order to reinforce Spanish presumptions of sovereignty along the Northwest Coast. Commanding the *Princesa* and accompanied by another ship, Martínez dutifully scoured the Gulf of Alaska and, confirming his worst fears, found the Russians well entrenched there. Worse, the Russians at Unalaska in the Aleutian Island chain corroborated reports about the profits British fur traders were making in Canton. Given the Russians' concurrent concerns about what this competition portended for the future of their own commercial interests, they apprised Martínez of a plan to occupy Nootka to pre-empt the British threat. In the event, this plan came to naught because, unknown to the Russian traders in Alaska, Empress Catherine had already abandoned the colonizing voyage in 1787 due to war priorities in Europe.[1]

After his first reconnaissance, Martínez recommended to Flores that he return to the north in 1789 and forestall both the British and Russians by establishing a presence at Nootka Sound using the rights from prior Spanish discovery as a basis. The full train of events at Nootka Sound are outside the scope of this study, but a few highlights may be laid out. Martínez reached Friendly Cove, the site of the Mowachaht village of Yuquot just inside the entrance to Nootka Sound, in early May. There he serially encountered not only several British ships but also agents of yet another national entry in the fur trade sweepstakes, Robert Gray and John Kendrick, American fur traders from New England. Meares was not planning to return to Nootka from China that year, nor would he as a consequence of subsequent events, but over the course of the trading season, four ships allied to his enterprises were there in force, not merely to trade but to formally establish a mercantile outpost. William Douglas, captain of the *Iphigenia Nubiana,* and Robert Funter, commander of the *North West America,* were a part of the original Meares concern. They were joined (in a sense, for at no time were all four ships tied to Meares at Nootka Sound concurrently) by Thomas Hudson, taking the place of Charles Duncan in command of the *Princess Royal,* and by James Colnett, former captain of the *Prince of Wales* but now in charge of the *Argonaut.* Colnett would prove to be the most important figure in the ensuing saga other than Meares and Martínez. The *Princess Royal* under Duncan and the *Prince of Wales* under Colnett were formerly in the exclusive employ of the Etches concern, doing business as the King George's Sound Company. When Colnett and Duncan had arrived in Macao the previous November (1788) after two years' worth of trading, they bore the news to the merchant community there that the Americans were now also on the Northwest Coast. In light of this threat, the voyage's supercargo John Etches (brother of Richard Cadman Etches, the enterprise's senior partner) and Meares decided to join forces. Duncan proceeded in the *Prince of Wales* to directly to Great Britain, carrying his chart of Juan de Fuca's strait that Dalyrmple found so interesting he had it published. Colnett was outfitted with a new ship, the *Argonaut,* and sailed back to the Pacific Northwest. He was placed in overall command of the 1789 joint effort, but his *Argonaut* was the last to arrive at Nootka Sound.

The new company was called the "Associated Merchants Trading to the Northwest Coast of America."[2] One key feature of this arrangement, and a critical one in the event at Nootka, is that the Etches brothers actually held proper licences issued in Great Britain that allowed their agents to trade within the zones otherwise controlled by the East India and South Sea Companies. As a practical matter, this meant that the *Argonaut* and *Princess Royal* no longer had to employ the subterfuge of sailing under Macao's Portuguese colours but could fly the Union Jack. The strategic thought behind the symbolism, as Colnett phrased it in his journal, went as follows: "It was thought advisable by both parties [Etches and Meares] to form a Junction of trade under the British Flag, each flattering himself [that we] would soon expel all other adventurers, and enable us to make returns adequate to [meet] expenses of outfit."[3] This nationalist gambit not only failed to scare off the Americans but had the added and unintended effect of inflaming Martínez and nearly led to war with Spain.

Meares's legacy ship, the *Iphigenia* under Douglas, was the first to encounter Martínez at Nootka Sound in May 1789. Martínez demanded to see Douglas's papers, which were in Portuguese, and, on the premise he had violated Spanish sovereignty and was trading without a licence, ordered the ship detained. The *Iphigenia* was eventually released and sailed from Nootka on June 2 on the promise that Douglas would proceed to China without conducting any further trade along the coast (a commitment that was not kept). A week later, the other original Meares vessel, the *North West America,* commanded by Funter, entered the harbour. A series of events similar to the seizure of the *Iphigenia* ensued, except that Martínez was determined to keep this smaller ship as a prize to conduct Spanish trade and exploration in the region. In mid-June, the *Princess Royal,* the first of the Etches-Meares ships under Hudson, was the next to arrive. Martínez scrutinized Hudson's papers but did not act on them, even though the ship was flying British colours. He transferred Funter's crew and their pelts to the *Princess Royal* with a warning to Hudson, emulating the previous directive to Douglas, not to trade in any harbour south of Prince William Sound (in the Russian sphere of influence). This too was ignored.

Colnett and the *Argonaut* anchored off the opening to Nootka Sound in calm waters on July 2, 1789, oblivious to all that had just transpired. Like his compatriots, Colnett was shocked to be meeting the Spaniards who rowed out to greet him, but not grasping the underlying dynamic, he allowed the *Argonaut* to be towed into Friendly Cove. Once there, he was alarmed to see two heavily armed Spanish warships. Colnett and Martínez exchanged their views – the former maintaining that he had permission not merely to trade in these waters but also to form a permanent colony there, the latter that he had orders to seize any British ships he might encounter within Spain's domain. The discussion turned into an argument, then an exchange of national insults, with swordplay imminent. When Colnett attempted to leave Martínez's cabin, the Spanish commandant had him arrested. A week later, Hudson and the *Princess Royal* returned to the entrance of Nootka Sound from a short trading cruise, hoping to meet up with their sister ship and Colnett, her captain, the leader of the overall expedition. Hudson and his ship were also captured. The *Argonaut,* now flying Spanish colours and with Colnett as a prisoner, sailed on July 14, 1789, for San Blas in Mexico under a Spanish prize crew. This was the same day that Alexander Mackenzie reached the Arctic and the Bastille fell in Paris. The *Princess Royal* followed at the end of the month.

All this took place under the watchful gaze of the American traders, Kendrick and Gray. Martínez was suspicious of the Americans too, but the British were the main threat to Spanish sovereignty in the Pacific Northwest and more broadly, so he left the *Lady Washington* and *Columbia Rediviva* alone. This did not escape the notice of the several British captains, but the freedom of maneuver the Americans enjoyed proved decisive in the final unwinding of the Nootka Sound affair. When the *Iphigenia* anchored in Macao on October 4, 1789, all Douglas could report to Meares was that the Spanish had been pre-emptory. Neither Douglas nor Meares had any information on the fate of the other three Associated Merchants vessels, which, unknown to them, were now under Spanish control. Thus it fell Gray, who arrived on November 17 in command of the *Columbia Rediviva* (in the course of the first American circumnavigation), to fill out the picture. Gray carried letters and other reports from the British principals at Nootka and had his own story to share. As Richard Nokes relates, the news "provided by the *Iphigenia* and the *Columbia* crews would be the primary British documentation that sent the government in London into crisis mode."[4]

Neither time nor space permits a retracing of Colnett other than to say that he, the *Argonaut,* and the *Princess Royal* were eventually released. Our story dictates that we follow Meares, who, after the final debriefing facilitated by Gray, hastened to Great Britain in high dudgeon, arriving in mid-April 1790. On April 30, he published his "memorial" to Parliament, which was formally received on May 13. Expressing outrage over the Spanish effrontery in seizing British ships and the destruction of his small outpost at Nootka, this pamphlet so inflamed public and official opinion it nearly impelled Great Britain into a war with Spain. Conflict was averted in October when the respective governments reached an accord, known commonly as the Nootka Convention. The key aspects of this instrument dictated restitution of Meares's property at Nootka Sound and free access to all parties in the waters adjoining the Northwest Coast. George Vancouver was eventually commissioned to effectuate this agreement in situ.

Meares's still more relevant travel narrative was published in late November or early December 1790. Titled *Voyages Made in the Years 1788 and 1789, from China to the North West Coast of America,* this book, written with great verve, galvanized the intellectual debate about the Northwest Passage already in the process of transformation because of Dalrymple's *Plan for Promoting the Fur-Trade,* published a year earlier. The visual centrepiece of Meares's *Voyages* was his "Chart of the Northern Pacific Ocean." This map was placed right after the list of subscribers who purchased the original quarto version of his account and the table of contents. In this image, the eye is drawn to the Northwest Coast of America. The inscription superimposed on the chart notes extravagantly that this area was "Explored in 1778 & 1779, by Captain Cook, and further Explored in 1788 & 1789, by John Meares."[5]

Meares, historian Stephen Bown tells us, "was persuasive, with a talent for the loose interpretation of prior discoveries and place names, and a powerful writer who never let truth get in the way of his version of the story." Glyn Williams also considers the portion of the Meares narrative dealing with the Spanish question "a mendacious account," but our interest lies not in sorting out that imbroglio but rather in the contents of an essay forming part of the book's prefatory section, titled "Observations on the Probable Existence of a North West Passage." In this regard, Barbara Belyea asserts that Meares became "the leading theoretician of the northwest passage," superseding Dalrymple because, having actually sailed to the Northwest Coast, "he could not be written off as a closet geographer." This last term was James Cook's famous put-down for speculative cartographers.[6]

Far less sober than Dalrymple's *Plan,* Meares's *Voyages* begins humbly enough, declaring in the preface that his three trading seasons on the Northwest Coast over the course of 1786–89 "were Voyages of COMMERCE, and not of DISCOVERY." His findings, "whatever novelty they may possess, or original information they may bestow, arose out of, and form, as it were, an incidental part of a commercial undertaking." This disclaimer did not affect the temper of his geographical assertions, some of them so extraordinary as to approach the ludicrous nature of the de Fonte legend. Meares was unabashed in declaring his underlying motive. The promotion of "British commerce," he wrote, was the object of his zeal. This interest, of course, perfectly matched Dalrymple's *Plan,* issued the year before, which publication Meares invoked "in supporting my opinion."[7] These two avid advocates for revisiting the question of the Northwest Passage created what in modern parlance would be called a positive feedback loop that led directly to the Vancouver expedition.

Meares's twenty-five-page essay on "the Probable Existence of a North West Passage" began with a review of Cook's third voyage. Misconstruing the Admiralty's guidance to Cook, Meares asserted that instead of beginning his survey at latitude 65°N "according to the language of his instructions," Cook fell in with land "in a much lower latitude." Of course, in actuality Cook proceeded in complete fidelity to the stated mission. He reached the Oregon coast in the spring of the year looking for a place to replenish wood and water after a long passage east and north across the Pacific from his staging grounds in Tahiti. After chiding Cook for not proceeding directly to 65°N, Meares observed that it was "too much regretted" that the great navigator did not spend more time exploring the Northwest Coast between 47°N and 56°N. Congratulating himself for having done precisely that, Meares averred that "there is every reason to lament that Captain Cook was himself prevented from such an examination as would have proceeded from him." This was an acknowledgment that foul weather had forced Cook far offshore until he reached Mount Edgecumbe near present-day Sitka, Alaska, at 57°N.[8]

Meares was cognizant of Samuel Hearne's "track across the continent," which seemed to "annihilate all hopes of a passage" in the lower latitudes. Notwithstanding the well-established fact that Hearne's trek very sensibly informed Cook's discovery agenda, Meares attempted to make the great navigator an implied supporter of the existence of a mid-latitude corridor to the east. He did this by suggesting that the reason Cook first struck North America at a lower parallel than 65°N was that "he did not think the route of Mr. Hearne so very conclusive as it has since been imagined." There is nothing in Cook's journal or the account of the voyage published afterwards to lend *any* credence to this assertion. Indeed, the opposite is true. Cook was derisively dismissive of theoretical geography and both he and the Admiralty took Hearne's findings as dispositive. Accordingly, Cook didn't begin his discovery phase until he approached Prince William Sound at 60°N. Continually alternating between criticizing Cook and trying to cast him posthumously as an ally, Meares asserted that from 45°N to Cook Inlet – indeed, all the way to 65°N – Cook "never saw" the true shore of the continent. In one of his more outlandish suppositions, Meares suggested that a "chain of islands" obscured "the real continent of America: For there is a ground for more than common conjecture, that [Nootka] Sound, Cook's River, and the whole coast hitherto seen, are part of a lengthened chain of detached islands." By this means Meares, on the basis of his own observations plus the discoveries of others such as Dixon, hinted at the possibility that there was *no* continental mainland immediately behind (that is, to the east of) Nootka and the Queen Charlotte Islands.[9]

Noting that Cook was ultimately stymied by the Arctic ice pack, Meares delved into that phenomenon as well. His great "Northern Archipelago" that he supposed ran twenty parallels of latitude was riven, he stated, by deep channels "wide and capacious." These openings were filled not only with whales and sea otters but also occasionally with "islands of ice." Explaining their appearance in his "latitudinal chain of islands," Meares argued that these floes "could never have been formed on the Western side of America" because of its "mild and moderate climate." That Meares could come to this conclusion after having spent a winter icebound in Prince William Sound is all the more astonishing. Attempting to reconcile the temperate climate of the Northwest Coast with the appearance of ice floating about in the Alaska littoral and the northern British Columbia coast, Meares concluded that "they received their formation in the Eastern Seas, and have been drifted by tides or currents through the passage for whose existence we are contending." Indeed, even the "icy fragments" at Cook Inlet and Prince William Sound were, Meares assumed, broken off from the solid ice pack Cook had encountered above the Arctic Circle and driven south by northerly winds.[10]

To Meares, Hudson Bay was a leading theoretical suspect for the ice found on the Northwest Coast. Even if Hearne's Coppermine River and the lands adjoining it seemed to interpose themselves between the bay and the Pacific, Meares disposed of that impediment by questioning whether the large body of water the HBC explorer had passed over on his return (Great Slave Lake) was "*fresh water*" or a southerly extension of the Arctic. If the latter, perhaps it was one and the same with "Lake de Fonte," which, as that hoary legend maintained, "communicates with the Northern Pacific Ocean" on the east side of his great archipelago, supposedly screening the continental mainland from an oceanic vantage. Hearne could not be fully trusted, Meares contended, as evidenced by the lack of authoritativeness in the HBC explorer's reckoning about the Coppermine outfall at 72°N. This had been called into question by Turnor's charts, Dalrymple's critique, and Pond's assertion (via the Ogden family) that the polar sea could be reached at 68°N. Extrapolating further from Pond, whom he characterized as "the respectable authority of the Canadian Merchants," Meares concluded that it was "by no means improbable that Cook's River may have some communication with this sea." Deeming Pond's "Canadian accounts" more reliable than Hearne, Meares asserted that it was "more than possible" that there was a "communication between Cook's River and the Southernmost part of Baffin's Bay, or the Northernmost part of Hudson's Bay into the Atlantic Ocean."[11] The important point to note with this text, for which purpose the word "communication" is key, is that given the freshwater nature of any river and the saline quality of Baffin and Hudson Bays, here Meares, like Dalrymple before him, subtly shifted the nature of any potential Northwest passages away from the version sought by Cook.

But to Meares a far better prospect lay further south – the fabled Strait of Juan de Fuca. In this case, unlike the northern district of Hearne and Pond, Meares was able to draw on personal experience and that of fellow maritime fur traders with whom he conversed at Nootka, elsewhere on the Northwest Coast, and in China. The de Fuca story, which Meares termed as one of "the ancient accounts," was found in his experience and those of his contemporaries to be so compelling "as to border on conviction." Recounting Captain Barkley's sighting of the legendary strait in 1787 that

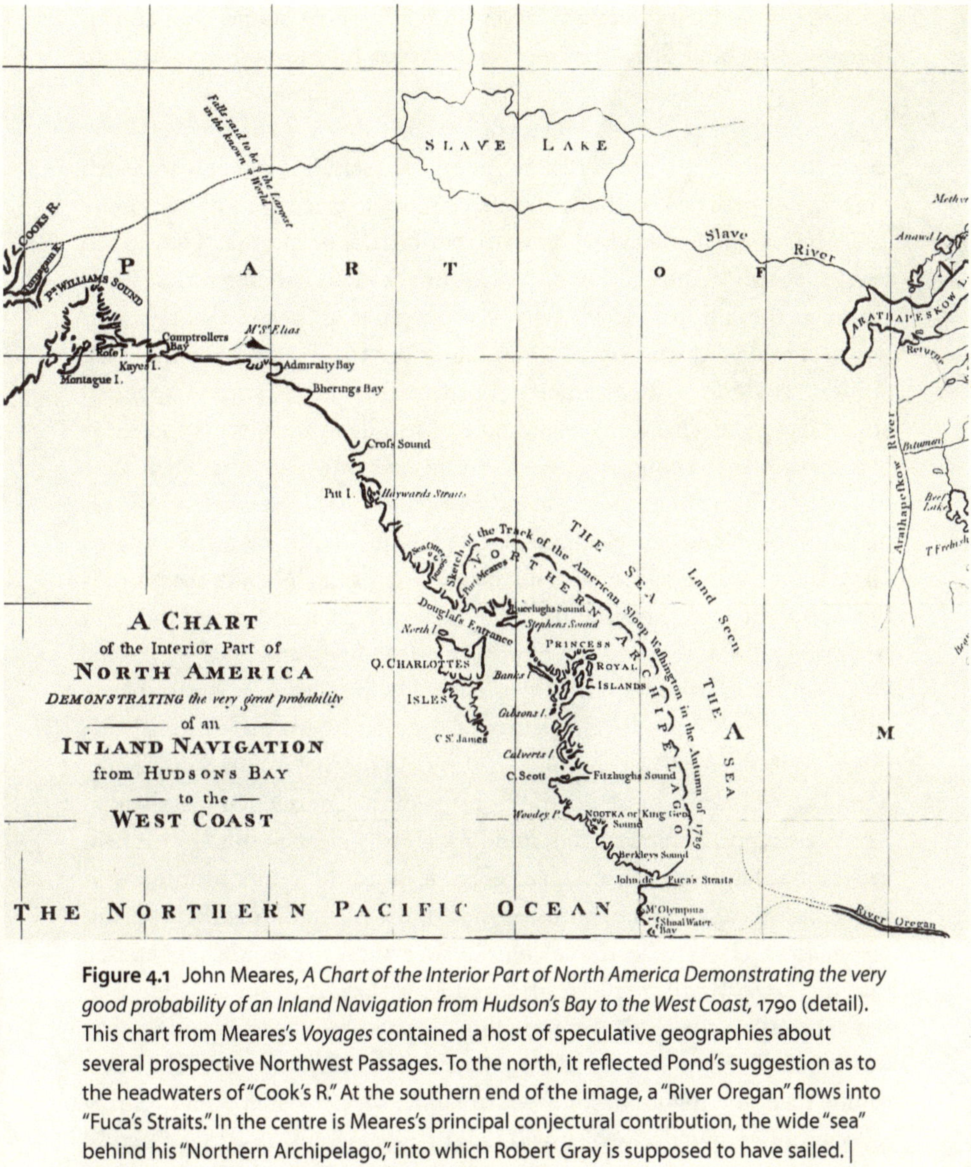

Figure 4.1 John Meares, *A Chart of the Interior Part of North America Demonstrating the very good probability of an Inland Navigation from Hudson's Bay to the West Coast*, 1790 (detail). This chart from Meares's *Voyages* contained a host of speculative geographies about several prospective Northwest Passages. To the north, it reflected Pond's suggestion as to the headwaters of "Cook's R." At the southern end of the image, a "River Oregan" flows into "Fuca's Straits." In the centre is Meares's principal conjectural contribution, the wide "sea" behind his "Northern Archipelago," into which Robert Gray is supposed to have sailed. | Washington State Historical Society, 1918.15.9.1

offered "a clear, uninterrupted horizon to the East," and Dalrymple's calculation of the seemingly manageable distance of 1,380 miles from the Northwest Coast and Hudson Bay, Meares stated that he would have pursued the quest of attempting to link the two but for "the destruction of our commercial enterprize by the ships of his Catholic Majesty." This last was a reference to Colnett's fracas with Martínez at Nootka Sound. Not coincidentally, Vancouver's instructions would point him to de Fuca's strait to assess the promise that Meares had laid before the public.[12]

As if any further incentive for an official British voyage of discovery were required, Meares also noted that an American, Gray, had not only transited the Strait of Juan de Fuca but also entered into "an extensive sea" (see Figure 4.1). After steering northward "upwards of 8 degrees of latitude," Gray exited this supposed Mediterranean of North America and re-entered the Pacific Ocean at the 56th parallel via what was imagined as de Fonte's strait. This was a complete distortion of Gray's early fur-trading foray in the *Lady Washington.* Meares was suggesting, in effect, that the American trader had circumnavigated what would soon come to be called Vancouver Island. Gray had done so such thing, and he was astonished two years later when Vancouver told him about what Meares had published. Nevertheless, Meares had stumbled onto an element of geographical truth by attributing to Gray's supposed voyage into an inland sea proof that "Nootka Sound, and the parts adjacent, are islands," one of many in his "Great Northern Archipelago" that screened the North American mainland from the Pacific Ocean in those latitudes. Historians have long puzzled over Meares's motivation for inventing the Gray story. It may have been mere propaganda intended to provoke British policy-makers. As for the insight that Nootka Sound was a bay on an island and not part of the continental mainland as Cook has supposed, the traders could have easily gleaned this key piece of geographical information after half a decade of steady communication with Indigenous People on the southwest side of the island during their pursuit of sea otters. Drawing on the calculus in Dalrymple's *Plan,* Meares believed that the *Lady Washington* had advanced to within 1,020 miles of Hudson Bay. The intermediate gap between Gray's easternmost station and the bay was "a question that must be left to the result of future discovery." The intervening medium might be "sea, river, or land," Meares stated. This ordering reflected his preference, which was that of all geographers of the practical or armchair variety, for the evolving image of the Northwest Passage.[13]

Meares concluded his summary of the southern district with an indirect but unmistakable criticism of Cook. Believing he had "unveiled" the true promise of the Northwest Coast previously dismissed by Cook, Meares asserted that his findings, and those of his contemporaries in the maritime fur trade, "will teach us also to pay some attention to the account[s] of former navigators." In a double allusion to Cook's journal and Barkley's exploratory breakthrough, Meares contended that the narratives that had once been considered "to be errors or fictions, now turn out to be real discoveries." In a vigorous attempt to sustain his enthusiasm for what the sighting of de Fuca's strait portended for the future of the Northwest Passage, Meares regaled his readers with information, also cited by Dalrymple in his *Plan,* that Spanish sailors had recently entered a strait at latitude 47°45'N. After a voyage of twenty-seven days they found themselves in "the vicinity of Hudson's Bay." This was a variation on the original de Fuca legend dating to 1592, in which the old Greek pilot was led, said Meares, "into a far broader sea, wherein he sailed twenty days, and arrived in the Atlantic Ocean."[14]

A report of the supposed Spanish "rediscovery" of de Fuca's strait (as originally conceived), roughly contemporaneous with Barkley's, came to the attention of Meares and Dalrymple via Sir John Macpherson, the governor general of India, which knowledge he was said to have secured passing through Cape Town. This fanciful claim, which should have been thought preposterous on its face, was superseded by the even more outlandish idea circulating among some Spanish sailors on the Northwest Coast that de Fuca's strait opened to an inland sea of such magnitude that, rather than merely approaching the vicinity of Hudson Bay, it facilitated a shortcut to the port of New Orleans on the Gulf of Mexico.

This last tale originated with the voyage of José María Narvaéz, a pilot assigned to the Martínez squadron stationed at Nootka Sound. Like the British whom he was attempting to thwart, Martínez was very intrigued by Barkley's discovery of de Fuca's strait and on June 21, 1789, he dispatched Narvaéz to assess its prospects. Narvaéz did so in command of a schooner called the *Santa Gertrudis la Magna.* This was the former *North West America,* built at Nootka by Meares the previous September but then captured by Martínez just two weeks earlier in one of the several contretemps playing out in that harbour. Narvaéz sailed into the strait about sixty-five miles and returned two weeks later. He reported finding the same opening discerned by Barkley, whose midpoint was 48°30'N. The strait measured from "side to side" twenty-one miles in width, but the more significant news conveyed by Narvaéz was that the passage "continued on inland for almost the same width as far as one could see" until it led to a division of its waters. With particular emphasis, he stated that "no horizon" was "visible in the east-southeast." This report fired Martínez's imagination. Perhaps, he thought, Narvaéz had discovered not only a possible connection to Hudson Bay to the northeast but also one to the Mississippi basin to the southeast. Thinking this vision would also startle geographers in Europe as much as it had himself, Martínez conveyed it to his superior in Mexico.[15]

The idea that de Fuca's strait led somehow to the Mississippi Valley should have been dismissed as patent nonsense. Indeed, when the Spanish viceroy, Count Revillagigedo, heard this idea, he said it belonged in the "sphere of dreams" equal to other mythologies.[16] Nevertheless, given the fevers raging through geographical and commercial circles, too much was at stake for Spain to simply cede matters to the ambitious British. Thus, in the summer of 1790, Martínez's successor, Francisco de Eliza, ordered Manuel Quimper to conduct a follow-up survey that added to topographical knowledge of the region, notably the existence of what is now known as the San Juan Islands.

In the summer of 1791, de Eliza himself commanded a second expedition into the strait, which resulted in the penetration of these islands by Juan Pantoja and Narvaéz plus a preliminary exploration of the Strait of Georgia. When Alejandro Malaspina, who reached Acapulco in 1791 during the course of his scientific voyage emulating Cook in the Pacific, heard about these voyages, he decided to have two of his highly trained young officers investigate further. Thus, in the summer of 1792, Dionisio Alcalá Galiano and Cayetano Valdés sailed north in the *Sutil* and the *Mexicana* to follow up the discoveries of Quimper, Eliza, and Narvaéz. There, as discussed more fully in the next chapter, they would meet, and briefly collaborate with, George Vancouver.

With sentiments that mirrored Spain's desire to avoid being caught flat-footed while some other nation made the geographical discovery of the age, Meares reached a narrative climax in his essay on the Northwest Passage by appealing to "the honour of this country to bring these researches to a conclusion." He argued that post-Cook "it has been a received opinion that it would be in vain to look for a passage" to Hudson Bay south of 67°N. But what if, he asked fervently, "the sea seen by Mr. Hearne be that very highest point? May not the Northern Archipelago, the Straits of John de Fuca, and Cook's River, all stretching to the North East ... be the very passages?" With these lines, we see the hazy shape of what would become the exploratory agenda of the voyage of discovery soon to be commanded by Vancouver. To Meares's thinking, the demographic dispersion of the "the Esquimaux" made the existence of a Northwest Passage a necessity. This, of course, was a thread of speculation that had intrigued Cook earlier, though he was more cautious than Meares in his geographical deductions. To Meares, these northern "channels," however constituted, provided "a safe and easy passage" for whales "and other huge marine animals" and were, perforce, "sufficiently deep and capacious for navigation."[17]

Noting that the maritime fur trade was still in its "infant state" and limited to merely "an archipelago of islands, forming a kind of barrier" to the North American mainland, Meares insisted that upon maturity it could be expanded to form "a commercial communication with the continent itself." Echoing Dalrymple's *Plan,* published the year before, Meares's prospectus envisioned a "chain of trade between Hudson's Bay, Canada, and the North West Coast of America" that would become "a great and very valuable source of commerce ... to our country." In this way, Meares also presaged the strategic vision that Alexander Mackenzie laid out in his *Voyages from Montreal,* published in 1801, which in turn prompted Thomas Jefferson to launch the Lewis and Clark Expedition. We know that Mackenzie studied Meares because he drew on the latter's assertion that Indians on the Northwest Coast decorated their canoes with shiny white studs made from human teeth. This was an observation that Mackenzie mistakenly attributed to Cook (an error William Clark replicated by plagiarizing Mackenzie). It actually came from Meares's book, not the great navigator's published journal.[18]

By far the most important part of Meares's geographical memoir in the main body of his book, as opposed to his prologue on the Northwest Passage emphasized here, was his account of a longboat excursion he commissioned in the summer of 1788. This venture investigated the opening Barkley had sighted a year earlier, which he presumed was the de Fuca strait of legend. In this regard, Meares was following the lead of Charles Duncan, who conducted a follow-up inspection in the weeks immediately after news of Barkley's remarkable discovery reached Nootka. Meares opined "that if Captain Cook had seen this strait, he would have thought it worthy of farther examination." Since Cook was, in fact, on the lookout for a harbour to secure wood and water, it is conceivable he could have nosed into the strait to a limited extent if it had not been obscured from his view when he was sailing north in 1778. However, given the primacy of the Hearne dictum that mandated starting the discovery phase at a more northerly latitude, it is extremely unlikely that Cook would have conducted the expedition that Vancouver did in response to the pleadings of Dixon, Pond, Dalrymple, and Meares. If Cook had ventured in, it is certain he would have been more exact in his description and exercised more restraint than Meares did in reporting on the exploits of the small squad drawn from his larger crew on

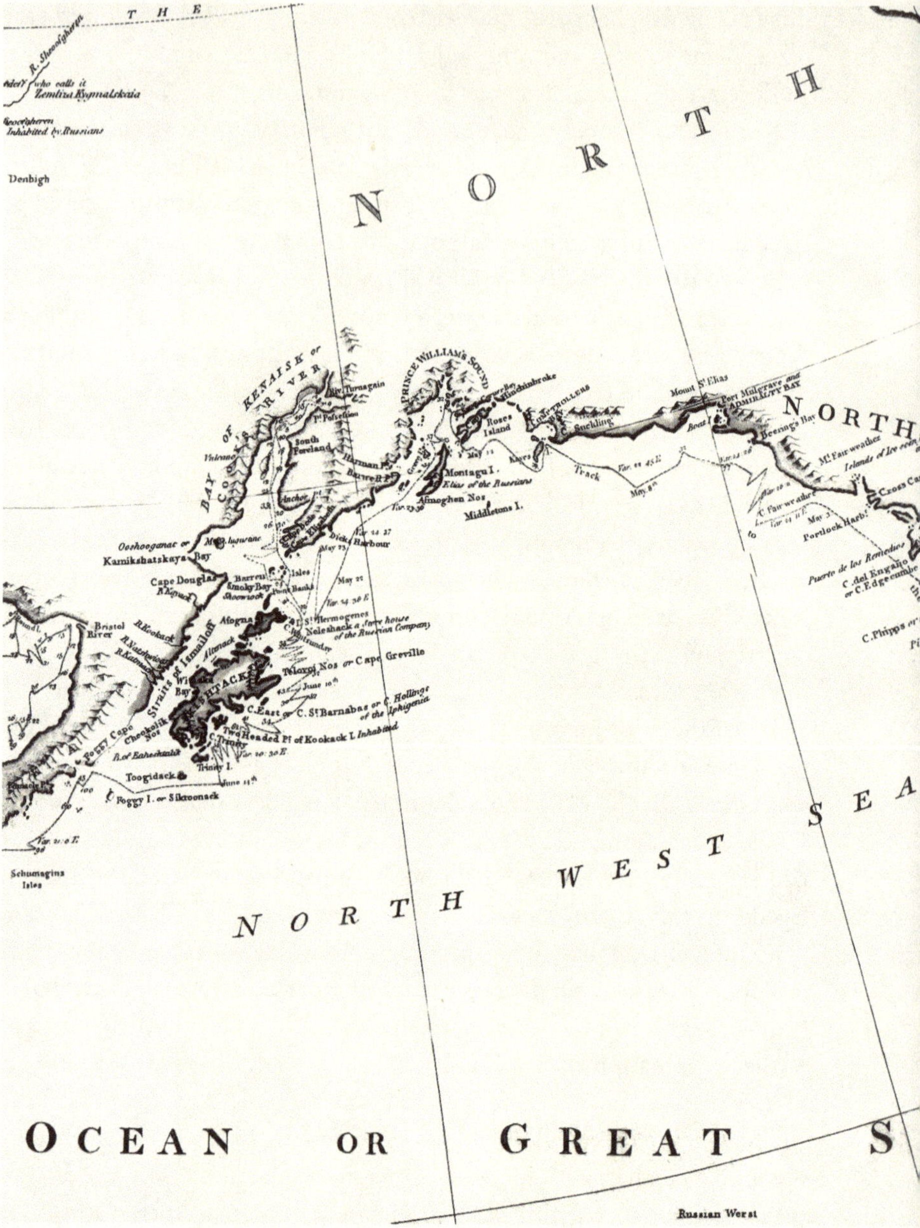

Figure 4.2 Henry Roberts, *Chart of the N. W. Coast of America and the N.E. Coast of Asia, explored in the years 1778 and 1779*, 1794 (detail). This inset from a very rare map (never previously published) shows the evolution of the Northwest Coast as a cartographic image in the decade after Cook's third voyage (compare with Figure 1.1 on page 26). Successive maritime fur traders began to reveal the true complexity of the coastline, including

the prominence of de Fuca's strait in the lower right-hand corner. This map was created by using the same copper plate upon which the original engraving was made in 1784. This image was the most accurate published representation of the Northwest Coast extant until Vancouver's charts were published four years later. | Washington State Library, Rare Map-14 912.798 Roberts, 1794 the *Felice Adventurer*. Meares, who did not actually participate in this "ocular demonstration," said his men sailed ninety miles into the strait before being turned around by the hostility of the Indigenous People to further incursion. Their last view of the eastern horizon before returning to Nootka appeared absent of land for forty-five nautical miles. Meares was enchanted by the possibilities this circumstance augured, musing in *Voyages:* "Such an extraordinary circumstance filled us with strange conjectures as to the extremity of this strait, which we concluded, at all events, could not be any great distance from Hudson's Bay."[19]

Glyn Williams has pointed out that the longboat's log actually showed a transect of only 30 miles,[20] but what would a mere 60 miles matter to a fevered imagination envisioning a clear path to the proximity of Hudson Bay, nearly 1,000 miles distant? Even in his own time, the exaggerated nature of Meares's claims was criticized by sensible reviewers after the appearance of his book, including Dalrymple, who was not the wholehearted ally that Meares believed him to be. Nonetheless, in the last of Dalrymple's four pamphlets published in 1789–90 discussing the Nootka crisis and geographical theory, *Spanish Pretensions Fairly Discussed,* he argued that although the accounts of de Fuca and de Fonte were "vaguely and imperfectly recited," they still needed to be authoritatively investigated until "the *reality* of *such Voyages* [was] positively *denied.*" He asserted that the original discoveries of de Fuca and de Fonte, though perhaps now "*outcast* and *abandoned*" by Spanish officials in Madrid, might "by adoption, become *English.*" With diligent follow-up, the "*true Right of Discovery*" might accrue to Great Britain.[21]

Thus, by the end of 1790, after half a decade's worth of intriguing reports from maritime and terrestrial fur traders, as amplified by Alexander Dalrymple in a torrent of published geographical analysis, promotion, and speculation, the authoritativeness of Cook's third voyage assessment of the Northwest Coast was undone (see Figure 4.2). In the words of Barry Gough, "all of a sudden the entrenched Admiralty calculations about Juan de Fuca's humbug ... had to be thrown out the window." This critique gained credibility, Barbara Belyea explains, because "the traders were both factfinders and apologists for the illusory passage."[22] Nevertheless, Cook, by virtue of the implied or real criticism directed toward his last voyage by Dixon, Dalrymple, and Meares, was being held responsible for not discovering what he was not commissioned to explore. His zone of discovery was the truly high northern latitudes near or above the Arctic Circle, not the more temperate parallels between 45°N and 57°N.

After further review by Dalrymple and others, Hearne's projection of the mouth of the Coppermine River, which was foundational to Cook's directive from the Admiralty, was judiciously repositioned from 72°N further south by four degrees of latitude, where it actually lies. The first transcontinental voyage of Alexander Mackenzie also seemed to confirm a more southerly coastline for the Arctic when he tracked the river bearing his name to salt water at 69°N. On the other hand, Mackenzie's expedition actually strengthened the case for Hearne's essential finding, which was not the latitude of the Arctic coast but the more substantial fact that the Atlantic portals for a Northwest Passage – Hudson and Baffin Bays – were separated from the Pacific Ocean by a vast quarter of North America. Further south, however, Dixon and Barkley, among others, punctured the linear solidity of Cook's cartographic projection for the Northwest Coast. Their voyages, as has been simplistically argued from that time to the present, undeservedly undermined the great navigator's reputation. In the wake of the maritime fur traders, the ancient legend of the Northwest Passage indeed took on new life in Dalrymple's pamphlets and Meares's book. But what emerged was an evolution of the classic image – not the direct saltwater link Cook sought but, at best, a Pacific analogue to Hudson Bay that merely shortened the distance between the two oceans. The fervour Dalrymple and Meares displayed helped lay the groundwork for a new formal exploration of those waters, but the search that ensued, led by Vancouver, did not, as is commonly contended by historians, finish the work that a supposedly fatigued Cook failed to accomplish on his last voyage.

5
The Complicated Origins of the Vancouver Expedition

IT HAS LONG BEEN an axiom in Pacific Northwest historiography that George Vancouver was dispatched to the Northwest Coast to *complete* the work James Cook *failed* to accomplish, and thereby ended the ancient quest for the Northwest Passage. The estimable W. Kaye Lamb, editor of the journals of both Vancouver and Alexander Mackenzie, writes: "Cook had been instructed to examine the north-west coast in the hope that he would find its western exit; Vancouver would soon receive similar instructions when he sailed to undertake his detailed survey of the Coast." Robert J. King states that Vancouver's coastal survey was an "attempt to settle the question of the existence of a sea passage to the North Atlantic." With the exploratory proposition thus posed, historians have been drawn to the conclusion, as Robin Fisher and Hugh Johnston write, that Vancouver "eliminated the possibility of a northwest passage." Barbara Belyea asserts forthrightly that "Vancouver would prove that a northwest passage did not exist." In 2012, Barry Gough stated that Vancouver, "by investigating each and every continental inlet to its disappointing head, disprove[d] the existence of a northwest passage in these latitudes." That same year, Freeman Tovell, Robin Inglis, and Iris Engstrand wrote that "Vancouver's meticulous survey ... confirmed that no navigable passage across the continent existed in the temperate latitudes."[1]

The assertions by Lamb and King are commonplace but wrong. The last four, if taken out of context, are misleading. As I demonstrated elsewhere at length and touch on lightly in these pages, Cook was *not* instructed to explore the Northwest Coast. His guidance directed him to refresh his provisions there before proceeding to his actual zone of discovery many degrees of latitude further north. Vancouver's instructions bore *no* resemblance to Cook's. As we saw in the previous chapters, the very notion of a Northwest Passage had undergone a dramatic revision in the decade after Cook's death. The original direct saltwater route was succeeded by both a set of conjoined bays that might "communicate" with one another in some unspecified manner, as well as a transcontinental network of lakes and rivers. Though the British Admiralty ultimately acceded to the idea of a new scientific voyage to the Northwest Coast – after much pleading by maritime fur trade interests, by Alexander Dalrymple, and especially as a consequence of the diplomatic requirements to settle the Nootka Sound affair with Spain – it never doubted the fundamental worthiness of Cook's third voyage or faulted him in any respect. If anything, as Gough notes, Vancouver was directed "to disprove the existence of a northwest passage" in the latitudes Cook bypassed, as opposed to finding it, and in that sense the Admiralty sought to reaffirm Cook's reputation.[2]

Vancouver fully endorsed that objective. As Glyn Williams observes, he "viewed his mission not as an attempt to find the passage, but as one to confirm his old commander's insistence that it did not exist." In the event, Vancouver occasionally found minor fault with some of Cook's cartography, but his esteem for the great navigator was constant. He shared his teacher's doubts about the practicability of even the truncated Northwest Passage of inland seas narrowing the gap between the Atlantic and the Pacific. Properly mentored during the last two Cook expeditions, Vancouver produced charts that matched those of the master. As J.C. Beaglehole notes, Vancouver conducted his three-year survey of the Northwest Coast "in detail impossible for Cook [and] was one worthy of Cook, in accuracy, completeness, and conservation of life."[3]

The Evolution of a Plan

What became the Vancouver expedition has a very complicated origin story featuring two false starts, although Vancouver's personal career trajectory was simple. He enlisted at the age of fifteen and sailed toward Antarctica with Cook as a midshipman. Emblematic of his youth, legend has it that he climbed out on the bowsprit when *Resolution* reached its southernmost station and exclaimed "*Ne plus ultra!*" – (in effect) no one farther.[4] After serving as a midshipman again on Cook's last voyage, Vancouver had several unremarkable postings and returned to England from the Caribbean in September 1789. Shortly thereafter he was posted as first lieutenant under another Cook alumnus, Henry Roberts, captain of the *Discovery.* This was a new ship, launched on December 19, 1789, not the one Charles Clerke commanded during the third voyage as a consort to Cook's *Resolution.* By the time Roberts boarded the *Discovery,* Dalrymple had already published his *Plan for Promoting the Fur-Trade,* scoping out the prospect of a new Northwest Passage in response to the geographical insights gleaned by Northwest Coast maritime fur traders George Dixon and Charles William Barkley, but Peter Pond's map had not yet been published, nor had news arrived about the Colnett/Martínez fracas at Nootka Sound. Thus, upon its commissioning, the *Discovery* was not destined for the Northwest Coast but rather the South Atlantic.

Prior to John Meares's trumpeting of the Nootka Sound controversy, the needs of the South Atlantic whale fishery loomed far larger in the calculations of the British government and the Admiralty than Dalrymple's visions relative to finding a shortcut to Hudson Bay from the Pacific. The southern whale industry had grown prodigiously at the conclusion of the American War for Independence and especially after 1786, when the royally chartered East India Company and South Seas Company agreed to allow independent whalers and sealers to hunt in the high latitudes of the southern oceans. Nevertheless, safe harbours were in short supply in that hemisphere. Spain foreclosed such opportunities in South America. When British merchantmen ventured into Spanish ports for repair or resupply, their ships and stores were confiscated. Cape Town had been closed to British traffic after the Dutch joined the Franco-American alliance against Great Britain in December 1780 (several months after John Gore and James King passed through on their way home at the conclusion of Cook's last voyage). The British attempted to take Cape Town by force in 1781 but that effort failed, leaving forlorn St. Helena as the empire's sole possession in the South Atlantic. Given the East India Company's pressing need for a South African way station for its vessels transiting to and from ports in India and China, the situation was acute. In 1785, the *Nautilus* conducted a survey of the southwest coast of Africa, from Luanda, Angola, southward to Lüderitz Bay on Namibia's Diamond Coast, looking for a harbour that could approach, if not duplicate, the serviceability of Cape Town.

The *Nautilus* returned to England the following year without a prospect in hand. With the requirements of the East India Company still unmet and the maritime hunters unrelenting in their pleadings, yet another African voyage was commissioned in June 1789. This was the same time as the controversy over the affray with the Spanish at Nootka Sound was reaching a crescendo. William Grenville, the British home secretary, drew up a new set of instructions to reinvestigate possibilities between Luanda and Lüderitz Bay and every part of the South Atlantic southwest of Cape Town (at latitude 34°S) between 42°S and 45°S, and as far west as longitude 48°W, just shy of the Malvinas/Falkland Islands, then under Spanish control. The urgency was such that Grenville envisioned this new mission going beyond those limits to "every part of that unfrequented Ocean" if need be. After depositing a report on that search at St. Helena, from which it could be conveyed to England via routine East India Company traffic, the operation called for a search of the coast of Southeast Africa beyond the limits of Dutch control as far north as the Portuguese settlement of "Mosambique," at 15°S. These instructions were delivered to Roberts, the *Discovery*'s captain, on January 1, 1790. Vancouver joined the expedition as first lieutenant six days later.[5] This is the unfamiliar origin of a voyage that is commonly misperceived as having to compensate for the shortcomings of Cook's last expedition.

The course of *Discovery*'s forthcoming voyage and Vancouver's career were all reset on February 10, 1790, when the Spanish ambassador in London notified the British government that several English ships had been seized at Nootka Sound. His Catholic Majesty claimed this bay by virtue of both centuries-old presumptions under the Treaty of Tordesillas with Portugal and a more recent exploration (the voyage of Juan Pérez in 1774) prior Cook's stop there in 1778. In response to this stunning diplomatic development, no less a figure than Joseph Banks was consulted as to the legitimacy of Spanish claims to prior discovery. Banks, after checking his sources, including Cook's third-voyage journal, asserted that Spain was not "entitled to the Whole Coast." Thus armed, British authorities, in the words of a government white paper issued after the Nootka Sound crisis was over, deemed it necessary "to prevent a Repetition of this extraordinary Conduct by the Spaniards, and to maintain the undoubted Right of This Country to an undisturbed Possession of such Parts of the North-West Coast of America, as had been originally settled or visited, for the Purposes of Trade, by the English." Home Office Undersecretary Evan Nepean sent a letter to the Admiralty expressing the government's response to the Nootka Sound affair, stipulating that it was "highly expedient that measures should instantly be taken for affording protection to such [British] Subjects as may have already proceeded to that part of the American Continent."[6]

The initial plan envisioned a British operation emanating from the new settlement at Botany Bay in Australia while Roberts, in command of the *Discovery*, pursued his original mission in the South Atlantic. At the end of February 1790, a revised strategy emerged. New, secret instructions, which Roberts was not to open until he had sailed from England, advised that the original exercise was to be postponed. His ship was to be "employed on a particular Service in a more distant part of the world." The *Discovery* was to make way to New South Wales, join a larger flotilla strengthened with a detachment of marines and a cadre of convict workers, and then proceed to the Northwest Coast of America. The objective was to "lay the foundation of an establishment for the assistance of His Majesty's subjects in the prosecution of the Fur Trade." This plan echoed an idea previously put forth by Richard Cadman Etches in the summer of 1788. In a letter to Banks, Etches cited the difficulties his fur-trading captains (Portlock and Dixon) had encountered. He stated that "their *Powers* of Government were not competent to the task – nor cou'd they form any Establishment with a certainty of it ever being prosperous, because they had not *Powers* to form any real government for the regulating the People, consequently all wou'd soon have been Anarchy and Confusion." Etches called for a small number of convicts with the requisite community-building skills to be placed at the disposal of his captains, forming the core of a permanent settlement or two that would help regulate trade. After that was accomplished, Etches envisioned that the naval lieutenant in overall command of the flotilla might conduct "a *regular survey* of the whole of the Coast" from Nootka Sound to "Cook's River."[7] This last item, of course, was a response to the traders' discernment of a more complicated coastline than was suggested by Cook's relatively straight line. This discovery agenda anticipated the one Vancouver would eventually pursue over several seasons, but it is important to note that nothing came of Etches's exploratory idea until Spain attempted to delimit British access to Nootka Sound and nearby waters.

Additional developments further complicated matters in the spring of 1790. First, British officialdom had learned the year before that Spain had dispatched Alejandro Malaspina to the Pacific to conduct a voyage of scientific discovery. This venture was a long-delayed response to Cook's third voyage and the intermediate answer by La Pérouse from France. When Malaspina's instructions were first drawn up (over the winter of 1788–89), the Northwest Coast was not on his agenda. He left Cadiz on July 30, 1789, in command of the *Descubierta,* with his companion, José de Bustamente, as captain of the *Atrevida.* These ships reached Mexico in late 1790, prior to a planned exploration of Polynesia. The following spring, however, they were redirected to the Pacific Northwest instead, partly as a consequence of the emerging Nootka Sound controversy.

The other factor was the news that the influential French cartographer Jean-Nicolas Buache de la Neuville, had suggested that the legendary voyage of Lorenzo Ferrer Maldonado, from the Atlantic to the Pacific through North America, was credible. Buache's address in Paris in November 1790 literally resurrected the Maldonado legend from the proverbial dustbin of history. It was supposed that Maldonado had sailed through Davis Strait near Greenland in 1588, reaching latitude 75°N before emerging on the Pacific side at or near 60°N, where he met a ship laden with gold, silk, and other precious cargo. Even more incredibly, this craft was staffed by seamen from Baltic ports who spoke Latin.

The Buache report about Maldonado, in retrospect as absurd on the face of it as the de Fonte legend, arrived in Mexico just before Malaspina's planned departure for the South Pacific. So instead of his planned route, Malaspina sailed north from Acapulco in May 1791. He was not eager about this new assignment, nor was his crew, for he had to stem the desertion of twelve seamen when they heard that their next destination was Alaska, not Hawai'i. Nevertheless, the excitement in European circles surrounding a revitalized Northwest Passage created an eventual Spanish interface with what would become the Vancouver expedition.

In terms of British planning, the intended fur trade/convict settlement was moved to an indeterminate destination further north, in the vicinity of the Queen Charlotte Islands, first explored and promoted by George Dixon, and away from Nootka Sound, where the Spanish seemed to be concentrating their presence. It was believed, based on intelligence gleaned from Dixon, that there was "a considerable River or navigable inlet into the interior Country" near latitude 51°N, which, should it have turned out to be true, was to be the site the site of the settlement. Afterwards, the *Discovery* was to conduct a thorough survey of the Pacific Slope from Mendocino to "Cook's River." If Malaspina's or any other Spanish ships were encountered, Roberts was instructed to avoid them and, like Cook before him, any Spanish settlements. But if the *Discovery* were to come across any British vessels that had been seized, Roberts was to demand immediate restitution on the premise, as his instructions read, that British subjects were entitled to the "free and uninterrupted intercourse with the whole of the American Coast lying to the North of Latitude of 40 degrees, except such parts only as may be actually settled and occupied by any other European power." Here the British government was drawing on a set of long-standing principles in international law (or law between European nations, at least), known collectively as the doctrine of discovery. According to the code governing this particular instance, prior discovery could lead to actual sovereignty only if early exploratory efforts were "perfected" by way of commercial development or other forms of settlement. In the absence of that, new lands were open to all comers.[8]

The *Discovery* was set to sail from England with these orders when the exploits of two British seafarers came to public attention and again altered the Admiralty's calculus. The first and most sensational occurred on March 14, 1790, when Lieutenant William Bligh, sailing master on Cook's last voyage but more recently captain of the *Bounty,* arrived in London. The need to respond to the mutiny that had occurred on his ship, and to reinstitute the original purpose of her voyage (which involved an agricultural experiment transplanting breadfruit from the South Pacific to the Caribbean), obviated the use of New South Wales and a convict-based settlement on the Northwest Coast. The focus of New South Wales now shifted to helping track down the mutineers. Then, within a fortnight of Bligh's return, John Meares arrived in London with what would later be shown to be an exaggerated claim of his "settlement" at Nootka Sound that Esteban José Martínez had expropriated, among other outrages perpetrated against British traders there. These views were memorialized in an inflammatory pamphlet Meares published later that spring.

The British cabinet took Meares's assertions at face value and, facing enraged public opinion over the affair, issued an ultimatum to the Spanish government seeking immediate redress. This step was buttressed by a full mobilization of the Royal Navy, including cancellation of Roberts's command intended for the Northwest Coast. This redirection, known as the Spanish Armament, was eventually defused with Spain's accession to the so-called Nootka Convention. This diplomatic protocol was worked out on October 28, 1790, between the British ambassador to Madrid, Alleyne

Fitzherbert (the Baron St. Helens, source of a famous place name put on the map of the Pacific Northwest by Vancouver), and the Spanish prime minister, the Count of Floridablanca. But in the short term, as a consequence of the international dispute Meares trumpeted, the *Discovery's* entire crew was redistributed to other ships of the line. For example, in May 1790, Vancouver was assigned to the *Courageous* as her third lieutenant. This ship conducted maneuvers off the European continent, during which service Vancouver was promoted to first lieutenant.

In the Nootka Convention, Spain yielded to British demands for access to and free trade within coastal ports north of California. A key requirement of this agreement was the full restoration of Meares's property (a small cabin and its fixtures) at the Mowachaht village of Yuquot in Nootka Sound, which, it is important to note, was to be effected in situ by representatives of both governments. When the *Discovery*'s mission to the Pacific Northwest was reconstituted in November and December 1790, one of the original components of the project, establishing a settlement in the maritime fur trade zone, was abandoned and replaced with the need to implement this provision.

This was not the only significant change. On December 20, Vancouver was called to the Admiralty's offices and advised that he was to become the *Discovery*'s captain, replacing Roberts. This development made it into the press shortly after the New Year. The *Times* of London reported on January 6, 1791 that Vancouver and the *Discovery* were to sail to the Northwest Coast. As for its former commander, the *Star* of London noted a week later that "Captain Roberts, we understand, is also to sail; ... but on a different destination from that of Captain Vancouver." His assignment reverted to the original mission intended for the *Discovery* off the coast of Africa. (Roberts later ended up being detailed to the West Indies, where he contracted yellow fever and died in 1796.) As for Vancouver, the January 12 edition of the *World* described his prospective voyage as one intended "to make an accurate survey of the North West Coast of America, and, if possible, explore the Straits of Juan de Fuca, and the Western Sea, which extends to the East, and *no doubt* has a communication with Lake Superior ... or with Hudson's Bay."[9]

In this last respect, we see the durability of the Northwest Passage as an *idée fixe,* and in the specific phraseology an echo of the essential aspects of Dalrymple's theoretical geography. The "Western Sea" was the previously discredited French cartographic ideal of the *Mer de l'Ouest* that the British maritime fur traders, principally Dixon and Barkley, had resurrected and that Meares had greatly popularized when his book appeared the previous November. The Northwest Passage, though not referred to as such, is dynamically posited in Dalrymple's mode as "a communication." As for Cook, in the torrent of press coverage touting the resuscitation of the *Discovery* voyage and the forthcoming survey of the Northwest Coast, he was never faulted for any previous shortcoming; indeed, the opposite was true. Both Roberts and Vancouver were considered credible expeditionary commanders precisely because they had served under the great navigator's tutelage.

The Vancouver expedition has been fundamentally misunderstood out of lazy deference to the Cook-as-flawed-navigator-on-his-third-voyage trope. Although Vancouver comprehensively explored seas and adjoining lands peripheral to Cook's last mission, his voyage was launched as a consequence of British interest in securing the right of her fur traders to roam freely in areas that may have been claimed by Spain but that, under the doctrine of discovery, it did not control. In that sense, Vancouver's discovery mission was subsidiary to his diplomatic assignment, and but for the actions of Esteban José Martínez at Nootka Sound, it might not have been launched at all. Since the purpose of this book is an analysis of the evolution of the Northwest Passage as a cartographic image after Cook's voyage, we shall henceforth limit any further description of Vancouver's diplomatic mission with his Spanish counterparts at Nootka Sound, except for purposes of clarifying the chronology of his discovery agenda.

Vancouver's Instructions

A recapitulation of the specific written guidance Vancouver received is essential for gaining an understanding of what his mission was, and was not, intended to accomplish. The first draft was drawn up by Admiralty board secretary Philip Stephens on February 11, 1791. There were few changes from the original draft but one noteworthy excision is worth highlighting: "The discoveries of Captain Cook & of the later Navigators [i.e., the maritime fur traders] seem to prove that any actual Communication by Sea, such as has commonly been understood by the name of a North West passage cannot be looked for with any probability of success." In the final version, after a general statement indicating that Vancouver's expedition was to be "undertaken for acquiring a more complete knowledge, than has yet been obtained, of the north-west coast of America," the following line, reflecting the evolution of the term's usage to signify a shortcut lessening the width of North America to be traversed, appeared. To wit: among Vancouver's "principal objects" was the procurement of "accurate information with respect to the nature and extent of any water-communication which may tend, in any considerable degree, to facilitate intercourse for the purposes of commerce, between the north-west coast, and the country upon the other side of the continent, which are inhabited or occupied by His Majesty's subjects." The last phrase was a reference to the fur-trading posts of the Hudson's Bay Company or the settlements on the St. Lawrence River.[10]

More specifically, Vancouver was told that "it would be of great importance if it should be found that, by means of any considerable inlets of the sea, or even of large rivers communicating with the lakes of the interior of the continent, such an intercourse, as hath been already mentioned, could be established." Mention of "large rivers communicating with the lakes of the interior" is an unmistakable reference to the speculative geography of Peter Pond. Toward that end, Vancouver was directed to conduct his survey so "as not only to ascertain the general line of the sea coast, but also the direction and extent of all such considerable inlets, whether made by arms of the sea, or by the mouths of large rivers, as may be likely to lead to, or facilitate, such communication as is above described." In order to be expeditious in his work, Vancouver was advised "not to pursue any inlet or river further than it shall appear to be navigable by vessels of such burden as might safely navigate the Pacific Ocean."[11] This provision was similar to the Admiralty's guidance to Cook before his third voyage on the necessity of avoiding attractive nuisances and staying focused on the mission in the high northern latitudes.

Vancouver was to pay "particular attention to the examination of the supposed straits of Juan de Fuca, said to be situated between 48° and 49° north latitude." This counsel was followed by a line of text clearly inspired by Meares that references the strait "through which the sloop Washington is reported to have passed in 1789," a reference to one of Robert Gray's ships. Once inland, Vancouver was advised that finding a "near communication between any such sea or strait, and any river running into, or from the lake of the woods, would be particularly useful."[12] Here we see an allusion to yet another geographical legend, the fabled "River of the West" that was long thought to reach the mid-latitudes of the Pacific coast from an undetermined spot in the center of the continent. This notion, like the *Mer de l'Ouest*, was a favourite idea in the French cartographic tradition, and for that reason took particular hold on the imagination of Thomas Jefferson.

The first edition of British cartographer Aaron Arrowsmith's pivotal map of North America, published in 1790, hints at a river winding its way west from the interior of North America, emptying into the Pacific in the vicinity of de Fuca's strait, near 48°N and 49°N. For this, Arrowsmith was clearly drawing on the Pond cartographic legacy. The 1784 map that Pond exhibited to the American Congress depicted a river emptying into the "South Sea" through "Flat head Indian Country" opposite the headwaters of the Missouri River. The map that he presented to British colonial officials in Quebec in 1785 showed the "West River" (that is, the River of the West) with its mouth on the Pacific shore near 47°N. Its headwaters originated in juxtaposition to an eastward-flowing tributary to "Lac du Bois" (Lake of the Woods), located near the wellspring of the Mississippi River and not far from the western-most point of Lake Superior. This cartographic image approximates the textual description found in Vancouver's instructions. Neither of Pond's last two maps, the one drafted in 1787 with the intention that it be presented to Empress Catherine or that published in London in 1790, conveyed any suggestion of a link between the "River of the West" and Lake of the Woods, but the first image had been implanted in the minds of geographers and so ended up on Vancouver's discovery agenda.[13]

If the de Fuca zone of discovery should prove unproductive, Vancouver was directed to the vicinity of "Cook's River," about which "there is the greatest probability that it will be found that said river rises in some of the lakes already known to the Canadian traders, and to the servants of the Hudson's bay company." This was a coded reference to the theoretical geography of Pond and the explorations of Samuel Hearne and his HBC successors. If Pond's image proved realizable, Vancouver was told that it would be "material to ascertain" the extent of "Cook's River" "with as much precision as the circumstances existing at the time may allow: but the discovery of any similar communication more to the southward (should any exist) would be much more advantageous for the purposes of commerce, and should, therefore, be preferably attended to, and you are, therefore, to give it a preferable attention accordingly."[14] Vancouver scrupulously followed this guidance, exhausting every possible recourse for an inlet or major opening in the latitudes south of 57°N during the summers of 1792 and 1793, before finalizing his work in the more northerly zone in 1794. These several elements of Vancouver's instructions clearly reveal that his voyage was not meant in any sense to correct

Cook's efforts, for unlike Cook's instructions, there is no hint that a transit to Hudson or Baffin Bay was even remotely possible.

Vancouver's Interface with the Explorers from Montreal

W. Kaye Lamb, in his annotation of Vancouver's journals, deemed it "surprising to find" that Pond's speculative connection to "Cook's River" from the lakes of the Athabasca district was still viable "as late as February 1791," when Vancouver's instructions were drawn up. Lamb reasoned that the news which arrived in London the previous December about Alexander Mackenzie's descent to the Arctic should have compromised Pond's theory.[15] An explanation lies in the visual power of Pond's maps, especially the one published in March 1790, in contrast to the merely verbal reports about Mackenzie's exploits. Besides, as Derek Hayes points out, we do not know the extent of information originating with Mackenzie that reached London. The North West Company itself never sent word to England, being more concerned about partnership agreements and trade monopolies. When we consider that news of Mackenzie's first expedition passed indirectly through several intermediating parties, compounded by distance and the complexity of North America's geography in that quadrant, the import of any such information was probably difficult to interpret in any event.

We know from Mackenzie's journal of his 1789 expedition that after the river he was expecting to take to the Pacific instead bent northward toward the Arctic, he learned from Indigenous sources that on the other side of the eponymous Mackenzie Mountains forestalling his westward path, there was yet "another large River." At the oceanic end of this stream, which ran "towards the Midday Sun," "a kind of large Beaver" was trapped and traded to the crews of large ships that visited "often." This was unquestionably a reference to the North Pacific sea otter fur trade, about which Mackenzie was previously aware. From the map one of his informants drew, Mackenzie deduced that the crude fort depicted was the Russian trading post Cook's men had encountered at Unalaska. Thus, he surmised, the river beyond the ridgeline of mountains to his west was the true source of "Cooks River," not the lakes of the Canadian North that Pond had postulated and whose geographic vision he was pursuing. This Indigenous report was probably a vague indication of the Yukon River, an arm of which (Apoon Pass) empties into Norton Sound, surveyed by Cook in the fall of 1778. It is also conceivable that this intelligence was a reference to the Fraser River, whose headwaters were much further south and which may be said, more properly, to run "towards the Midday Sun."[16]

After Mackenzie returned to his post at Lake Athabasca on September 12, 1789, the snow began to fall and the rivers and lakes of the Canadian North froze. In fifteen months' time, what he learned, or believed he learned, had been distributed to geographers in London. But relative to Lamb's query about why Mackenzie's discovery did not eliminate Cook Inlet from Vancouver's research agenda, nothing the Scotsman believed in real time, or was conveyed through the fur-trading network to officials in London, would have made the Admiralty certain of our modern normative understanding of the region's geography. After all, Mackenzie titled the 1789 journal a "Voyage ... to the Pacific Ocean." He termed the river now named after him merely "River Disappointment" – his trip pursuant to Pond's vision discovered nothing, so far as he was concerned. Yes, he had reached an ocean, but in real time he was not certain it was the Arctic. Indeed, it was not until 1794 (three years after Vancouver sailed), when Arrowsmith published the second version of his epochal map of North America, that geographers sensibly concluded that Mackenzie had reached the same body of water that Hearne had spied hundreds of miles to the east.[17]

The conduit for Mackenzie's hazy northern insights was Philip Turnor, the HBC surveyor who was posted at Cumberland House in northern Saskatchewan, near a rival North West Company fort. After spending the winter of 1789–90 on Lake Athabasca, Mackenzie returned to his company's field headquarters, Grand Portage, west of Lake Superior. On his way, he encountered Turnor at Cumberland House. The latter recorded on June 23, 1790: "Mr Mckensie says he has been at the Sea, but thinks it is the Hyperborean Sea but he does not seem acquainted with Observations which makes me think he is not well convinced where he has been." Mackenzie reached Grand Portage on July 13. After relating his discoveries to the partners, he wrote to his cousin Roderic, still in the Athabasca country: "My Expedition is hardly spoken of." Mackenzie's colleagues obviously felt nothing had been proved by the venture. A new partnership agreement negotiated that summer dominated everyone's thinking instead. Turnor's report of Mackenzie's exploits, conveyed through HBC channels, reached Dalrymple by the end of 1790 and spread from there. Daines Barrington wrote John Douglas, editor of Cook's second- and third-voyage accounts, specifying that Dalrymple had told him that "Mackenzy found Cook's Rivr to turn N: instead of the W: and that he hath proceeded to 70 degrees." On his way back to Athabasca in late August 1790, Mackenzie saw Turnor again at Cumberland House. We do not know what was discussed, but one year later, Turnor headed to the Athabasca country himself, with Peter Fidler as an associate, "to settle some dubious points." Vancouver was already on his way to the Northwest Coast by the time Turnor and Fidler conducted their more scientific survey. Their fieldwork informed the 1794 version of Arrowsmith's authoritative map, which showed Mackenzie's track to the Arctic, replacing the image found on the 1790 version. The latter map, under the sway of Peter Pond's speculative projection, showed the rivers and lakes of the Athabasca country flowing into Prince William Sound and "Cook's River."[18]

One other consequence of the Turnor/Mackenzie encounters bears mentioning: the Scotsman determined to become more practised in astronomy and the related arts of surveying (training put to good use on his 1793 voyage). Mackenzie spent the winter of 1791–92 in Britain, where, in addition to honing his navigational skills, he read Dalrymple's thesis and Meares's speculations and learned of Vancouver's mission. But even after his voyage to the Arctic, Pond's vision still weighed upon him. Before leaving the Athabasca country on the voyage that in the summer of 1793 took him to the same tidewater surveyed concurrently by Vancouver, Mackenzie dispatched a note to Roderic. It said in part: "I send you a couple guineas. The rest I take with me to traffick with the Russians"[19] – which is to say, when Mackenzie ventured forth on his second expedition, he was still expecting to find the headwaters of "Cook's River" across the range of mountains that had thwarted him several years earlier.

In his journal, Mackenzie professed a desire to pursue the north fork of the Peace River (today's Finlay River) that would take him, so he thought, to a mountain portage that might interface with the westward-flowing river (the Yukon), which he learned about after returning from the Arctic in 1789. However, his Indigenous guides in 1793 vehemently recommended against this course, arguing that the north fork of the Peace dissipated in the mountains and did not lead to a "great river that ran in any direction." (This is true: the Yukon's source lies in the lakes near Whitehorse, nearly 300 miles to the northwest of the Finlay's headwaters across an intractable mountain landscape.) Following their advice, Mackenzie took the south fork of the Peace (today's Parsnip River), which led to a "carrying place" (a portage). After one day's travel, it linked to a "large river" flowing south. This was the Fraser River, the headwaters of which Mackenzie was the first to explore and which, combined with a cross-country dash to the west, took him to

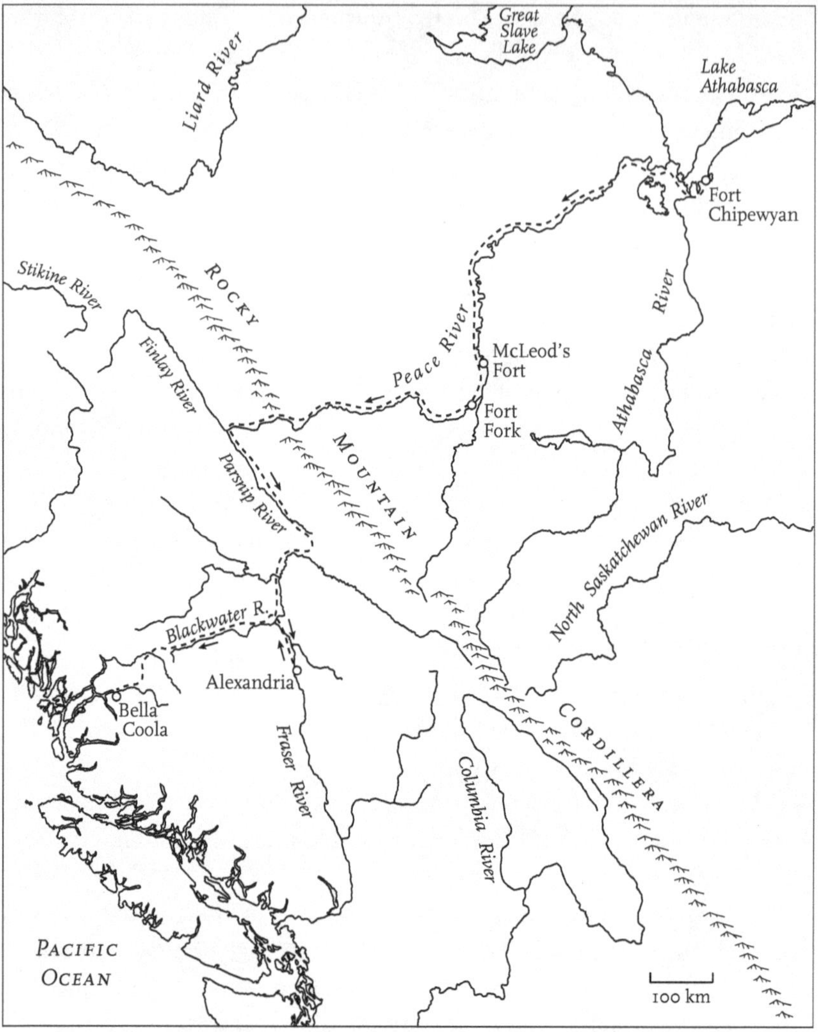

Figure 5.1 Alexander Mackenzie's route westward to Pacific tidewater, 1793. | Based on Barry Gough, *First across the Continent,* 113; redrawn by Eric Leinberger

Pacific tidewater near Bella Coola on the central British Columbia coast, hundreds of miles south of Cook Inlet.[20]

After returning to Upper Canada (present-day Ontario) on September 10, 1794, and thus without any knowledge of Vancouver's expedition, which was still in progress, Mackenzie wrote John Graves Simcoe, the provincial lieutenant governor headquartered at Navy House on the Niagara River. In this letter, he discussed his ascension of the Peace River and then, after crossing over a "Height of Land which is only 700 yards" wide, his arrival at a small stream contributory to a larger river that "the Natives call Tacoutch Tesse." This river carried him downstream "with great velocity." Learning that it took "too much a Southern Course to bring me to the sea as soon as I expected," and that it contained many formidable rapids, he abandoned it in favour of an alternate route that took him to tidewater at a higher latitude. To Mackenzie, the large river he forsook seemed to be the fabled "River of the West; a Branch if not the whole of which I take it to be." By the time his published account appeared in 1801, he was able to link this stream to the explorations of Gray and Vancouver on a river that was by then named the Columbia.[21] In time, the river Mackenzie discovered in 1793 was shown not to be the Columbia but a different stream, followed by a bitterly disappointed Simon Fraser, after whom it was named. The conclusive determination of the relationship between these two great rivers was not settled until 1811 by David Thompson, another fur trader/explorer in the employ of the North West Company.

Thus, even though Mackenzie's first expedition would seem to have excluded Cook Inlet from Vancouver's purview, the sketchy nature of his report, compounded by Turnor's deprecations, precluded the Admiralty from considering it an authoritative finding. Accordingly, Vancouver's final instructions included "Cook's River" among the "considerable inlets" that he was to investigate, in addition to ascertaining "the general line of the sea coast." The Admiralty also seems to have anticipated Vancouver's Cook-like perfectionism. They feared that mapping the dizzying detail of the long and complicated Northwest Coast might take longer than the two years nominally allotted to the task. In the event, Vancouver took three years to complete the job. After the voyage, he confessed that he conducted it "with a degree of minuteness far exceeding the letter of my commission or instructions." Vancouver acknowledged that his conduct may have bordered on disobedience, but, exercising the general grant of authority allowed to previous explorers, he, like his mentor Cook, pursued matters to his own standard of performance.[22]

Although Cook was criticized for missing features he was not looking for, the supposedly conclusive Vancouver voyage failed to discern several notable aspects of the Northwest Coast for which he was specifically tasked. He missed the outfalls of the Fraser, Skeena, Nass, and Stikine Rivers. He might never have sent a ship to explore the Columbia River, except for the accidental encounter with Robert Gray off the coast of what is now the state of Washington. Gray, who was prominently mentioned in Meares's book, apprised Vancouver of the opening he had discovered and named after his ship, *Columbia Rediviva,* in May 1792. Since the Columbia fit the description outlined in his instructions – a river that was "navigable by vessels of such burden as might safely navigate the Pacific Ocean"[23] – Vancouver detached Lieutenant William Broughton, in command of the *Discovery*'s consort, the *Chatham,* to explore it in the fall of 1792. Broughton proceeded as far as the appropriately named Vancouver, Washington, after the conclusion of the expedition's first summer in the Pacific Northwest surveying Puget Sound and the waters surrounding Vancouver Island (described in the next chapter). Vancouver later admitted that he could have been even more assiduous in the survey of the Northwest Coast than he was, but, like Cook before him, he was sensitive to the main mission and concentrated on the big picture, which meant looking for any sign of an analogue to Hudson Bay. Like Cook, Vancouver could also read a landscape. Having discerned Admiralty Inlet, Puget Sound, plus the Georgia Strait and its adjacent channels to the north creating the insular effect for Vancouver Island, he knew that the ever-present coast mountain ranges rimming these waters would prevent any ocean-going craft from reaching deep into the continent toward Hudson Bay.

6
Vancouver's Maze

THOUGH GEORGE VANCOUVER's expedition might have originated for reasons apart from Captain Cook's third voyage, this chapter and the next will focus on the question of the great navigator's thoroughness or lack thereof, and the corresponding evolution of the Northwest Passage as a cartographic image. Vancouver sailed from Great Britain on April 1, 1791 (as noted later in the voyage with full appreciation of that date's derisive association). The command ship, *Discovery*, held 100 men, including several principals in the work of discovery: Lieutenants Zachary Mudge, Peter Puget, and Joseph Baker, and sailing master Joseph Whidbey, all of whom had served with Vancouver in the Caribbean. The consort brig *Chatham*, commanded by Lieutenant William Broughton, held 45 men, notably master James Johnstone, who also figures prominently in the voyage's exploratory endeavour.

Vancouver proceeded first to Hawai'i. He observed that "Captain Cook's chart of the Sandwich islands presented little field for any improvement that could occupy the several winters we were likely to pass in their vicinity."[1] Leaving Hawai'i in March 1792, he aimed the *Discovery* and the *Chatham* for the North American mainland. Approaching the Northwest Coast, Vancouver was sarcastically effusive about the exploration he was now about to undertake. Informed by the narrative speculations of Alexander Dalrymple and John Meares, including the latter's maps, plus the long tradition of the French school of cartographic projection and its *Mer de l'Ouest*, he wrote that his "curiosity was much excited to explore the promised expansive mediterranean ocean, which by various accounts, is said to have existence in these regions." This Doubting Thomas of explorers struck the Pacific Coast near Point Cabrillo at latitude 40°N, proceeding north from there. During this run, Vancouver noted the absence, as he wrote in his journal, of any of the "large rivers and capacious inlets that have been described as discharging their contents into the pacific, between the 40th and 48th degree of north latitude."[2] Like Cook before him, Vancouver was spying this coast for a harbour to facilitate the re-provisioning and refitting of his ships. Unlike Cook, he actually did miss something he was tasked with charting – the soon-to-be-discovered Columbia River, the mouth of which he sailed past unaware on April 27, 1792.

Two days later, Vancouver encountered fur trader Robert Gray at sea off the coast of the present-day state of Washington. He was curious about Gray's exploration of the de Fuca strait and the inland sea to which it led, being under the mistaken impression that the American had done this on the basis of what Meares had written. Gray denied it all, having barely nosed into that narrowing of water, but he did share his theory about where to find the "Great River of the West," the outflow of which he thought he had detected a year earlier. Gray was now waiting for the weather to clear in order to explore it more fully. Vancouver seems to have been almost oblivious to this riverine prospect. He had been misled by Meares, who several years earlier charted and bestowed the name "Cape Disappointment" on a headland south of his meet-up with Gray. (Meares hoped this coastal promontory might be the opening to a great river. He concluded, mistakenly, that the cape was the northern gateway to merely another oceanic inlet, thus his disappointment.) After securing Gray's intelligence, Vancouver concluded that if the bay south of Cape Disappointment was truly the mouth of the Great River of the West, then it "must be a very intricate one, and inaccessible to vessels of our burthen, owing to the reefs and broken water ... in its neighbourhood."[3]

In its constituent parts, Vancouver's assessment was partly correct and partly mistaken. The Columbia River, as we know, could carry ships of the size of the *Discovery* and the *Chatham* inland for well over a hundred miles, but first one needed to be able to cross its bar, where two sizable hydraulic flows met one another with great force and surface turbulence. Later that fall, Vancouver would fail to cross this barrier with the *Discovery*, but Broughton was able to take the *Chatham* across that watery divide.

For now, the British and American explorers proceeded in opposite directions, though they would meet again later in the year at Nootka. Vancouver postponed following up on Gray's river in the short term because Charles William Barkley's discovery of the de Fuca strait, and its promotion as the new gateway to the interior and "communication" with Hudson Bay by Dalrymple and Meares, loomed larger in British maritime strategic thought. By definition, any mid-latitude river would continually narrow as one proceeded upstream to its headwaters in a remote corner of the continent, whereas an inlet a couple dozen miles wide could theoretically lead to an even larger body of salt water. Proceeding north, in the waters off Destruction Island, Vancouver referenced Dalrymple and his pamphlet the *Plan for Promoting the Fur-Trade,* published three years earlier. At latitude 47°37′N, Vancouver noted that "Mr. Dalrymple informs us" to expect an opening at 47°45′N, and "alledged that the Spaniards have recently found an entrance ... which in 27 days course brought them to the vicinity of Hudson's bay." Entering into the legendary de Fuca strait, Vancouver saw the "opening of considerable extent" previously described by Barkley and Meares, though he added that it did not seem to be a part of "any opening into the mediterranean sea."[4]

Vancouver enjoyed taking discreet jabs at Dalrymple as much as Cook had, maybe more so. Indeed, at this juncture in his account, Vancouver volunteered a defence of his mentor, noting (without mentioning names) that "theoretical geographers" had placed, over the course of decades, if not centuries, many "arms of the ocean, communicating with a mediterranean sea, and extensive rivers, with safe and convenient ports" north of Cape Mendocino. He decried the fact that these cartographic images were "not derived from any source of substantial information." Rather, he asserted, such ideas had "been adopted for the sole purpose of giving unlimited credit to the traditionary exploits of ancient foreigners, and to undervalue the laborious and enterprizing exertions of our own countrymen, in the noble science of discovery." Cook was virtually synonymous with scientific discovery. The criticism could only have been aimed at Juan de Fuca, Admiral de Fonte, and Dalrymple. Vancouver added to his derision of those who would mock Cook by specifying the reason behind the recent revival of interest in supposed "communications" between waters of the Pacific and the centre of North America – the pecuniary interest of "modern traders," referring to the likes of Meares and other fur men.[5]

Proceeding into the de Fuca strait, Vancouver scanned the expanse of the waterway trending to the southeast that both British fur traders and Spain's José María Narváez had glimpsed. Advancing inland further than any European before him, he perceived the northern tip of today's Kitsap Peninsula and a division of the water – a relatively narrow channel to the south southwest that he named Hood Canal after a vice admiral in the Royal Navy, and Admiralty Inlet to the east, after the navy's governing body (see Figure 6.1). From this vantage, he also detected a strong ebb tide.

Circumstances were suggesting to Vancouver that, in the argot of his mission, there was "extensive inland communication" further to the south. He chose to enter Hood Canal first, on May 12, 1792, the same day Gray tucked himself inside the bar of the Columbia River, which he named after his ship. Vancouver quickly discerned what has since been often noted as Hood Canal's distinguishing characteristic – the "great depth of the water." (Not coincidentally, Hood Canal is home today to the naval base at Bangor, part of the now consolidated Naval Base Kitsap. It is the principal submarine port for the United States Pacific Fleet.) He said that the adjoining countryside on the eastern slope of the Olympic Mountains was "delightfully situated," but something was amiss. The combination of water, landscape, and climate suggested that this district should be as well populated as Nootka Sound, but it wasn't. Vancouver was nowhere near the demographer that Cook was, yet he was perspicacious enough to discern the telltale evidence of smallpox among some of the Indigenous People he saw. In the understatement of the century, he stated that smallpox was "a disease which there is great reason to believe is very fatal amongst them."[6]

Vancouver determined that Hood Canal was a narrow cul-de-sac, and after mapping it he backed out. He then commenced an exploration of the adjoining Admiralty Inlet and points south, including where present-day Seattle is located, followed by "Puget's Sound," a multi-fingered bay south of The Narrows, near present-day Tacoma. (The image of the body of water and adjoining terrain known today as Puget Sound, without the possessive "s," has migrated northward over the more than two centuries since Vancouver's time. The modern understanding has replaced his sense of Admiralty Inlet, which ran from Vashon Island to the Strait of Juan de Fuca. Today's Admiralty Inlet is now understood to be merely the narrowing between the Olympic Peninsula and Whidbey Island.) The survey teams that Vancouver sent out (and occasionally led) were blessed with uncommonly balmy weather during late May 1792. As a mariner, Vancouver found this watershed "abounding

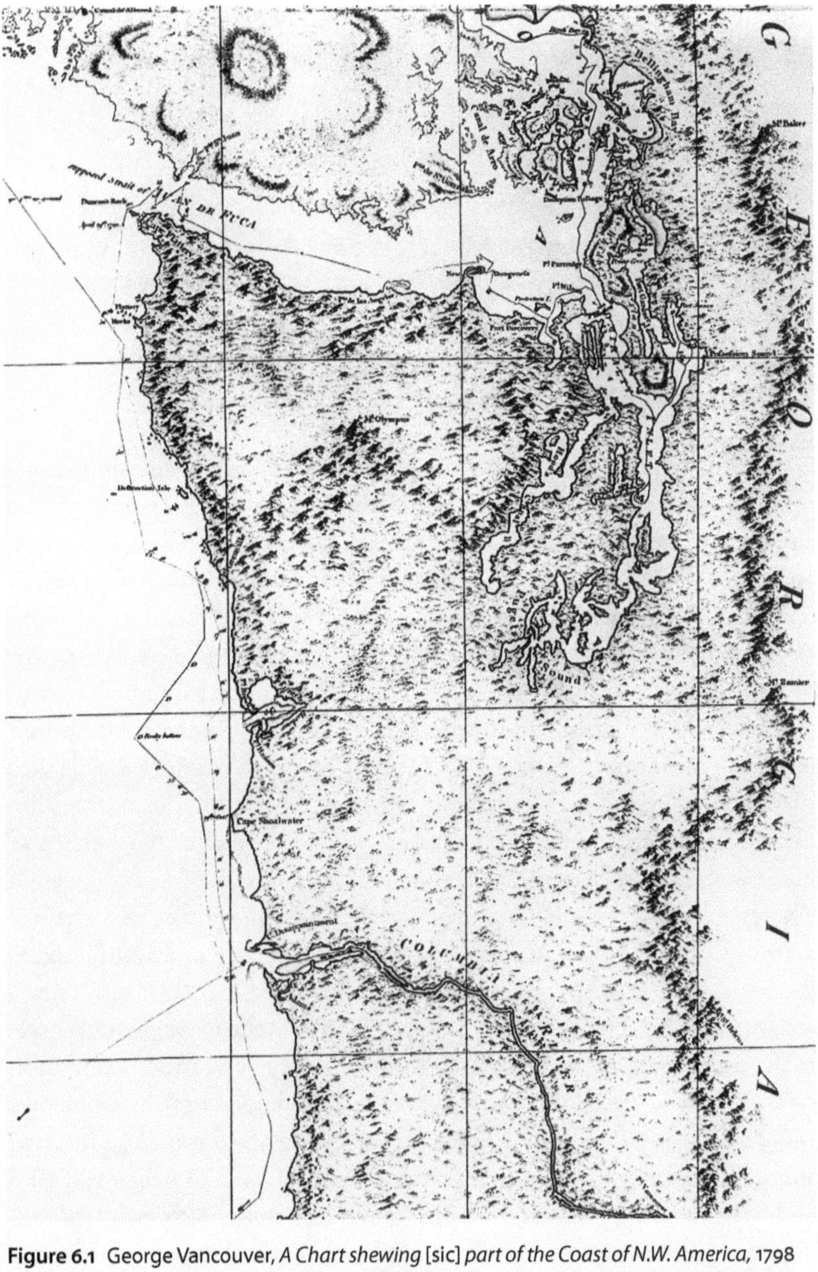

Figure 6.1 George Vancouver, *A Chart shewing* [sic] *part of the Coast of N.W. America*, 1798 (detail). This inset from the larger work, published in the atlas to Vancouver's *A Voyage of Discovery to the North Pacific Ocean and Round the World, 1791–95*, offered the first reasonably accurate view of waters inland from the Strait of Juan de Fuca. Note that the original cartographic construction of "Puget's Sound" was limited to the multi-fingered bay at the southern extremity of "Admiralty Inlet." | Washington State Historical Society, 1911.5.4

with materials." The shores were stacked with timber that promised to yield "thousands of the finest spars the world produces." He continued:

> To describe the beauties of this region, will, on some future occasion, be a very grateful task to the pen of a skilful panegyrist. The serenity of the climate, the innumerable pleasing landscapes, and the abundant fertility that unassisted nature puts forth, require only to be enriched by the industry of man with villages, mansions, cottages, and other buildings, to render it the most lovely country that can be imagined; whilst the labour of the inhabitants would be amply rewarded, in the bounties which nature seems ready to bestow of cultivation.[7]

Notwithstanding his pleading for some future "panegyrist," Vancouver's romantic description shows the evolution of nature writing. There is no such comparable passage in Cook's more empirical mode of the Enlightenment. Stylistically, Vancouver's remarks prefigured Meriwether Lewis's ruminations relative to the settlement and commercial prospects of the area where the Willamette River joins the lower Columbia near Portland, Oregon, the site of the US Pacific Northwest's second most populous metropolitan area.

Puget and Whidbey were dispatched from the ships anchored off Restoration Point (opposite West Seattle's Alki Point) to conduct the detailed survey of the sound that Vancouver later named after the former to commemorate his "exertions." "Puget's Sound" drained an attractive landscape, but it was not itself, nor did it lead to, a great inland sea of Dalrymplian or Mearesian proportions. Combining Puget and Whidbey's insights with those derived from his own tour along the southeastern edge of Puget Sound as far as Eld Inlet near Olympia, Vancouver concluded that two large "snowy mountains" and the adjoining foothills connecting them enclosed the basin to the south and east. This obviated the prospect of any further "communication" in those directions. Vancouver named the largest of these eminences Mount Rainier (after the now obscure Peter Rainier, a commodore and later admiral under whom Vancouver served during the Spanish Armament) and St. Helens (after the British diplomat who negotiated the Nootka Convention that spurred his mission).[8] Admiralty Inlet and Puget Sound proved not to be the North American Mediterranean long envisaged by French cartographers as recently repopularized by Meares and Dalrymple, but mapping their perimeter is arguably Vancouver's most significant contribution to geography.

Vancouver turned his gaze and the ships north, where the topographic pattern, in a sense, repeated itself. At first a "variety of branches" in that direction seemed "most capacious, and presented an unbounded horizon." During Cook's third voyage, a prospect like this would have been greeted by some as the long-wished-for passage to the other side of the continent, but Vancouver's journal conveys no comparable sense of anticipation or excitement. Neither he nor any of his officers started calculating the distances to Baffin or Hudson Bay, as happened on Cook's third-voyage deck. Instead, Vancouver reflected upon the forthcoming rigours and "numberless dangers and unpleasant situations" he and his crew were likely to face in the next phase of "ascertaining the continental boundary." In the short term, this meant completing the survey of the coastline through Possession Sound and Saratoga Passage into Skagit Bay, but it also prefigured experiences still further north during the course of the summer. On June 1, off Camano Island, Vancouver proceeded up a deepwater passage, but this soon "finished in the manner common to all we had hitherto examined," meaning the southern extremities of Admiralty Inlet/Puget Sound. Mount Baker, another of the "remarkably high snowy mountains" (named by Vancouver when he initially proceeded into de Fuca's strait) came back into view. Now fully sensing the logic of the topography before him, Vancouver concluded that a whole chain of mountains "continued, so as to connect the whole in one barrier along the coast." These massive peaks, the principal elements of what in time would be called the Cascade Range, to Vancouver represented the "probability" that they were a southerly extension of the several chains Cook had observed in 1778. This included the Coast Range of British Columbia, the Fairweather and St. Elias Ranges of Southeast Alaska, culminating in the Chugach Mountains that ring Prince William Sound.[9]

After spending two weeks examining the narrow passages east of Whidbey Island through a "tedious inland navigation," Vancouver was ready for "a new field of enquiry." However, an unmistakable trend had already been established as he began to realize that he did not have the luxury Cook had in being able to bypass the complexities of the Northwest Coast. A mere six weeks of investigation "incontestably proved, that the object of our voyage could alone be accomplished by very slow degrees."[10]

It was at this point that Vancouver fastened on the inevitability of how best to deploy the *Discovery* and the *Chatham* going forward. The ships were obviously too big for close coastal surveying, so the practice employed in Admiralty Inlet and Puget Sound became the template. He would anchor the ships in some convenient, visible, and safe anchorage, and then send out the small launches and cutters, under such figures as Puget, Whidbey, and Johnstone, from which the real work of the survey was done. This was a key decision, and after three years it resulted in a remarkable feat of geographical comprehension and physical endurance.

Proceeding into a stretch of water he named the Gulf of Georgia (after Britain's monarch), Vancouver, for the second time, failed to sight the estuarial outfall of a major river, later named the Fraser. This watercourse meets salt water to the south of the great city now named after him. Even though it was then at full flood because of the spring freshet, Vancouver missed it. Like Cook, he was not, it can be safely stated, a river man – or at least not a very good one. He saw evidence of the runoff, such as water discoloration created by silt and stray logs floating about or washed up on the bank, but he could not discern the Fraser's mouth. Vancouver was operating at a distance from the shore, and was otherwise screened from the river by many small islands and shoals in the river's delta. He concluded that any waterway here could carry only canoes, not ocean-going craft the size of ships. So, faithful to his mission like Cook before him, and pursuant to his instructions, he simply moved on. It is no small irony, given the standard interpretation of Cook's supposed lack of thoroughness in these latitudes and its corollary of Vancouver's having to finish the work his mentor left undone, that Vancouver missed both of the Pacific Northwest's two major rivers. During the balance of the voyage, Vancouver himself vigorously pursued only the one "river" Cook discovered – Cook Inlet. It was only in the fall of 1792, after his first season of exploration in the lower latitudes encompassing the Strait of Juan de Fuca and Vancouver Island and his work as British commissioner at Nootka, that Vancouver explored the Columbia River. But even then, the watercourse had already been pointed out by Robert Gray, and the actual survey of the stream was delegated to Broughton and the crew of the *Chatham*.

Partnering with the Spanish

As much as Vancouver was attracted to the salubrious qualities of Hood Canal, Admiralty Inlet, and "Puget's Sound," he consistently deemed the waters of today's Inside Passage and its adjoining terrain "a dreary and comfortless region."[11] His northerly course over the last half of June 1792 led to his second greatest contribution to geographical comprehension – the insularity of the island that bears his name. (Meares had speculated that Nootka Sound was part of an island, a rare instance where his conjecture proved true.) Toward this end, Vancouver was aided by his serendipitous encounter with two Spanish ships captained by men detached from Alejandro Malaspina's five-year-long scientific voyage of discovery to the Pacific. Dionisio Alcalá Galiano on the *Sutil* and Cayetano Valdés on the *Mexicana* had sailed from Spain with Malaspina's original two ships and accompanied him north in the fruitless search for Maldonado's mythical exit into the Pacific. After that frustrating experience in the summer of 1791, Malaspina returned to Acapulco and heard new reports from Nootka-based captains Esteban José Martínez and José María Narvaéz about where the strait "re-discovered" by Barkley might lead. He persuaded Spanish colonial officials that *his* men, with their superior training and instruments, were best able to carry out whatever exploration the suddenly revivified de Fuca legend required, and, if necessary, follow any "passage" leading east. The locally available *Sutil* and the *Mexicana* were identified for this duty, and because of their leadership by Galiano and Valdés, this venture is often construed as a detachment from Malaspina's voyage.

Malaspina himself was eager to move on with the main elements of his mission and did not want to waste still more of his time conducing another "farrago of nonsense," as Glyn Williams calls the Maldonado legend. Malaspina conducted his North Pacific venture in 1791 "in a mood of resignation rather than enthusiasm," as Williams put it. He proceeded to latitude 56°N before curving toward the continent. He spotted a cleft in the mountain-rimmed coast at 59°15′N – Yakutat Bay – which somewhat resembled the opening described by Maldonado. After anchoring, he loaded two small boats with two weeks' worth of provisions to search for the mythical Strait of Anián. A few hours later, the bay shoaled and "the thunderous sound of large blocks of ice calving from a glacier could be heard," Williams writes. Soon, Malaspina and his companions reached the toe of Hubbard Glacier, vanquishing one more dream of a Northwest Passage. After that predictably unproductive turn, Malaspina returned to Mexico for the winter, where colonial officials apprised him of British explorations toward the east end of de Fuca's strait, but more particularly the presumptively promising voyage of Narvaéz. It was in this context of Spain's desire to direct yet another investigation of the Northwest Coast that Malaspina detached Galiano and Valdés from his direct command while he and José de Bustamante sailed west toward the Philippines in March 1792. After entering de Fuca's strait and transiting the San Juan Islands, the *Sutil* and the *Mexicana* met Vancouver on June 22, 1792, in the Gulf of Georgia near the 49th parallel.[12]

Within the reigning ethos of the Enlightenment's Republic of Letters, the captains from both Great Britain and Spain shared information, both geographical and diplomatic. For example, it was from Galiano and Valdés that Vancouver learned that Juan Francisco Bodega y Quadra was already in harbour at Nootka Sound awaiting his counterpart's arrival in order to implement the terms agreed to by diplomatic convention as precipitated by the Meares affair. For his part, Vancouver sketched out what he had already examined that season, the most significant point being that the waters to the south (i.e., Admiralty Inlet/Puget Sound) did not lead, as Narvaéz had once intimated, toward the proximity of the Mississippi Valley.

Galiano and Valdés reciprocated with some coastal details in that vicinity. More importantly, they offered to join the British ships in a survey of the Strait of Georgia and waters to the north separating what would turn out to be the east side of Vancouver Island from the continent (see Figure 6.2). Both partners concentrated on the latter because of its more complex coastline and strategic value. All four ships, with intermittent meet-ups along the way, sailed into Desolation Sound and the mainland inlets near the 50th parallel. In a detached venture, James Johnstone conducted a small-boat transit of the more constricted Seymour Narrows and the strait that now bears his name into Queen Charlotte Strait at the northeastern end of Vancouver Island. Johnstone's survey was gratifying news to Vancouver, who would otherwise have had to spend a lot of time backtracking out of a dead end to get to Nootka. The strong currents in those narrows held the promise of open ocean further north.

At this juncture, when the insularity of Vancouver Island was established, the Spaniards decided to end the discovery partnership. Galiano and Valdés became concerned that they were retarding the progress of the British expedition. After bidding Vancouver farewell (till they met again at Nootka), the Spanish captains separately followed the British through Johnstone Strait into Queen Charlotte Strait. Once there, they took a divergent track. Galiano and Valdés surveyed the remaining extent of the island's northeast coast (later sharing those results with Vancouver) by proceeding out to sea via Goletas Channel before rounding its northern tip. They then sailed south directly to Nootka. Accordingly, they must be credited as the first Europeans to circumnavigate Vancouver Island.

On July 13, Vancouver left his "Spanish friends at anchor" near present-day Campbell River, British Columbia. Proceeding ahead of them, the *Discovery* and the *Chatham* transited the narrows into Queen Charlotte Strait, and then the swelling waters of Queen Charlotte Sound. There, the British ships coasted terrain laved by the main ocean in pursuit of the mission's larger ambit, a story we shall return to. As for the Spaniards, a quick summary may be offered here. Malaspina, after dispatching Galiano and Valdés northward into de Fuca's strait, proceeded across the Pacific to the Philippines, an intention that his Iberian colleagues shared with Vancouver when they first met him. Writing his final report after returning to Spain in 1794, Malaspina deemed the two years (1791–92) that his ships spent along the Northwest Coast chasing legends an insult to science and "modern navigation." He prophesied posterity's amazement that the myths of de Fuca, de Fonte, and Maldonado should have gained any credibility in an age that considered itself enlightened.[13] Even before Vancouver returned to Europe after the last of his three comprehensive seasons of exploration, Malaspina had already concluded that the Northwest Passage, in either its classic form or the new one, was a dead letter.

After parting ways with the gentlemanly Spanish sailors, but before turning his own ships toward Nootka and the rendezvous with his diplomatic counterpart, Vancouver conducted one more coastal foray to the north. Off the northern tip of Vancouver Island in Queen Charlotte Strait, he briefly paused for reflection. There, at approximately the 51st parallel, he took in a broad "view to the north" and the six remaining degrees of latitude between his station and the coast where James Cook had closed in on the continent fourteen years earlier. Having already vindicated his mentor's suspicion about the reliability of the de Fuca legend, he knew that the mainland to the northeast was the only remaining prospect for the western portal to de Fonte's old Northwest Passage, or Dalrymple's new, truncated version. The scene was largely a continuation of what Vancouver had seen since early May. The continental shore was "bounded by a mountainous country, irregular in the height of its eminences, and some of them capped with snow." There seemed to him little opportunity for "extensive inland navigation" through what looked like an "insurmountable barrier." Vancouver knew this much about the work ahead: the coastline "seemed wholly composed of rocky islands and rocks, and presented in their examination a very laborious task, to ascertain the continental boundary."[14]

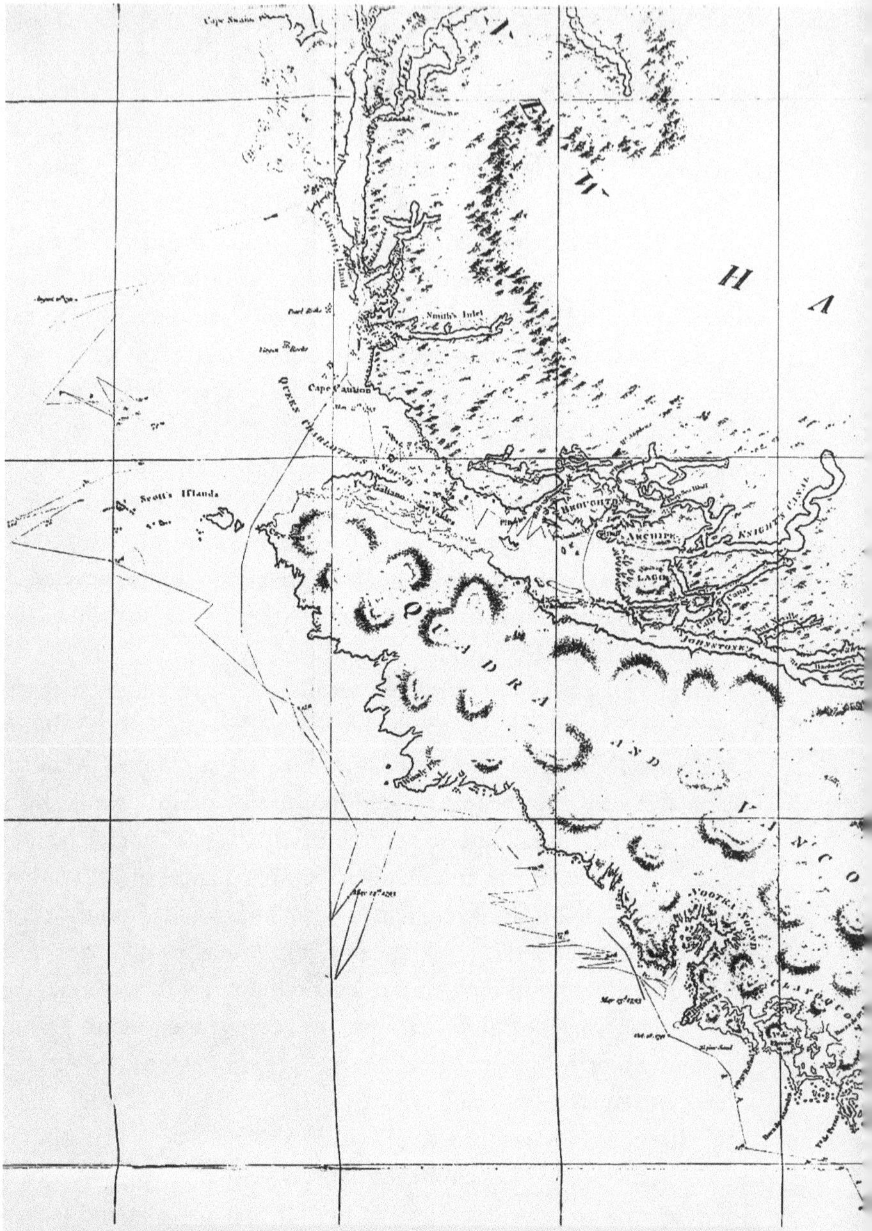

Figure 6.2 George Vancouver, *A Chart shewing* [sic] *part of the Coast of N.W. America*, 1798 (detail). This inset from Vancouver's chart was the first to depict the insular nature of what he billed as "Quadra and Vancouver's Island" in honour of his diplomatic counterpart, plus the famed Inside Passage. | Washington State Historical Society, 1911.5.4

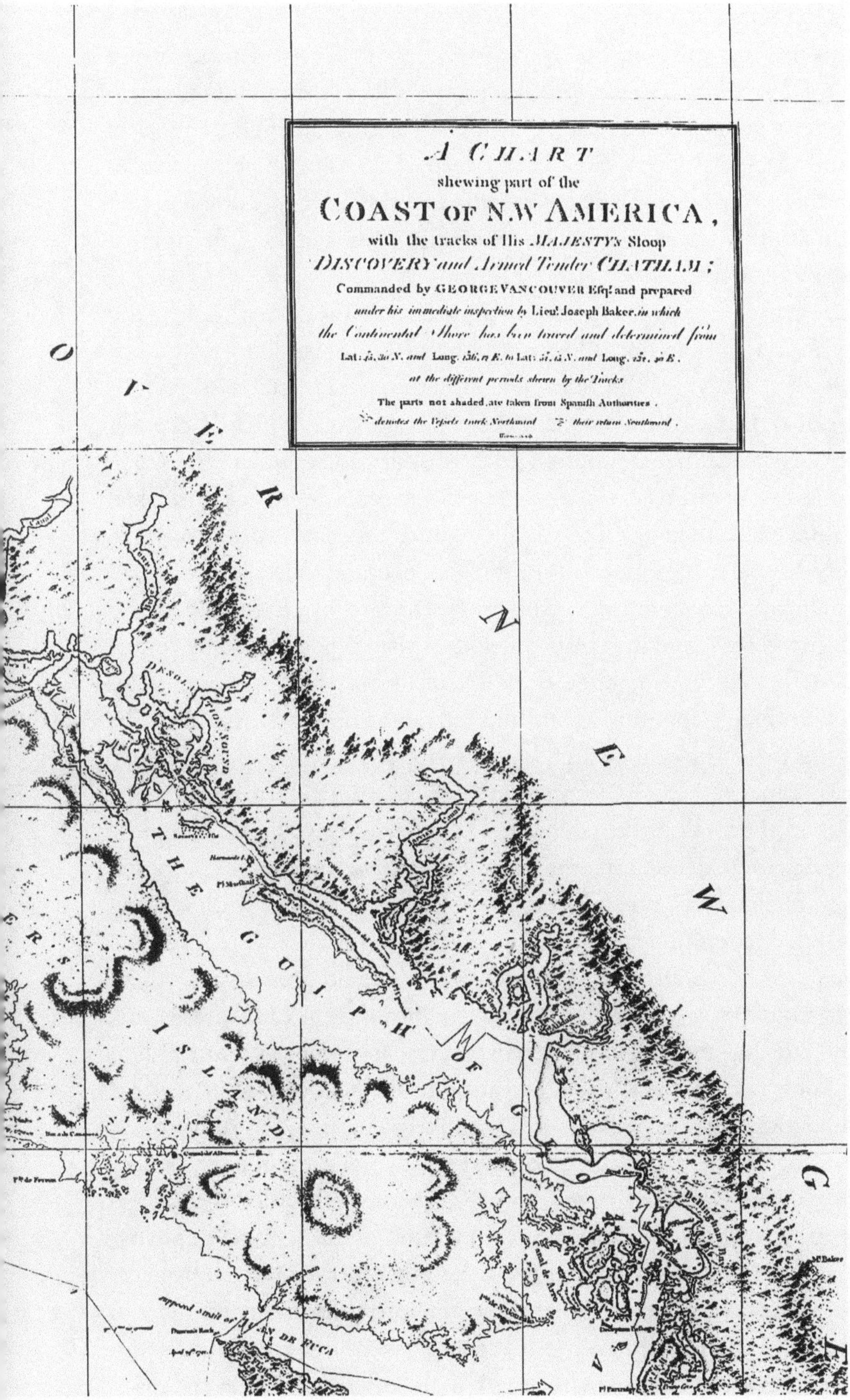
A CHART
shewing part of the
COAST OF N.W AMERICA,
with the tracks of His MAJESTY's Sloop
DISCOVERY and Armed Tender CHATHAM;
Commanded by GEORGE VANCOUVER Esq.r and prepared
under his immediate inspection by Lieu.t Joseph Baker, in which
the Continental Shore has been traced and determined from
at the different periods shewn by the Tracks
The parts not shaded, are taken from Spanish Authorities.
denotes the Vessels track Northward
their return Southward.
DESOLATION SOUND
THE GULPH OF GEORGIA
Bellingham Bay
Mt Baker

This last phrase is compelling evidence of how Vancouver perceived the particulars of his mission. His tactics varied from Cook's second-voyage search for *Terra Australis Incognita,* but in a way, he was emulating his mentor's larger strategy. That is, rather than crisscrossing the high latitudes of the South Pacific as Cook did – each sweep eliminating the prospects for any southern continent – Vancouver settled on the procedure of methodically probing this way or that while eliminating any and all pathways to a new Mediterranean.

Vancouver's gaze across Queen Charlotte Sound toward Fitzhugh Sound and his forecast of the difficulties ahead proved perspicacious, for on this next leg of the voyage, the *Discovery* ran aground. Disaster was avoided because at the time of the incident there was no swell, which Vancouver dubbed a "providential circumstance."[15] This, of course, was a term Cook had famously used on the *Endeavour* voyage when he struck the Great Barrier Reef, and then again in the Aleutians on his last expedition when he dodged some rocks hidden in the fog. An incoming tide, aided by some hard pulling on the *Chatham*'s cables by its crew, refloated the ship.

Vancouver was now in waters previously visited post-Cook by several maritime fur traders, notably George Dixon, from whose ventures place names like Queen Charlotte Sound and Fitzhugh Sound were derived. The northern extremities of the latter, Vancouver proved, terminated in fjords: Burke Channel and Dean Channel. In form, they resembled Hood Canal accessed off the Strait of Juan de Fuca. Dean Channel would be Alexander Mackenzie's route to tidewater the very next year. For Vancouver's purposes, however, Fitzhugh Sound and its arms were yet another labyrinth offering no navigable "communication" to the interior of the continent.

Vancouver had hoped that Fitzhugh Sound would provide a northern outlet to the ocean as Johnstone Strait had on the inside of Vancouver Island, thereby reducing the length of the voyage, but its complexity forced him to postpone a comprehensive investigation to the next year. With Bodega y Quadra awaiting him at Nootka, Vancouver determined in mid-August "to abandon the northern survey of the continental shore for the present season." Summing up his first year's work, Vancouver recorded that he had "the greatest reason to be satisfied with the result of our summer's employment, as it had by the concurrence of the most fortunate circumstances enabled us finally to trace and determine the western continental shore of North America, with all its various turnings, windings, numerous arms, inlets, creeks, bays, etc. from latitude 39°5′ ... to latitude 52°18′."[16] Vancouver, with descriptive and apt phrasing, had taken Cook's straight-line depiction of the Northwest Coast and replaced that image with one closely resembling a maze.

Vancouver's diplomatic activity at Friendly Cove is outside the scope of this study. We need only report that he got along famously well with his Spanish counterpart that summer, and then again in California later in the year, when the British ships were resupplied in Monterrey. Indeed, Vancouver likened Bodega y Quadra (whom he referred to simply as "Quadra" in his journal, thus normalizing his name in Canadian historiography) to Magnus von Behm, the governor of the Russian Far East who had been so helpful during the later stages of Cook's final expedition when it visited Kamchatka. Both men, Vancouver averred, were "actuated by the same motives of benevolence, and governed by principles of similar magnanimity."[17] Behm Channel, now Behm Canal in Southeast Alaska, was named and charted in his honour.

To memorialize their cordial relations and Spanish assistance with the cartography of the Inside Passage, Vancouver originally denominated the island that conventionally now bears his name alone to include that of Bodega y Quadra as well. This usage soon fell out of favour. Vancouver sent Zachary Mudge back to England aboard a fur trade vessel with "the most important parts of my journal, with a copy of our survey of this coast."[18] By way of China, Vancouver's dispatch reached the Admiralty's offices in London on June 10, 1793. Fifteen years after Cook sailed along the Northwest Coast, it was now unmistakably clear to the lords, or armchair geographers like Alexander Dalrymple, that after one season's work, any opening into what Vancouver called the North American "Mediterranean" had been delimited to merely five degrees of latitude, between Fitzhugh Sound and Mount Edgecumbe. The prospects for even the revised image of the Northwest Passage were shrinking rapidly.

Surveying the River of the West

Departing Nootka Sound in the fall of 1792 to conduct an authoritative survey of the California shore, Vancouver had only a single aspect of Northwest Coast geography still in mind: the "river and a harbour discovered by Mr. Gray in the Columbia between the 46th and 47th degrees of north latitude." Vancouver had encountered Gray in April northbound into the Strait of Juan de Fuca, but Bodega y Quadra later also met the American trader when he subsequently arrived at Nootka Sound. Gray entrusted the Spanish navigator with a sketch of his findings, which Bodega y Quadra shared with Vancouver in turn. Initially, Vancouver intended to explore Gray's opening with both ships. On October 19, he reported: "With a pleasant gale and fine weather we coasted along this delightful and apparently fertile part of New Georgia," his name for what would come down through history as the Oregon Country. At four in the afternoon, he noted "having nearly reached cape Disappointment, which forms the north point of entrance into Columbia river, so named by Mr. Gray."[19] Much to Vancouver's repute, he generously and without hesitation credited the American in his published account with the discovery of the river. This he did despite the technical assessment of William Broughton (after he had crossed the bar in the *Chatham*) that Gray had done nothing more than enter a bay into which several rivers emptied – the Columbia among other small ones originating in the coast ranges on either side of the estuary.

Approaching the bar, Vancouver signalled for Broughton to cross first, which he succeeded in doing. Vancouver struggled in getting the *Discovery* to follow because for "some little time" he was unable "to distinguish the entrance into the river, the sea breaking in a greater or less degree from shore to shore." As the *Chatham* proceeded inland seemingly, as Vancouver phrased it, "in a fair channel," the larger *Discovery* found its itself "in three fathoms, and as the water was becoming less deep, and breaking in all directions. ... I hauled to the westward in order to escape the threatened danger." Vancouver had the benefit of an ebbing tide so the change in course was effected quickly by the carrying power of the great river behind him, even though "a very heavy swell ... rolled from the westward directly on the shore." Once safely out in deep water and away from "an irregular and dangerous sea," Vancouver spent the night anchored off the coast.[20]

At daybreak, Vancouver could see the *Chatham* safely inside the bar, but from his vantage, even with a calm sea and fair weather, he was wary of "the heavy cross swell" and the line of breakers that ran "from shore to shore" across the breadth of the river's mouth. Facing an extremely dangerous zone that would in time become known as the "Graveyard of the Pacific," he assessed the advisability of trying to cross the bar again. After two days of soundings and further reflection on his "providential escape" from the nearly disastrous first try, he wrote in his journal: "My former opinion of this port being inaccessible to vessels of our burthen was now fully confirmed, with this exception, that in very fine weather, with moderate winds, and a smooth sea, vessels not exceeding four hundred tons might, so far as we were enabled to judge, gain admittance." Here Vancouver was referencing his Admiralty instructions as to the navigability of any opening he might find. He was also echoing Cook's famous dictum from the *Endeavour* voyage off the Great Barrier Reef about how the navigator must always walk a line between "*Timorousness* and want of Perseverance" versus "*Temerity*" to be found in "want of conduct." This was a passage that Vancouver was assuredly familiar with, and he must have sensed that his career would be judged according to its calculus. Before sailing away, he reported "this opening in the coast to be much more extensive than I had formerly imagined."[21] From his station off the coast, Vancouver resighted and finally named Mount St. Helens in the eastern distance after the British diplomat Baron St. Helens.

On the morning of October 21, "all hopes of getting into Columbia river vanished" as an offshore gale built up in the darkening sky. To Vancouver, a rapidly falling barometer gave "every appearance of approaching bad weather," so he cut his losses. The *Discovery*, he said, "weighed and stood to sea," bound for California.[22] Factored into Vancouver's thinking was his great trust in Broughton's ability to conduct a survey of the river. A previously arranged meet-up with the *Discovery* at their designated rendezvous – Monterrey Bay – was accomplished on November 22. Then and there, Vancouver learned the particulars of Broughton's excursion up the Columbia. These insights were then incorporated into his journal as a paraphrased addendum to the first-year discovery record, carefully keeping the account in his voice as opposed to his adjutant's.

Vancouver introduced the Columbia River narrative by observing that his sense of Broughton's zeal and energy were "found to have been worthily placed." Broughton's investigation of the estuary was "greatly retarded by the shallowness of the water." This circumstance "induced Mr. Broughton to give up the idea of removing the Chatham further up the inlet, the examination of which he determined to pursue in the boats."[23] Thus, with the *Chatham's* cutter and launch and a week's worth of provisions, Broughton aimed for the head of the inlet and the river that flowed into it (see Figure 6.3).

In this regard, Broughton made a credible technical determination that goes a long way toward explaining Lewis and Clark's confusion when they reached the lower Columbia River, and perhaps more particularly the scholarly debate surrounding when and where Thomas Jefferson's captains first saw the ocean. That is, Broughton did not consider the mouth of the

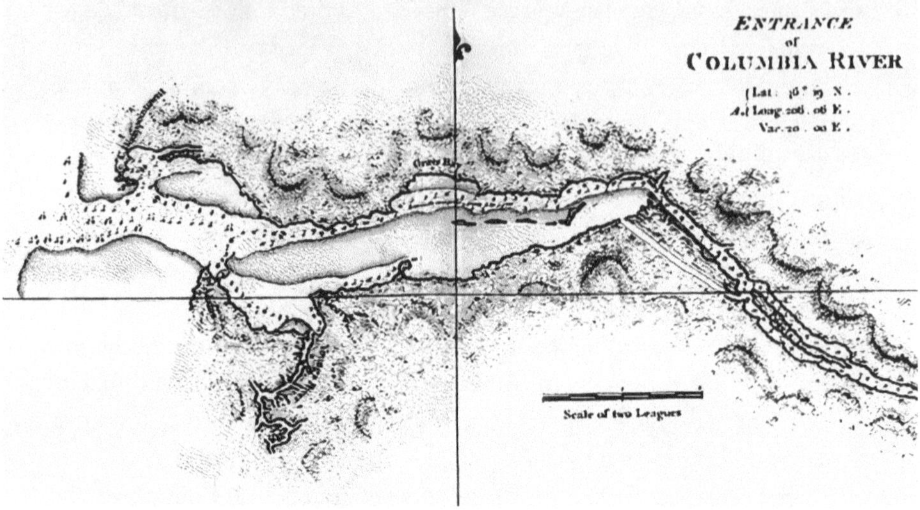

Figure 6.3 George Vancouver, *Entrance of Columbia River,* 1798. Included as an inset within Vancouver's larger chart of the Northwest Coast's mid-latitudes, this small chart reflected Lieutenant William Broughton's understanding of the river's mouth. | Washington State Historical Society, 1911.5.4

Columbia to be the bar between the capes where a riverine flow meets the ocean's waves. Informed by his experience in the *Chatham,* he considered that indent off the Pacific a shallow bay into which several local rivers and the larger Columbia all flowed. The Columbia itself, in his mind, terminated where it flowed into this bay, which point Broughton determined to be near the present-day town of Skamokawa, Washington, in the vicinity of Pillar Rock. This, of course, is precisely where William Clark penned his famous line "Great joy in camp we are in *View* of the *Ocian.*"[24] Given the curvature of the earth, one cannot see the ocean from Pillar Rock, if by ocean we mean an imaginary line running from Cape Disappointment south to Cape Adams on the Oregon side. In that sense, Clark was wrong. But if, as Broughton's trained eye told him thirteen years earlier, Clark was not viewing the ocean proper but an inlet off the ocean, then the American explorer was not mistaken at all.

Also of note, when Broughton's small boats reached the confluence of the Willamette River with the lower Columbia – a spot he called Belle Vue Point – "a very distant high snowy mountain" that was "beautifully conspicuous in the midst of an extensive tract of low, or moderately elevated, land" came into view. He christened this eminence Mount Hood, after the same admiral for whom Vancouver had earlier named a fjord in the Puget Sound lowlands. But more to the point regarding the evolution of the image of the Northwest Passage, Broughton stunningly concluded that Mount Hood "seemed to announce a termination to the river." Broughton, like Vancouver, and for that matter James Cook, was not a good judge of rivers. Indeed, few Englishmen of that generation could be. Oceans, not rivers, were the system of watery conveyance upon which their commercial culture was built. (It was the North Americans, like Mackenzie and Lewis and Clark, who were river men). Despite a rapid downstream current against which it was difficult for the crew "to make but little progress," Broughton concluded that he was close to the head of the Columbia River.[25] For this reason, mistaken as it was, combined with the exhaustion of his supplies, Broughton terminated the survey at a spot he named Point Vancouver, east of present-day Washougal, Washington.

Broughton's last view upstream from Point Vancouver suggested to him that "the river seemed to take a more northerly direction,"[26] and in truth from that spot there is about a twenty-degree northeasterly turn on a tangent leading to Beacon Rock, a landmark later named by Lewis and Clark. In the immediate aftermath of Vancouver's voyage, British map-makers would interpret Broughton's text literally. After Alexander Mackenzie's 1793 discovery of what he believed to be the headwaters of the Columbia River (though it was actually the Fraser), these cartographers depicted the Scotsman's "Tacoutch Tesse" as the northern fork, if not the main stem, of the Great River of the West. Nicholas King's map, copied from British charts for use by Lewis and Clark in the field, replicated this naive cartographic image. He complemented it with a completely conjectural southern fork of the Columbia that met Mackenzie's northern branch precisely at Broughton's upriver mark (see Figure 8.1, page 174.) The American explorers corrected this image in the fall of 1805 when they discovered that the northern fork of the Great River of the West converged with the southern branch – the Snake River, which they had descended – 150 miles upstream from Broughton's last vantage point, near present-day Pasco, Washington.

Broughton estimated via dead reckoning that he had travelled 100 miles upstream from the *Chatham*'s anchorage inside the bar, which is not far off the actual distance of 127 miles from Cape Disappointment to Point Vancouver. It took Broughton seven days to transect that distance, and, with provisions for only two or three more days, he concluded that there was no point, as recorded in Vancouver's paraphrase, to "any further examination, and was reconciled to this measure, because even thus far the river could hardly be considered as navigable for shipping." It is important to recall that Broughton was on the Columbia in October, the time of year when the river had its lowest flow. Edward Bell, the *Chatham*'s clerk, was on this sojourn, and from his unpublished log we learn that the considered opinion in the boats was that the Columbia ran about another hundred miles into the interior of the continent. In other words, this mini-expedition had travelled what they thought was perhaps about half the river's course. In phrasing that evoked Vancouver's instructions, and Dalrymple's speculative geography that preceded them, Bell concluded that the Columbia "*might* communicate with some of the Lakes at the opposite side of the Continent." Given the river's purported length, this seems contradictory on the face of it. Bell thought that such an interior connection was "not impossible, nor do I conceive it altogether improbable." Drawing on the ancient geographical theory of equipoise – the mirroring of geographical features – he asked rhetorically "why a River on this side of America shou'd not run up so high, as one on the opposite side,"[27] a reference to the St. Lawrence.

Bell's observation perfectly anticipated the third version of the Northwest Passage – a system of interconnecting rivers emerging from the heights of the continental interior – just as Vancouver was demolishing the second one, which hypothesized a large inland sea shortening the distance between the Pacific and the Atlantic. Mackenzie's discovery of a great southward-flowing river in 1793, and the image of it that he popularized in *Voyages from Montreal* in 1801, guided Jefferson's vision for what became the Lewis and Clark Expedition.

Notwithstanding Vancouver's gentlemanly deference to Gray's priority of discovery, in his journal he adopted Broughton's definition of where the river began, or, as it were, where it ended its downstream run. Vancouver postulated that his adjutant had sensibly executed his own claim of discovery on the basis "that the subjects of no other civilized nation or state had ever entered this river before." Citing "Mr. Gray's sketch," which he had secured from Bodega y Quadra, Vancouver added that the American trader neither "saw, nor was ever within five leagues [fifteen miles] of, its entrance." Quoting Broughton directly in the body of his own journal, Vancouver asserted that Gray "entered the river, or rather the sound, and had named it after the ship he then commanded. The extent Mr. Gray became acquainted with on that occasion is no further than what I have called Gray's bay."[28] In time, these lines of text would be added to the arsenal of argument that Great Britain deployed in its mid-nineteenth-century negotiation with the United States over the proper division of the Oregon Country.

Broughton returned to the *Chatham* twelve days after departing upriver. He sailed across the bar in tandem with a fur-trading vessel, the *Jenny*, that had arrived in the meantime and that he found anchored in the lee of Cape Disappointment. On its exit, Broughton reported that the *Chatham* was pitched about "incessantly" and "the sea broke several times over her from stem to stern."[29] These and other tales penned by Vancouver but based on Broughton's experiences, including a man being thrown overboard but miraculously rescued, formed the first and emblematic imprint of the bar of the Columbia upon the literary imagination of the western world.

7

Defining the Labyrinth, Cook's Legacy, and the Delusion of the Northwest Passage

After wintering in California and Hawai'i, George Vancouver returned to the Northwest Coast in the spring of 1793. He resumed his survey where he had left off the previous year, to the northward of Vancouver Island. William Broughton had parted company with Vancouver at Monterrey, California, and, via Mexico, returned to Great Britain to update officials there on the progress – or, more accurately, the lack thereof – in the negotiations with Juan Francisco Bodega y Quadra at Nootka. Lieutenant Peter Puget succeeded him at the helm of the *Chatham.*

Upon resuming work in his discovery zone, Vancouver wrote: "I consider myself answerable only for the certainty of the connection of the continental shores." "Agreeably, however, to my former practice [in a reference to his efforts the previous year] we kept the starboard shore on board, as the most positive means of ascertaining the continental boundary."[1] Unknown to each other, Vancouver and Alexander Mackenzie missed each other by two days shy of seven weeks in the waters of Dean Channel, an arm of Fitzhugh Sound, on the central coast of British Columbia. Vancouver deemed this stretch of the "continental shore" off Dixon Entrance an "uninteresting region" because it "afforded nothing further worthy of notice." By that he meant that the plethora of inlets (along what is unquestionably one of the most complex shorelines in the world) all "terminated as usual" (see figure 7.1). Vancouver discerned other constants. "The continental shore was composed of a range of lofty mountains covered with snow," which gave every evidence of being a continuation of the chain of peaks James

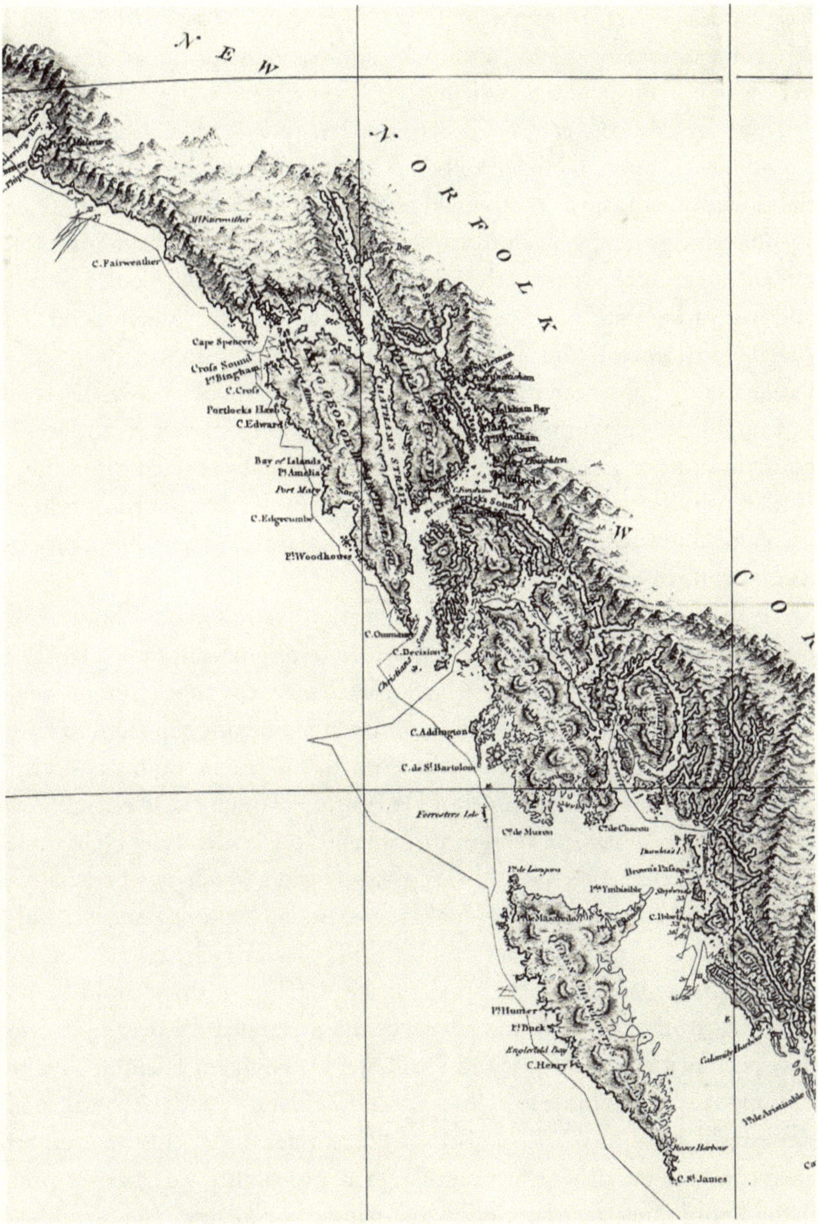

Figure 7.1 George Vancouver, *A Chart shewing* [sic] *part of the Coast of N.W. America*, 1798 (detail). Taken from the most geographically comprehensive chart in Vancouver's *A Voyage of Discovery to the North Pacific Ocean and Round the World, 1791–95,* showing the Pacific coast from San Diego to Kodiak Island, this inset depicts the labyrinthine nature of the coastline of northern British Columbia and the Alaska Panhandle. | Washington State Historical Society, 1911.5.4.17

Cook had sighted further north. At the water level, the seaboard was lined by "innumerable rocky islets and rocks, above and below the surface of the sea," which made landings difficult.[2]

In a way, Vancouver was far better suited to this mission than Cook would have been. Cook liked to be on the move, covering wide swaths of the globe's oceanic surface; he would have considered Vancouver's pointillist agenda boring and, by his third voyage certainly, perhaps beneath him. But after delineating the coast in the low to mid-50s bypassed by Cook a decade and half earlier, Vancouver was ready to "determine the reality of the discoveries attributed to the labours of Admiral de Fonte." In August 1793, he wrote from a spot near the present-day city of Ketchikan, Alaska: "The surrounding country consisted of a huge mass of steep, barren, rocky mountains, destitute of soil; whose summits were perpetually covered with snow." With unvarying predictability, each arm of every inlet terminated at its head with a patch of land from which "these mountains rose in nearly perpendicular cliffs from the water's edge."[3]

To Vancouver the geographical implications were several. The streams flowing into these channels were deemed "too insignificant to be dignified by the name of rivers, and in truth scarcely deserve the appellation of rivulets." In a reference to the legendary continental opening supposed to have existed in these latitudes, Vancouver scorned "the late prevailing conceits" about "the pretended discoveries of De Font, De Fonta, or De Fuentes," as assisted by the "*hypothetical projectors*," an allusion to Alexander Dalrymple and John Meares. After two seasons of exploration, Vancouver concluded, "our *practical labours* have thus far made it totter." He threw one last alliterative jab before departing for the California coast, which he would survey as far south as latitude 30°N. Thus far, he concluded, there could be no "doubt as to the extent or the fallacy of the pretended discoveries said to have been made by De Fuca, and De Fonte, De Fonta, or Fuentes." By the time the 1793 season came to a close, Vancouver had surveyed the continental shore from 30°N to 56°N – a total of 1,700 miles. From this he "derived great consolation in the reflection that, in all probability, we had overcome the most arduous part of our task, and that our future researches would be attended with less disappointment and fatigue."[4]

After spending the early months of 1794 in Hawai'i, Vancouver was rested and, though bothered by a chronic debilitating illness, ready to cross the North Pacific for the conclusive phase of his survey of "the coast of Northwest America." He viewed the balance of his mission thus: "Having ascertained satisfactorily that there was not any extensive navigation eastwardly, between the 30th and 56th degrees of north latitude, on this side of the American continent, I was led to believe, that if any *did exist,* it would most probably be found communicating with Cook's river." Of course, in retrospect we know that the Columbia River, via Broughton's excursion and assessment, was in fact the principal corridor providing access to the interior of North America, as Lewis and Clark and North West Company fur trader and explorer David Thompson proved within a generation. If the Columbia was not precisely an "extensive navigation" for ocean-going ships, which was Vancouver's metric, it was still a significant gateway of another sort that served as the main corridor of Hudson's Bay Company operations for the first half of the nineteenth century. With regard to "Cook's river," however, Vancouver stated that he "entertained no doubt of penetrating to a very considerable distance."[5] In this he would be disappointed.

Vancouver wanted to complete his "researches in the course of the present season" (1794), but if not, "we should at any rate ... reduce the unexplored part of the coast within very narrow limits." Since "the examination of Cook's river appeared to me to be the most important, and I did not doubt would prove the most laborious part of our task in the ensuing campaign," he determined to head there first, rather than resume where he had left off the previous summer at latitude 56°N. Vancouver wanted "a certainty in the accomplishment of that object," so he set a course for 60°N that would put him "in readiness to commence the pursuit the instant the spring was sufficiently advanced to render our endeavours practicable." After investigating "Cook's river," the intention was to resurvey Prince William Sound and then proceed south to link up with the end point of the previous year's survey in the Alaska littoral.[6]

Vancouver planned to reach the Alaskan subcontinent via the fog-shrouded coast of the eastern Aleutians that had been screened from Cook's view by rain and hazy conditions in 1778. By April 5, he had the heights of Kodiak Island in view. Conditions were frosty. Vancouver noted that "the thermometer rose to 35." He did not truly discern the complexity of the Alaska Peninsula and the adjoining islands any more clearly than Cook had, but this zone was a geographical sideshow compared with plumbing the depths of "Cook's river." To Vancouver, it was "important that not a moment should be lost" on peripheral concerns because the spring weather "had now put on a very favorable appearance." Bending past the Barren Islands, the *Discovery* entered Cook Inlet on April 12, 1794. (The *Chatham* had separated from the command ship during the sail north from Hawai'i, though Vancouver was confident his consort would soon catch up.[7])

Taking advantage of tides and favourable winds, Vancouver proceeded toward the vicinity of present-day Anchorage, Alaska. He had the "very cheerful" benefit of cold but clear weather. This afforded "an excellent view of the surrounding region, composed, at a little distance from the river, of stupendous mountains, whose rugged and romantic forms, clothed in a perpetual sheet of ice and snow, presented a prospect, though magnificently grand, yet dreary, cold and inhospitable." This was the same terrestrial prospect that had led Cook to quickly conclude that he was in a dead end, not on his way through the Northwest Passage. With the benefit of time that Cook did not have in 1778, Vancouver probed more deeply into the shoal-infused inlet with "large pieces of floating ice" abounding at full tide. Indeed, the ice drifted about the *Discovery* in such profusion as to precipitate "apprehensions for the ship's safety." The floes were "nearly as hard and ponderous as the solid rock," several of which were driven against the ship's bow, "occasioning such shocks as to awaken our fears, lest the hull of the vessel should sustain some material injury." Cook had not faced this problem because he entered the inlet over a full month later in the year. Though Vancouver had never sailed into the very high latitudes of the Atlantic, he likened his circumstance to that of "a Greenland ship." Indeed, some Russian fur traders visiting the *Discovery* were so concerned for their personal safety that they refused Vancouver's offer for them to spend the night aboard.[8]

Since Cook's time, the Russian merchants had moved their zone of influence from the Aleutians, where Cook found them, onto the shores of the Alaskan subcontinent, even into the northern borderlands of the Tlingit near Yakutat Bay. The major figure in the Russian trade at the time of Vancouver's appearance was the famous Aleksandr Baranov, working for what was then called the Northeastern Company. Formerly the Golikov-Shelikov concern from the Cook era, the company provided the foundation for the future crown-chartered monopoly known as the Russian-American Company, for which Baranov was the future and long-term governor. Baranov had arrived in Unalaska in 1790, and as chief manager moved the company's headquarters to Kodiak Island in 1791, where he was in the spring of 1794 when Vancouver arrived at Cook Inlet. Baranov later told owner Grigorii Shelikov that he had wanted to meet with Vancouver but was unable to, offering his understanding that the Englishmen "were friendly and fair with our men."[9] For his part, Vancouver was under the impression that Baranov was avoiding him. This may have been true, but for good reason. The year 1794 was a very tumultuous one in the Russian trade. The workforce was discontented, and compounding the intensifying rivalry between Russian groups (which Vancouver detected), an ever-increasing number of Yankee and British traders were visiting Alaskan waters. Now the British Royal Navy was there.

Instead of Baranov, the Northeastern Company's Egor Purtov (identified as George Portoff in Vancouver's journal) was the Englishman's principal contact, both at Cook Inlet and then later in Yakutat Bay. Purtov provided Vancouver with geographic information that essentially snuffed out the last remaining hope for a subarctic opening into the interior of the continent. Vancouver was told that he was not on "a river, but in an arm of the sea" that ran inland merely ten miles further from Cook's Point Possession (approximately where the two arms of the inlet could be said to meet), in water "scarcely navigable for boats," let alone his large ship. Vancouver also learned that the heights at the eastern extremity of Turnagain Arm formed a narrow strip of land separating it from the northwesterly extension of Prince William Sound. And if that information was not sufficiently discouraging to British interests in the region, the Russians used "every endeavour to impress us with an idea, that the American continent and adjacent islands ... belonged exclusively to the Russian empire."[10]

Vancouver, as a man of Cook, had to verify the geography for himself. Ship's master Joseph Whidbey was dispatched from the *Discovery*'s anchorage near Point Possession to conduct a preliminary survey of each arm of the inlet. Whidbey's small boats proceeded further, Vancouver noted pointedly, "than the boats from the Resolution and the Discovery had penetrated in the year 1778." Knik Arm was assessed first and, according to Vancouver, Whidbey "found all the shores round to the northward composed of compact low land; and unless the branch took a very sharp turn to the east or S.E. it had every appearance of terminating not many miles beyond the extent of his examination, in a spacious bason or harbour." That quotidian word from the 1792 and 1793 surveys – "terminating" – and its variants, took on a somewhat different tone here at Cook Inlet in the spring of 1794, for Vancouver was now in the latitude of not only Cook but also Peter Pond. The next line in Vancouver's account is revealing. Whidbey's report "disappointed my expectations, as it was not easily reconcilable with the idea we had formed of the interior distance to which we should be led by waters of this extensive opening." Vancouver was making the same mistake his mentor had – one probably reinforced, in his case, by a recollection from two years earlier about the sizable bay into which the Columbia River flowed. "The presumption that our progress would speedily be stopped, became by this information very strong," he wrote. The implication was that the *Discovery* should have been able to sail deeper into the continent in the manner that Broughton's expedition pursued the Columbia. Like his teacher, but with the temporal latitude Cook never had in these waters, Vancouver insisted that Whidbey's theory "remained yet to be proved." Accordingly, he "determined to persevere in my former intention" of a more extensive sail. Weighing anchor at the next flood tide, he "steered for the bason or harbour described by Mr. Whidbey."[11]

Vancouver had learned well. Having come this far, he was not going to let himself be second-guessed by the likes of Dalrymple and other enthusiasts for a Northwest Passage as had happened to Cook posthumously. But Whidbey's surmise proved correct. Probing even deeper with his ship than Whidbey had gone in the small boat, Vancouver sailed the *Discovery* into a small cove (near the Eagle River), and from that vantage, the "shores, in every direction in which we had seen them, had uniformly appeared to descend gradually, from the mountains to their termination at the water-side." Knik Arm was thereby ruled out as an inland corridor, which seemed a vindication of Cook. But Vancouver was to more fully discern something that had misled the great navigator when Vancouver determined that the water nearly to the head of the "bason" was "little, if at all, fresher than that of the ocean, although at the distance of near 70 leagues [210 miles] from the sea."[12] To explain this discrepancy, we must bear in mind that Cook was visiting at a time (late May/early June) when both the floating ice had fully melted and the glacial runoff was at its peak, thus giving the inlet a freshwater overlay. Vancouver, arriving two months earlier, hereby found unmistakable evidence that Cook's opening was a saltwater inlet and not a river.

Whidbey's probe of Turnagain Arm also proved the verity of Russian intelligence. As Vancouver recorded, this "arm, (for it was now proved to be no longer intitled to the name of a river)," terminated in a circular basin "surrounded by high steep barren mountains, covered with perpetual snow." Dodging ice and finding that a thirty-foot ebb tide almost completely drained its upper extent, "Mr. Whidbey considered it unsafe to be navigated, and undeserving any further examination." Upon Whidbey's return, Vancouver deemed the "termination of this extensive inlet" an "extraordinary and unexpected" event. As with Knik Arm, Vancouver again conducted

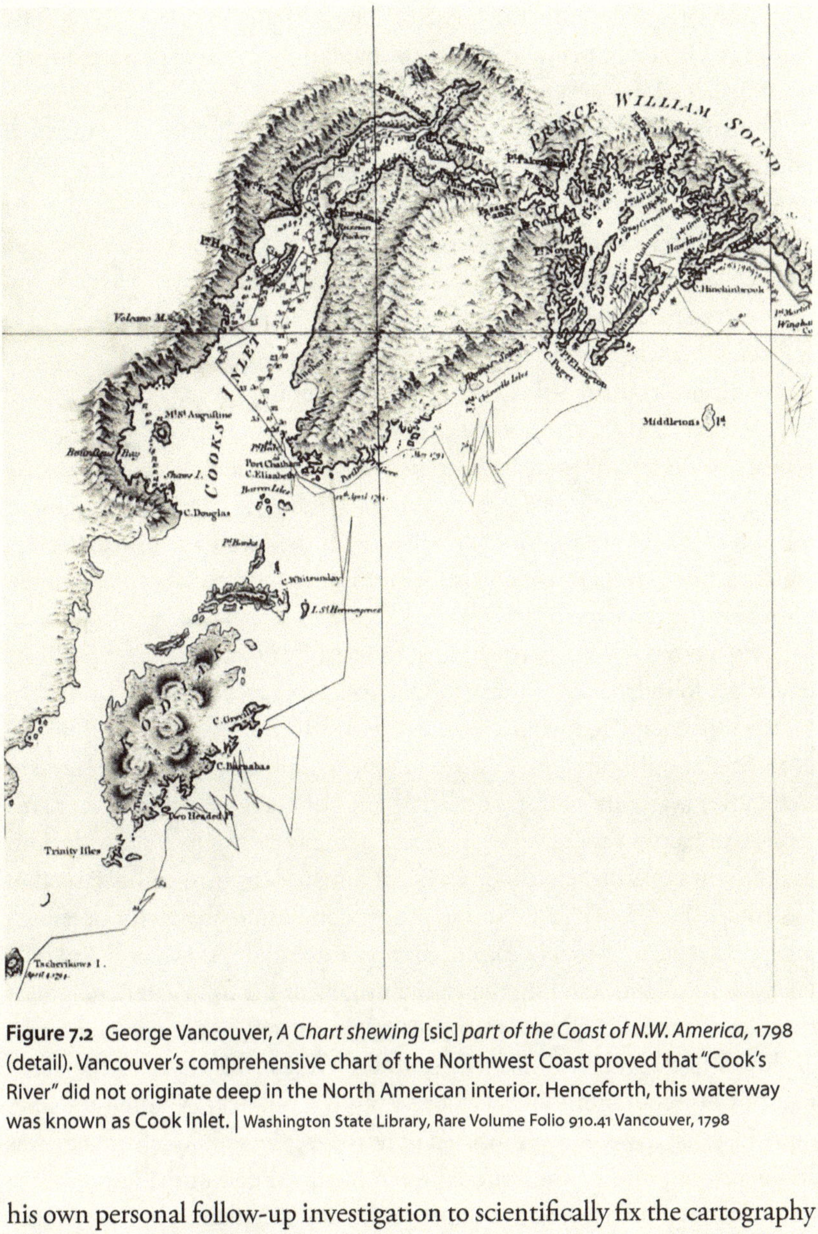

Figure 7.2 George Vancouver, *A Chart shewing* [sic] *part of the Coast of N.W. America,* 1798 (detail). Vancouver's comprehensive chart of the Northwest Coast proved that "Cook's River" did not originate deep in the North American interior. Henceforth, this waterway was known as Cook Inlet. | Washington State Library, Rare Volume Folio 910.41 Vancouver, 1798

his own personal follow-up investigation to scientifically fix the cartography of the upper extent of the inlet for his chart, rather than to prove Whidbey's principal conclusion right or wrong. Vancouver saw the upper limit of Turnagain Arm encircled by "stupendous mountains covered with snow." This was followed by the now common scene and textual passage from his account of the three-year survey: "This we had found to be the case with the connected ridges on the coast of North West America, whose lofty summits formed in many instances the bases only of the still more stupendous detached mountains."[13]

Vancouver had reached the tipping point not only for his voyage but perhaps also for Europe's centuries-long quest for a navigable route above or through North America. Given its terrestrial boundaries and how the water was constituted chemically, Vancouver concluded that what had heretofore been called Cook's *River* could no longer be considered such. He wrote on May 6, 1794: "I shall therefore distinguish it henceforth as an *inlet*." He followed this with a commentary alternately praising Cook and criticizing him: "Thus terminated this very extensive opening on the coast of North West America, to which, had the great and first discoverer of it, whose name it bears, dedicated one day more to its further examination, he would have spared the theoretical navigators, who have followed him in their closets, the task of ingeniously ascribing to this arm of the ocean a channel, through which a north-west passage existing according to their doctrines, might ultimately be discovered."[14]

A few comments are in order. First, it must be noted that these few lines of text are foundational to the orthodox historiographic interpretation that Vancouver completed what Cook had failed to do. Notwithstanding the evolution of the Northwest Passage as a cartographic image, or the fact that Vancouver was looking for a different kind of corridor at a different latitude than Cook, Vancouver's experience actually proves how perspicacious his tutor was. With less time, and without help from the Russians that Vancouver had, Cook came to the proper conclusion about this opening's prospects as a commercial shortcut for ocean-going ships. His only real mistake was not understanding the glacial origins of the freshwater content of the inlet. Nor could Cook have anticipated the imaginative use that terrestrial explorers like Pond and Mackenzie would make of his error, to the detriment of his reputation. Furthermore, even if Cook had explored the inlet one more day, as Vancouver insisted in retrospect he should have, there was still a vast amount of coastline between Mount Edgecumbe and Nootka Sound that could have, and did, prove ripe for speculation by "theoretical navigators" such as Dalrymple. Finally, as Robin Inglis points out, historians have chosen to read Vancouver's text as a criticism of Cook but it could just as easily be viewed as a forlorn wish that his former commander had "dedicated one day more" to the effort in order to mitigate the negative impression that had resulted. Not that Cook was wrong in his ultimate determination of the practicability of this waterway; rather, in retrospect, his inadvertent use of innocent nomenclature such as "River Turnagain" (broadened by others to "Cook's River") was used against him.[15] Similarly, Vancouver would have been chagrined to have seen what use was made of his journal to criticize his mentor. (See Figure 7.2.)

If not deflation, then a sense of anticlimax infused Vancouver's journal from this point on. That the extent of Cook Inlet was so quickly determined freed up a lot of time for the balance of that season's exploratory agenda. The good news, as Vancouver phrased it, was that "the prospect of concluding our survey of the coast during the present season was greatly improved." This did not mitigate, however, the troubling "mortification from the reflection, that our opinions respecting the extent to which these waters were likely to lead, had been so extremely erroneous."[16] "Cook's river" never having been fully described by Pond, we are at a loss to know just what Vancouver was expecting to see along its upper reaches, unless he meant sailing all the way to Great Slave Lake as depicted in *Gentleman's Magazine*. Both Dalrymple's *Plan for Promoting the Fur-Trade,* and, correspondingly, Vancouver's instructions from the Admiralty, were far more heavily invested in what might be called the zones of de Fuca and de Fonte further south than they were with the Pond/"Cook's River" nexus.

Nevertheless, some words are literary signposts in travel literature, and "mortification" is one of them. It strikes at two of the most important premises underlying exploratory accounts: the heroic significance of the enterprise and the prescience of the principal observer. No explorer wants to be made a fool of or proved wrong, which is why the typical traveller's narrative showed an almost preternatural knowledge of future outcomes. This awareness, of course, was based on the retrospective casting of their journals, wherein surprises are rarely recorded and odious outcomes can be edited out, especially prior to publication. (The only other time in Vancouver's four-year voyage that he uses the term "mortification" was when he was surprised to find that Dionisio Alcalá Galiano and Cayetano Valdés had preceded him into the Gulf of Georgia.[17]) Vancouver's explanation for his predicament was what he called "the fallacy of analogous reasoning, which the human mind is too apt to render subservient to some favorite hypothesis, and too frequently to adopt, even under the appearance of manifest contradictions."[18] This phrasing appears to be a nod in Cook's direction regarding the primacy of actual observation over speculation in general, and specifically the seductive idea that Hudson Bay in the Atlantic basin was in some fashion mirrored by a comparably sized inlet off the Pacific.

The *Chatham* finally caught up with the *Discovery* on May 7 and both ships exited Cook Inlet in tandem. It was at this juncture, on their way out to the ocean, that Vancouver bemoaned not being able to meet with Baranov, someone he deemed a well-travelled man who could contribute to "the science of geography." A reliable source like Baranov stood in contrast to some lower-ranking Russians who informed Vancouver that a branch of Cook Inlet "communicated with an immense lake to the north-east, in which were whales, seals, sea otters, and a variety of salt-water fish." This was a geographical fantasy worthy of de Fuca or de Fonte. Vancouver dryly observed: "This incident will afford sufficient proof how little dependance is to be placed on information assumed from persons but ill qualified to answer our questions if understood, and with whom, for want of language, our communication was necessarily very uncertain and imperfect." He offered this episode as evidence "on how slight foundation the theories of mediterranean seas, and of a north-west passage, may often rest."[19]

In an ironic comment echoing his mentor's outlook when leaving the same body of water in 1778, Vancouver closed his discussion by noting that, notwithstanding their very early start to the 1794 season of discovery, the expedition "had yet accomplished little in comparison to the extent of coast we had yet to survey." He set a course for Prince William Sound, reversing the track Cook had taken sixteen years earlier. The scene there in late May and early June, he wrote, "seemed to wear a much more wintry aspect than the countries bordering on those more northern inland waters we had so recently quitted." Inevitably, Vancouver harked back to Cook's more cursory survey of this inlet in May 1778, a somewhat fractious time during that voyage given the great navigator's skepticism of his junior officers' bounding optimism that Prince William Sound was the gateway to a passage above the continent. In contrast to Cook, Vancouver pointedly noted that the "principal object" of his expedition was "*the survey of the shore of the continent.*"[20]

Toward that end, Vancouver dispatched Whidbey and Johnstone in separate shore boat surveys of every appendage of the complex Prince William Sound shoreline. Encountering a settlement of the seemingly ubiquitous Russian fur traders in one of these northern inlets, Johnstone asked, with a view toward expediting his work, whether they possessed a "chart of the Russian discoveries in these regions." This inquiry was consistent with Vancouver's desire to meet with Baranov when they were at Cook

Inlet. As best as could be explained to Johnstone across the language barrier, there was no chart, but a "Mr. Smyloff, who commanded the galiot, was well informed." Here was a proverbial small-world moment, for this was the same Gerasim Izmailov whom Cook had met at Unalaska in the fall of 1778, and to whom he had entrusted vital correspondence. Unfortunately, Izmailov was now "absent on an excursion," and he was indeed a peripatetic figure.[21] We know from Russian records that, post-Cook, Izmailov captained the ship that first took Grigorii Shelikov to Kodiak in 1783, and in 1786 he pioneered his firm's expansion into Prince William Sound. A few months hence, he would sail from Okhotsk to Kodiak Island with a shipload of passengers (convict settlers and Orthodox churchmen), animals, seed grain, and tools, as well as correspondence from Northeastern Company headquarters intended for Baranov.[22]

Unknown to Johnstone, at that very moment Izmailov was heading for the British. Vancouver related after their meet-up: "This gentleman having heard of our arrival in Cook's inlet, he had gone thither for the purpose of paying us a visit, but had been disappointed of his object by our departure thence." In his zeal to catch up with these visitors, Izmailov reached Prince William Sound over the narrow "isthmus" separating it from Cook Inlet. Here was very practical confirmation of the intelligence gleaned previously that the eastern extent of Turnagain Arm was proximate to Prince William Sound. (Today a combined rail-highway tunnel links these bodies of water.) Vancouver aptly characterized Izmailov's role, noting that he "presided over the maritime part of this mercantile company" – and thus was able to share both geographical hints about the adjoining coastline to the southeast and insights into his firm's operation. He told Vancouver that his company's easternmost settlement was the one Johnstone had stumbled upon, established the previous year. Vancouver also learned that the "Russian government had little to do with these settlements." All the fur ventures in the North Pacific were "independent mercantile companies," employing about 400 Russians. Without doubt, the sharpest, somewhat ironic disclosure was Vancouver's, who admitted: "I have to regret very much our want of knowledge of the Russian language, because the results of all our inquiries concerning the object of that [nation's] power in extending its immense empire to these distant and inhospitable shores, became from that cause in a great measure nugatory, vague, or contradictory."[23] If, as some historians assert, Cook was negligent for not taking a Russian speaker with him on his third voyage, what must be said about Vancouver, who sailed in his wake with knowledge of such a need? Cook was not the only British navigator to sail to the North Pacific unprepared for this contingency.

Having, through the efforts of Whidbey and Johnstone, "completed the examination of the whole of Prince William's sound, as it respected the boundary of the continent," Vancouver re-entered the Gulf of Alaska. Plying to the southeast, he took in the view of "the still connected chain of lofty mountains, whose summits are but the base from whence mount St. Elias towers, majestically conspicuous in regions of perpetual frost."[24] Mount St. Elias reaches 18,008 feet in height, and, given its prominence as a benchmark in the narratives of both Cook and Vancouver, it is flanked by peaks named after both British explorers.

In Yakutat Bay, the *Chatham*'s Peter Puget encountered ten Russian fur traders and a legion of Alutiiq hunters hailing from as far away as Kodiak Island. This was the pioneering effort of the Northeastern Company into that bay, led by the same Egor Purtov who had met the British expedition earlier at Cook Inlet. It offered further evidence of both the continual eastward colonization of the Russians, who in Cook's time had ventured no further than the easternmost island in the Aleutian chain, and the vassalage the Alutiiq were held in. (Purtov was also in the vanguard of the firm's move into Sitka Bay two years later, an expansionary effort that culminated in a two-year war with the Tlingit starting in 1802.) Purtov's men provided the British sailors with a supply of halibut, and on their eventual separation, he gifted Puget with a luxuriant sea otter pelt and a three-person baidarka, a large kayak-like vessel. The Vancouver expedition's roving encounter with Russian expansionism prefigured the contest of sovereignty claims fuelled by commercial competition that would play out in the waters of the northeast Pacific and European capitals over the next three decades.[25]

As the expedition proceeded southward along the coast of Southeast Alaska, magnificently conspicuous Mount Fairweather came into view. Gazing about, Vancouver noted that it was part of range of snow covered peaks running in a southeasterly direction from Mount St. Elias; the latter still visible even though it was now 150 miles away. As he conducted the ships toward Lituya Bay and then Cross Sound, Vancouver marvelled at "the length of time we had been in sight of these very remarkable lofty mountains." They gave every evidence of being an undivided range the entire length of the Northwest Coast. Entering Cross Sound, he tackled the last portion of the continental shore that Cook had bypassed in his rush to reach latitude 65°N in 1778. Cross Sound, Vancouver reminded readers of his journal, had been named by Cook. Within the larger context of Cook's supposed inaccuracy upon which his own voyage was founded, Vancouver stated that "in common justice to the abilities of that able navigator, it must be observed, that he has given a much more accurate description of it, than from the transitory distant view he had of it, might have been reasonably expected."[26] Going still further in his defence of Cook, he maintained that whatever navigational errors were made along the Northwest Coast, such as niggling mistakes in latitude and longitude, none would have occurred "had circumstances permitted Captain Cook to have approached nearer to these shores on passing by them."[27]

Vancouver was confirming that from afar Cook had discerned that Cross Sound branched "into several arms, the largest appearing to take a northerly direction," principally referring to what Vancouver called Glacier Bay. (Lynn Canal, a neighbouring channel to the east linked by Icy Strait, which has the present-day city of Skagway at its head, is still larger.) Glacier Bay had a "great number of small, though hard pieces of loose ice" floating at the entrance, "some of which, at first sight, occasioned considerable alarm, from their strong resemblance to sea-beaten rocks, just level with the surface of the water." An exploration of Glacier Bay found that it terminated in "compact solid mountains of ice, rising perpendicularly from the water's edge, and bounded to the north by a continuation of the united lofty frozen mountains that extend eastward from mount Fairweather." These glaciers, of course, were the source of the floating ice chunks encountered where Cross Sound met Icy Strait. Modern ice scientists have determined that the glaciers in this inlet had started receding by the time of Vancouver's appearance, coming as it did near the end of the Little Ice Age and its period of maximum glaciation between the years 1740 and 1750. The pace of recession has accelerated since the start of the industrial age.[28]

Proceeding deeper into the labyrinth, Vancouver dispatched Whidbey on the most extensive of all the small-boat surveys conducted during the three-year investigation of the Northwest Coast. The examination of Lynn Canal, named by Vancouver after his birthplace, showed that "both sides of this arm were bounded by lofty stupendous mountains, covered with perpetual ice and snow." From Whidbey's report, Vancouver correctly deduced that these peaks were a continuation of the same "lofty ridge of snowy mountains so repeatedly mentioned, as stretching eastwardly from mount Fairweather, and which, in every point of view they had hitherto been seen, appeared to be a firm and close-connected range of stupendous mountains, for ever doomed to support a burthen of undissolving ice and snow." He took pains to point out that the head of Lynn Canal was a noteworthy geographical marker in that its calculated longitude of 135°W "approaches nearer to those interior waters of the continent, which are said to be known to the traders and travellers from the opposite side of America, than we had found the waters of the North Pacific penetrate in any former instance." To Vancouver, this was "Arathapescow lake, as laid down in Captain Cook's chart"[29] – Samuel Hearne's name for Great Slave Lake as drawn on the master map completed after the third voyage, theoretically only 320 miles away.

Hearne was profoundly wrong about how far west he had been, as was Pond, who explored the Athabasca district more extensively in the decade after Cook. Great Slave Lake is actually 700 miles from Lynn Canal. But without regard to the accuracy of this lake's placement, there was a larger truth at play about the Northwest Passage, even in its revised form. As Vancouver put it, given "the close connection and continuation of the lofty snowy barrier, so frequently before adverted to, trending south-eastward, and nearly parallel to the direction of the continental shore, little probability can remain of there being any navigable communication, even for canoes between such waters and the North Pacific Ocean, without the interruption of falls, cataracts, and various other impediments."[30] This assessment was both a recapitulation of his true mission, not the one historians construed, and a practical reassessment of topographic realities in the wake of fur trader criticism of Cook and cartographic hyperbole. Its reference to canoe travel perfectly anticipated the experiences of Alexander Mackenzie (occurring almost simultaneously), Lewis and Clark, and Simon Fraser.

By late July 1794, Vancouver, in a Cook-like process of elimination, had reduced the prospect of a Northwest Passage to a little more than one parallel of latitude. The last redoubt lay at the head of the inlets running south from where Lynn Canal met the Icy Strait, a watercourse that would be named Chatham Strait after his consort. Vancouver employed his standard tactic of anchoring the large ships while the shore boats scoured the inside (eastern shore) of Chichagof and Baranof Islands. When the rowers completed their work in late August and returned to the ships at a station Vancouver called "Port Conclusion" (in Frederick Sound on the opposite side of the island holding present-day Sitka), Whidbey and Johnstone "communicated the glad tidings of their having effectually performed the service, and attained the object that had been expected from this expedition." Vancouver added, appropriately, that the "laborious nature" of his voyage could be better and more properly conceived by looking at his charts, as opposed to reading his text. The intellectual and physical effort that went into the completion of his comprehensive survey (from San Diego, California, to Anchorage, Alaska) was "a service that had demanded our constant and unwearied attention, and had required our utmost abilities and exertions to bring thus to a conclusion, could not, after the indefatigable labours of the three preceding years, fail of exciting in the bosoms of our little community, sensations of a nature so pleasing and satisfactory, that few are likely to experience in the same degree, who were not participators in its execution; and to the imagination of those alone, must I refer the happiness we experienced on this interesting event."[31]

Whidbey and Johnstone, by encircling Admiralty Island and surveying the inlets leading eastward off Stephens Passage (named after the secretary of the Admiralty board), found, as Vancouver phrased it, "the shores of the continent ... bounded by a continuation of those lofty frozen mountains," just as they had been "from Cook's inlet to this station."[32] In this moment, Vancouver and his men effectively closed the book on the nature of the coastline that some had thought Cook had inadvertently mischaracterized, but that, in a fundamental way, he had not. In this fashion, Vancouver's expedition was methodologically akin to Cook's second and third, in that he had discovered nothing by proving that, in this case, a "communication" to the interior of North America navigable by ocean-going ships did not exist.

The end of the survey prompted a spontaneous celebration, in no small measure because the crew in Whidbey's shore boat excursion had survived a series of skirmishes with the Tlingit, likely brought about because the Englishmen mistook Tlingit eagerness to trade for hostile intentions. (By this time, the intercultural environment was quite contentious, as opposed to Cook's time on the Northwest Coast. Vancouver's naturalist, Archibald Menzies, reported that in 1792, merely twelve years after John Gore and James King had returned Cook's ships to England, thirty trading vessels were spotted along the Northwest Coast.) Relating insights found in Whidbey's journal, Vancouver recorded that "it is not possible for language to describe the joy that was manifested in every countenance, on thus meeting their comrades and fellow-adventurers, by which happy circumstance, a principal object of the voyage was brought to a conclusion." Cheers, hugs, and "hearty congratulations" abounded in recognition of the "laudable pride that been entertained by both parties, in having been instrumental to the attainment of so grand an object."[33]

Vancouver related a humorous anecdote recounting the travels of Whidbey's and Johnstone's "little squadron," which, after they met up, "proceeded to a cove ... where they took up their abode for the night. In the course of the evening no small portion of facetious mirth passed amongst the seamen, in consequence of our having sailed from old England on the *first of April,* for the purpose of discovering a north-west passage, by following up the discoveries of De Fuca, De Fonte, and a numerous train of hypothetical navigators." Alexander Dalrymple would have been counted among the latter, and to an extent John Meares, even though he was an actual navigator, because of his inventions and exaggerations. Vancouver considered the moment when the boats returned to the ships at anchor at the tip of Baranof Island the "finishing stroke" of the discovery phase of his expedition. With "the principal object ... of this voyage having at length been completed," he was now of a mind to state, somewhat naively as it turned out, that "the precision with which the survey of the coast of North West America has been carried into effect, will remove every doubt, and set aside every opinion of *a north-west passage*, or any water communication navigable for shipping, existing between the North Pacific, and the interior of the American continent, within the limits of our researches."[34]

This became an important sentence in the history of the Northwest Passage, and especially of Cook's connection to that quest, because Vancouver's instructions *never* referenced his mission as a search for the Northwest Passage, but rather the assessment of several prospects that might offer a vague "communication" or opening analogous to Hudson Bay that was navigable into the depths of the continent. Thus we see here a bit of literary grandiosity on Vancouver's part, though it may have been added to the manuscript by his brother. Vancouver died before his journal was published in late August or early September 1798, and we know its finishing touches were applied by John Vancouver.

Even if one discounts that theory, Vancouver's post-expeditionary disputes with two prominent figures may have tempted him to embellish his contributions to geographical understanding with an oblique but still recognizable jab at Cook. (He put the on-board naturalist, Joseph Banks's protege Archibald Menzies, under house arrest crossing the Atlantic on the last leg home, and had former midshipman Thomas Pitt, later Lord Camelford, flogged and sent home from Hawaiʻi on another ship because of disciplinary infractions. Pitt would publicly challenge Vancouver to a duel and otherwise humiliate him on the streets of London.) Up to this point in his account,

Vancouver was content to simply deride "all speculative opinions respecting the existence of a *hyperborean or mediterranean ocean* within the limits of our survey,"[35] but at this pivotal juncture, where the voyage account reaches its narrative crescendo, he introduces for the first and only time the notion that *he* has vanquished the ancient idea of the Northwest Passage. Several generations of historians have turned this sentence into proof that Vancouver finished what Cook had "failed" to do, when in fact it was merely gratuitous bravado by Vancouver or perhaps his brother.

This can be verified independently. Vancouver made one last call at Nootka Sound on the way home. There, on September 8, 1794, he sent a letter to Admiralty board secretary Philip Stephens that was entrusted to a fur trader in the hope it would reach London before he did. Coming at the conclusion of his three-year survey, this missive can be viewed as an unvarnished recapitulation of his efforts. Requesting "the honor of communicating to my Lords Commissioners of the Admiralty," Vancouver stated, in a textual nod to his instructions, "it is with the greatest satisfaction I am at length able to inform them of our having finally traced and determined the continental boundary of N.W. America ... through all its various turnings and windings," at least those that were large enough to admit his shore craft. "During this investigation," he continued, "we have never been able to penetrate beyond the barrier of the lofty mountains, which covered with eternal frost and snow, [and] extend nearly in a connected chain along the western border of the continent ... though in some instances considerably farther removed from the ocean than in others." He noted having entered "Cook's River" the previous April in "boisterous and severe cold weather," but by May 6 had deduced that "the inlet by no means constitutes a river." To substantiate this point, Vancouver documented that his men had gone "a few leagues" farther than William Bligh did for Cook, and established that the *Resolution*'s sailing master had in fact reached the furthest navigable extent for ships on Knik Arm. Beyond that station, the inlet shoaled and was completely "dry at low tide." Turnagain Arm "terminated in a similar manner."[36] In this correspondence with his superiors (and formerly Cook's), Vancouver never made any claim to have disproved the existence of a Northwest Passage, even the truncated version popularized by Dalrymple.

In a private letter written a few weeks later and addressed to a friend, John Sykes, father of one of his midshipmen, Vancouver declared that he had arrived at Nootka "in high health and spirits having *finally determined* the nonexistence of any water communication between this & the opposite side of America ... beyond all doubt or disputation."[37] Here Vancouver trumpets his accomplishment a bit more than he did in his letter to Stephens, though not as much as he (or his brother) did in the published journal. The concept of a "water communication" reaching the "opposite side of America" could be read as being consistent with an inland western sea that narrowed the distance from the Pacific Ocean to Hudson Bay. On the other hand, such phrasing was open to being interpreted as his having foreclosed any corridor *directly* connected with the Atlantic basin. But then, no sensible geographer, armchair variety or otherwise, including Dalrymple, truly believed that this image of a Northwest Passage was realistic after Hearne's findings became known in the mid-1770s. If, as historian Barry Gough suggests, Vancouver was somehow disappointed in not finding the Northwest Passage, even in its post-Dalrymple truncated form, the scant documentary evidence for that assertion revolves solely around his expectations for the inland extent of "Cook's river."[38]

Archibald Menzies also sent a revealing letter home from Nootka Sound, to his patron Banks. He wrote that "Cook's River" was "pursued to its outmost extremity, and You will no doubt, Sir, be equally surprized as we were, when I inform You that it is only an Inlet of the Sea, similar to many other great Inlet[s] on this coast – falling short, not only, of those great expectations conceived of it, but even, of the extent assigned to it."[39] These remarks, combined with Vancouver's on the same subject, suggest that "Cook's River" loomed larger in the reckoning of prospects for a Northwest Passage for those actually conducting the expedition than did the Admiralty instructions for the voyage, which emphasized openings much further south. This speaks to the durability of the image Pond had been able to convey about his vision of the inland lake and river system emerging at Pacific tidewater over the last half of the previous decade, culminating with the *Gentleman's Magazine* report published in 1790 during the formative stage of what became the Vancouver expedition.

Vancouver's Summary Reflections

After stops in California, Chile, and St. Helena, Vancouver reached London on October 17, 1795. He prepared a brief for his Admiralty superiors three days later, in which he borrowed from the sentiments expressed in his journal and the letter sent from Nootka Sound. From his "very extensive and accurate examinations," he had "removed every doubt and set aside every opinion of a Northwest Passage within the limits of our searches."[40] There was now little time available to Vancouver before his death on May 12, 1798, at the age of forty-one. (Historians have speculated that the cause of death was a severe thyroid deficiency, or, since he came down with malaria during his stint in the Caribbean, that exposure may have led to a chronic renal condition known as nephritis, or Bright's disease.) While preparing his journal for publication, Vancouver was also fending off public accusations from Menzies, Pitt (now Lord Camelford), and others that he had behaved intemperately during the voyage.

As noted earlier, John Vancouver finished the manuscript journal for the press and covered it with a letter addressed to the King, which was included in the publication issued that summer. The tone of this letter and some of its phrasing provides evidence that the text touting George Vancouver's role in vanquishing the idea of a Northwest Passage in juxtaposition to Cook's mission came not from the commander's pen but was rather a kind of family apologia. First, John Vancouver gave Cook's final voyage very short shrift by citing as its principal accomplishment "the near approximation of the Northern Shores of Asia to those of America." His brother George, on the other hand, had the greater accomplishment, having proven "the complete certainty that, within the limits of his researches on the Continental Shore of North West America, NO INTERNAL SEA, OR OTHER NAVIGABLE COMMUNICATION whatever exists, uniting the Pacific and Atlantic Oceans."[41]

The title George Vancouver applied to his account is probative – *A Voyage of Discovery to the North Pacific Ocean ... to ascertain the existence of any navigable communication between the North Pacific and North Atlantic Oceans* – in contrast to Cook's last title, *A Voyage to the Pacific Ocean ... to Determine ... the Practicability of a Northern Passage to Europe.* In his introduction, Vancouver readdressed the circumstances that had brought his venture into being. Fur trade "adventurers," he charged, navigated the Northwest Coast without "proper astronomical and nautical instruments," and with only commercial interests in mind, "they had neither the means, nor the leisure, that were indispensably requisite for amassing any certain geographical information." They cavalierly filled "the blanks in the charts of Captain Cook" with cartographic images and accompanying texts, which often "contradicted each another." In Cook's wake, the traders saw "a coast apparently much broken by numberless *inlets*, which they had left almost intirely unexplored." Their findings "gave encouragement once more to the hypotheses; and the favorite opinion that had slept since the publication of Captain Cook's last voyage, of a north-eastern communication between the waters of the Pacific and Atlantic oceans, was again roused from its state of slumber, and brought forward with renovated vigour."[42]

Vancouver's implication was that there was nothing intrinsically faulty with Cook's work along the coast; relatedly, he also carefully employed the term "communication" instead of "passage" to describe the inter-oceanic link that he was looking for. The legends of de Fonte and de Fuca had been dredged up, he charged, by "the enthusiasm of modern *closet philosophy*, eager to revenge itself for the refutation of its former fallacious speculations," a coded passage aimed at Alexander Dalrymple and his theory about *Terra Australis Incognita*. Vancouver objected to an even worse calumny against Cook's memory, the suggestion that he was "amongst the *pursuers of peltry*," a criticism aimed in the direction of John Meares and his ilk, who "dared even to drag him [Cook] forward in support of ... visionary conjectures." Vancouver allowed that Cook himself would have coped with such "animadversions" through the force of "his own judicious opinions founded as they were on the solid principles of experience, and of ocular demonstration, uninfluenced by any prejudice, and unbiassed by any pre-conceived theory or hypothesis."[43]

For Vancouver the "profundity" of Cook's judgment alone was sufficient "to obtain credit with the public," and he took care to point out that at the conclusion of Cook's third voyage, no follow-up survey was envisioned. But then, at the end of April 1790, "intelligence was received that the Spaniards had committed depredations on different branches of the British commerce on the coast of North-West America, and that they had seized on the English vessels and factories in Nootka sound," cancelling the *Discovery*'s mission to "southern regions." Vancouver established that it was not until after the "disputes between the courts of London and Madrid" were settled that it was "deemed expedient" to send a British official to Nootka Sound to accept "restitution," with the additional element of conducting "an accurate survey of the coast" from Cook Inlet to the tip of California at latitude 30°N.[44] In this respect, Vancouver's narrative conveys the notion that his voyage was one originating in the demands of statecraft and not discovery, and certainly not as a corrective of Cook as evidenced, in part, by the addition of the low latitudes of the California coast, which Cook had never visited, to the research agenda.

In the concluding sections of Vancouver's account, the scope of his mission and accomplishments was revisited one last time. His was "a plain narrative" of fact and observation without "the obtrusion of specious opinions, to hypotheses, however ingenious," another salvo aimed at Dalrymple and Meares. Indirectly, Vancouver presciently addressed the twentieth-century critique of J.C. Beaglehole, who propounded the thesis, which became an orthodox view among historians, that a supposedly tired Cook had overlooked several important opportunities for discovery during the staging phase of the third voyage in the South Pacific. According to Vancouver, only a few "small spaces" remained "unexplored in the Pacific or Indian oceans," and they were "too insignificant to be an object of enterprize." On the other hand, there were "parts of the coasts, both of Asia and America, which would yet afford employment for the labourers in the science of DISCOVERY." But then the narrative ends on a grandiose note, with Vancouver asserting that his voyage to the Northwest Coast was on par with Cook's second expedition, and at the same time had somehow completed Cook's third-voyage mission. His proof that there was no "navigable communication between the waters of the Pacific and Atlantic oceans" was an accomplishment equivalent to "the hypothesis which gave as a counterpoise to the globe *a southern continent*, which the indefatigable diligence of Captain Cook completely subverted."[45] However, here again we must note that the commander's brother, John, finished the concluding sections of the draft manuscript and so the authorship of this text is in doubt.

Personal or family boastfulness aside, the entirety of Vancouver's journal more consistently conveys the idea that the various reports of straits or channels to the western Mediterranean he was directed to explore were groundless from the beginning. In the words of historian Glyn Williams, Vancouver's voyage "enabled him to contradict and humiliate those who, in their eagerness to find a northwest passage, had dared to impugn the achievements of his old commander." Thus it turned out, as Williams summarizes, that "Cook had been wrong on the location of the continental shore, but was instinctively right about the wider question of the existence of a northwest passage."[46] In that sense, then, Vancouver did not so much finish work Cook had failed to accomplish; instead, he clarified misconceptions that had crept into the evolving image of the Northwest Coast as created by the fur traders, most notably John Meares.

In an addendum to his narrative labelled "Notes and Miscellaneous Observations," Vancouver pondered one last item of relevance to the Cook third-voyage nexus. Without any apparent means of "communication" with the interior of the continent, how could one explain "the similarity observed in the race of people inhabiting the shores of Hudson's bay and those to the northward of North West America?"[47] The demographic dispersion of the Inuit was the same problem Cook had puzzled over, and, as sailors, neither seems to have been able to countenance the notion that this northern people could have traversed the Arctic basin without access to navigable water, seasonally or otherwise.

Aftermath

Vancouver suffered the same posthumous fate as Cook in two respects. In the most obvious instance, Vancouver, like Cook, was subject to post-voyage criticism for his supposedly arbitrary treatment of the crew. More central to the theme of this book, even after Vancouver succeeded in the mission of disproving the Dalrymple/Meares idea of a western analogue to Hudson Bay, the Northwest Passage as an *idée fixe* simply shifted shape into its third iteration, from a saline solution to a network of freshwater bodies that could span the continent and spur commerce with Asian markets. One of the first reviews of Vancouver's published account appeared in the *Annual Register* for 1799. It criticized him for not exploring the Columbia River in greater detail, suggesting (correctly, it turned out, with regard to the Fraser River) that if one opportunity had been missed perhaps others had been too. The review concluded: "On the whole, we must be allowed to repeat, that the prospect is considerably lessened, but, that it is by no means yet proved, that a N.W. passage does not exist."[48]

What explains the hardiness of this and other geographical delusions? Historian David Mackay argues that before Cook demolished the pastoral fantasy that had previously impelled British interest in the mythical southern continent, "one could say that armchair geographers and the more optimistic discoverers read what they wanted to see into new lands, informed by the prevailing Arcadian tropes. They employed prophetic geography which saw only opportunities. No dark, tortured mind hypothesized a frigid southern continent in high, unreachable latitudes." As for the Northwest Coast after Cook but before Vancouver, "no one in the eighteenth century accurately predicted the ... formidable, northern coast of British Columbia and Alaska. No one suggested that all those coastal indentations were simply that – dead ends which provided no channels to the East."[49]

Beaglehole once described the durability of the Northwest Passage thus: "If men wanted a new strait from the Atlantic into the Pacific, then they themselves must cut through the land" because "no fleets would ever disembogue from the wide Strait of Anian."[50] As described earlier, the visionary elements of this terrestrial version of the passage were first seeded by Pond but its strategic import was formalized by Mackenzie. In *Voyages from Montreal* (1801), Mackenzie cited "Vancouver's minute survey," which revealed the Columbia River as the only "level country in all the Southern extent of continental coast from Cook's entry," that is, Cook Inlet. In several oft-overlooked lines of text coming just prior to his description of how British commercial interests might dominate the North American fur trade, Mackenzie wrote: "The discovery of a passage by sea, North-East or North-West from the Atlantic to the Pacific Ocean, has for many years excited the attention of governments, and encouraged the enterprising spirit of individuals. The nonexistence, however, of any such practical passage being at length determined, the practicability of a passage through the continents of Asia and America becomes an object of consideration." Adding Vancouver's serial contributions to his 1793 expedition to the Pacific by way of a chain of lakes and rivers from the Canadian interior, Mackenzie asserted that "the non-existence of a practicable passage by sea, and the existence of one through the continent, are clearly proved." Mackenzie concluded with a line that by all accounts galvanized Thomas Jefferson's attention as the new president of the United States. The Scotsman referred to the Columbia River, using Dalrymplian nomenclature, as the long-sought "line of communication from the Pacific Ocean," its low banks serving also as "the most Northern situation fit for colonization, and suitable to the residence of a civilized people."[51]

The premise behind Mackenzie's confidence was the (mistaken) belief that the large river whose headwaters he discovered and followed south, before cutting over to Pacific tidewater, was the Columbia, whose lower extent Broughton had charted for Vancouver. The river he actually encountered and then abandoned, pursuant to the information gleaned from the local Sekani and Shuswap people about formidable rapids in his projected path, was the Fraser. But if Mackenzie did not know that, in either 1793 or 1801, neither did Jefferson in 1803. At Jefferson's behest, Lewis and Clark ventured up the Missouri to find a hypothetical southern branch of the Columbia in order to obviate the potential of British/Canadian encirclement.

Because of the laborious nature of hauling canoes and goods over portages, this third version of the Northwest Passage would prove no more practicable than the classic form Cook sought, or the adaptive fallback scheme of Dalrymple that Vancouver demolished with Cookian thoroughness. When Meriwether Lewis returned to St. Louis in the fall of 1806, he wrote to Jefferson and bravely suggested that he and Clark had "discovered the most practicable rout which dose exist across the continent by means of the navigable branches of the Missouri and Columbia Rivers." This was true only under the most flexible definition of "practicable" because, as Lewis faithfully reported, between the navigable portions of those great streams there lay a "passage by land of 340 miles," 140 of which were constituted by "tremendious mountains which for 60 mls. are covered with eternal snows."[52]

Instead of the idealized vision of an interlocking river system requiring minimal portaging imagined by Jefferson at the outset of the expedition, Lewis's report was the very negation of the idea of a Northwest Passage to the Pacific and Asian markets. This view was cemented with the publication of the journals of Lewis and Clark in 1814. Thus, between 1778 and 1806, three different versions of the Northwest Passage failed to materialize: over the top of the continent (Cook); a shortcut through the continent (Dalrymple/Vancouver); and across the continent (Pond/Mackenzie/Lewis). In time, the Hudson's Bay Company proved that the Columbia could be made to work as a riverine highway, but its utility in that regard was short-lived and never approached the "Passage to India" that had captivated Western civilization from the sixteenth century onward. "The most daunting implication" of these truths when they set in, Glyn Williams writes, was the realization that if a Northwest Passage "existed it must lie far to the north among the ice-choked channels of the Arctic Ocean."[53]

In time, the ancient quest would revert to what it was before Cook. In 1818, forty years after Cook's voyage in search of the passage, and with the Napoleonic Wars in Europe concluded, the British Admiralty began to redeploy its new surplus of officers in yet another multi-generational quest for an Arctic link between the Atlantic and the Pacific. Over the course of the nineteenth century, this alternately quixotic and tragic effort resulted in the mapping of the bleak Canadian north, but it was not until 1905 that the Norwegian Roald Amundsen, after several icebound seasons, successfully navigated the last unexplored leg of the Northwest Passage. The 1969 voyage of the oil supertanker *Manhattan* transecting the Northwest Passage seemed to realize the dream of mariners since the fifteenth century of a viable northern route. In reality, the one-off *Manhattan* voyage simply proved that the Northwest Passage remained effectively impracticable. It was not until 2013, with the full onset of global warming in the decades since the *Manhattan,* that a bulk cargo ship joined the nineteenth-century explorers and adventurers who followed Cook into the Arctic.[54]

8

The American Response, or, the Lewis and Clark Expedition as Diplomatic History

THE VOYAGES OF Captains James Cook and George Vancouver to the North Pacific were demonstrably sponsored by the British government, the latter principally as a tactical response to Spain's presumed infringements on the coastal fur trade. But what became known as the Lewis and Clark Expedition was even more consciously conceived as a strategic element of national statecraft. The central figure in this development was Thomas Jefferson, who served successively as member of the Continental Congress, minister to France, George Washington's secretary of state, and president of the United States. He was also a member, and later president, of the American Philosophical Society (APS), an association of learned naturalists headquartered in Philadelphia, which, like the Royal Society vis-à-vis Cook's voyages, was vital to the origin of Lewis and Clark's venture.

In the eulogy for Meriwether Lewis that Jefferson crafted for inclusion in Nicholas Biddle's edition of the expedition's journals, he provided what was in essence a genealogy of the events that led to the launching of the captains into the American West. Jefferson cited two stillborn efforts prior to Lewis and Clark's success, but there were actually two others. With the ink barely dry on the treaty that Benjamin Franklin, John Adams, and John Jay finalized with Great Britain on September 3, 1783, Jefferson wrote to George Rogers Clark, leader of the Kentucky militia that had fought the British in that sector. Dated December 4 of the same year, his letter referenced rumours circulating in Philadelphia that the British were planning an expedition "from the Missisipi to California." He added: "They pretend it is only to promote knoledge. I am afraid they have thoughts of colonising into that quarter." Falling within the larger context of Anglo-American mutual mistrust and the resultant wrangling over the force and meaning of many provisions in the so-called peace treaty, Jefferson read the situation as necessitating an immediate response. In so doing, he prefigured the moment when, as president, he read Alexander Mackenzie's *Voyages from Montreal.* He told Clark that "some of us have been talking here in a feeble way of making the attempt to search that country," adding weakly that "I doubt whether we have enough of that kind of spirit to raise the money." Getting to the point, Jefferson asked Clark: "How would you like to lead such a party?" Even after posing the proposition, he confessed that he was "afraid our prospect is not worth asking the question."[1]

Clark discerned that he was being asked to volunteer for this mission, which he stated was "extreamly agreable to me could I afford it." Clark, who would spend the latter part of his life in financial straits, told Jefferson that if Congress authorized the "proposition," he would "take pleasure in lending all the aid in my power as an Individual." The new American government, organized under the Articles of Confederation adopted in 1781 and struggling on every front, was in no position to underwrite such a venture. But Clark, experienced frontiersman that he was, passed along a piece of tactical advice to Jefferson that made a strong impression, warning the future president that explorers travelling in "Large parties" would "allarm the Indian Nations they pass through."[2]

Jefferson's overture to Clark came after the unauthorized accounts of Cook's third voyage were published, but before the official Admiralty narrative was issued in June 1784. His initiative preceded the wave of geographical speculation about the Northwest Passage that arose after the first maritime fur traders followed Cook's track to the Northwest Coast between 1785 and 1791. Therefore, exploration west of the Mississippi had more to do with diplomatic considerations than either scientific or commercial ones. But the story soon turned on another aspect of Cook's final voyage, and intersects with this book's prologue.

The American-born, Dartmouth-educated John Ledyard was a corporal of marines on Cook's *Resolution.* As related earlier, Ledyard, after failing to organize a fur trade venture to the Northwest Coast with sponsors in America's Atlantic port cities, travelled to Europe to attempt the same there. In the course of this journey met Jefferson in Paris. Jefferson reported that after failing a second time "to engage a mercantile company in the fur-trade,"

Ledyard settled on what might be called the Cookian vision of walking around the world. In the Lewis eulogy, he described "John Ledyard of Connecticut" as someone with "a roaming disposition," who in 1786 "was now panting for some new enterprize." According to Jefferson, it was he who then "proposed to him to go by land to Kamschatka, cross in some of the Russian vessels to Nootka sound, fall down into the latitude of the Missouri, and penetrate to and thro' that to the U.S."[3] The comprehensive Cook/King third-voyage narrative had been published by this time, and we know that Jefferson was familiar with it, but his comprehension of northern Pacific geography and maritime commerce was contorted if not naive as well. The Russians had indeed spread eastward from their base in Kamchatka into the Aleutians, as Cook and Ledyard saw with their own eyes, but by Vancouver's time, nearly a decade later, the Russians had barely reached Alaska's Cook Inlet and Prince William Sound, let alone Nootka Sound on the central British Columbia coast.

Jefferson stated that Ledyard "eagerly siesed the idea" (impliedly his own) "and only asked to be assured of the permission of the Russian government." Jefferson was in a position to help with this by interceding with the diplomatic corps in the French capital, and did so on behalf of a fellow American who seemed determined to conduct such a quest, which, in its culminating phase, was the reverse of what he had proposed to Clark three years earlier. Writing in retrospect, Jefferson believed that Empress Catherine's "permission was obtained" but that she later "changed her mind" and had Ledyard evicted from her domain after he reached eastern Siberia. Apparently having forgotten about his earlier overture to Clark, Jefferson considered the Ledyard saga the first failed "attempt to explore the Western part of our Northern continent." But before leaving the subject, he projected Lewis's fate upon Ledyard, asserting that Ledyard's attempt at circumambulation "broke down his constitution" and impaired "his bodily strength" even if "His mind however remained firm." Recommitting to a journey to seek the source of the Niger River, Ledyard, according to Jefferson, "ended his career and life" in Egypt.[4] In fact, Ledyard died of natural causes, a circumstance that contemporary accounts made plain. It would seem that Jefferson was attempting to destigmatize Lewis's fate by suggesting that his end was not an uncommon one for ambitious explorers.

Following the adoption of the new US Constitution and George Washington's election to the presidency in 1787, the very first diplomatic crisis the new government faced was, *mirabile dictu,* the Nootka Sound controversy. The American fear was that if provoked to war, Great Britain, from its base in Canada, would strike at Spain's holdings in New Orleans, or at the Louisiana territory more broadly. This placed the Washington administration in an awkward position. If the British asked the United States for permission to cross its territory, refusal might mean war with the former mother country, but if the United States acquiesced, it would likely mean conflict with Spain. Faced with this dilemma, Washington consulted his cabinet, whose leading light was the more-famous (of late) Alexander Hamilton, secretary of the treasury. The secretary of war was Henry Knox, and the secretary of state was John Jay, whose brief term ended when he was replaced by Jefferson on March 22, 1790. Jefferson's absence from the scene of these deliberations probably explains his lack of working knowledge about another, largely unreported attempt at western exploration.

Spain and Great Britain averted war via the Nootka Convention negotiated in 1790, quite happily for the United States, which rode out the storm as a neutral power. Before the terms of the convention were agreed to and publicized, however, Knox wrote Josiah Harmar, commander of the western army in Ohio, on December 20, 1789. He asked him to devise a "practicable plan for exploring the branch of the Mississippi called the Missouri, up to its source and all its southern branches, and tracing particularly the distances between the said branches and any of the navigable streams that run into the Great North River which empties itself into the gulf of Mexico."[5] The "Great North River" is an archaic term for what now is commonly referred to as the Rio Grande.

Washington's cabinet seems to have settled on the idea of a scouting expedition that could simultaneously reconnoitre British trading interests and related placements on the Missouri and Spanish control of the Louisiana borderlands. Given the contested diplomatic context, and the extraterritorial nature of this venture, Knox told Harmar: "You will easily see that this object cannot be undertaken with the sanction of public authority." Knox believed that "an enterprizing Officer" accompanied by a non-commissioned soldier "well acquainted with living in the woods, and perfectly capable of describing rivers and countries, accompanied by four or five hardy indians perfectly attached to the United States would in my opinion be the best mode of obtaining the information suggested." Because of attendant risks, Harmar was to consider sending out two such parties in staggered starts. That way, it would be "highly probable that one if not both would succeed and return." Knox stated that he was not able to make commitments, but he vowed to see that such men were "liberally rewarded on their return," terms strongly suggestive of the espionage that was at play. In phrasing reflective of the naïveté surrounding Jefferson's overtures to George Rogers Clark and John Ledyard, Knox thought that the only extraordinary equipment necessary to complete this mission were compasses, pencil, and paper.[6]

On February 20, 1790, Harmar replied that he wanted to consult with Arthur St. Clair, governor of the Northwest Territory, at its capital in Marietta, Ohio, before committing to the mission. But at first blush, he told Knox, "it will be a very difficult dangerous undertaking." In his query to St. Clair, written the same day, Harmar asked for counsel but confessed that the proposal seemed "too adventurous." Moving matters along, he settled upon Lieutenant John Armstrong as a potential candidate and directed him to consult directly with St. Clair. After doing so, Armstrong reported to Harmar in June that the governor was of the opinion that this "was not only a difficult task, but one that in his [St. Clair's] opinion could be executed in the character of a trader only." St. Clair's was wise advice and, of course, it fully reflected the nature of Alexander Mackenzie's two forays into the wilderness and Indian country that bracketed this time frame.[7]

According to Armstrong, there were additional difficulties with Knox's plan, starting with the likely Spanish opposition to such an incursion and the dubious notion of Indians "well attached to the United States." Armstrong professed to "know not where they are to be found," reflecting the western reality that was soon to result in the Northwest Indian wars with the Miami, Shawnee, and Delaware Indian nations, including Harmar's forthcoming defeat near present-day Fort Wayne, Indiana. With candid phrasing, he told Harmar that Knox's plan was "a business much easier plan[n]ed then executed," and if called upon to conduct it, he wanted to be much better equipped than proposed, especially given the "hostile disposition of the Indians." Harmar reported to Knox on June 9, 1790, that he had tapped Armstrong "to endeavor to carry into effect your Secret orders," including a consultation with St. Clair, who "was of opinion that it was neither prudent nor practicable to undertake it."[8] Thus ended the third American attempt to explore the Trans-Mississippi West.

The second false start in Jefferson's typology, though in truth the fourth overall, was sponsored by the APS. In 1792, now that he was secretary of state, Jefferson proposed to the organization that it conduct a "subscription to engage some competent person to explore" the western part of the continent. In his vision, this expedition would ascend the Missouri River, cross the Rockies, then descend "the nearest river to the Pacific." Since it was not known at the time that Robert Gray would be the first, that year, to cross the bar of a long-rumoured river, Jefferson was likely referring to the legendary "Great River of the West," later the Columbia. (News of Gray's discovery of the river named after his ship, the *Columbia Rediviva,* did not reach Atlantic America until August 1793.) As Jefferson explained, "it was proposed that the person engaged should be attended by a single companion only, to avoid exciting alarm among the Indians."[9]

Here Jefferson was repeating the advice George Rogers Clark had given him a decade earlier. In the event, André Michaux, a French botanist and author, secured the appointment. In the Lewis eulogy, Jefferson disclosed that the young Meriwether Lewis had applied for this mission. Before Michaux was approached, Caspar Wistar, one of the APS's leading lights, sounded out a local botanist, Moses Marshall, about his interest in the subscription Jefferson was initiating, but the latter never pursued the matter.

The actual subscription agreement used to raise money for the venture provided greater specificity as to what was envisioned. The designee was "to explore the interior country of North America, from the Missisipi along the Missouri and Westwardly to the Pacific ocean, or in such other direction as shall be advised by the American Philosophical society." Upon return, the society's agent was to "communicate ... the information he shall have acquired of the geography of the said country, it's inhabitants, soil, climate, animals, vegetables & minerals, & other circumstances of note." This was the standard catalogue of the Enlightenment's areas of inquiry popularized by the Admiralty's publication of Cook's instructions in the volumes that documented his travels. The APS traveller was to be paid one-fourth of the sum raised "on demand," with the balance payable upon a successful mission, defined as reaching the Pacific, or an unspecified proportion of that sum if he did not. All proceeds from the publication of an account of the expedition would accrue to the explorer. This venture attracted the interest of no less a figure than President Washington. He told Jefferson in January 1793: "If the Subscription is not confined to the members of the Philosophical Society I would readily add my mite to the means for encouraging Mr. Michaud's undertaking."[10]

Not unexpectedly, Jefferson drafted Michaux's instructions, in late April 1793. Michaux was advised that "the chief objects of your journey are to find the shortest & most convenient route of communication between the U.S. & the Pacific ocean, within the temperate latitudes." The Missouri

River had long since been established in geographers' minds as the gateway to the West. Accordingly, Michaux was advised: "As a channel of communication between these states & the Pacific ocean, the Missouri, so far as it extends, presents itself under circumstances of unquestioned preference." For emphasis, Jefferson told Michaux that the Missouri was "a fundamental object of the subscription, (not to be dispensed with)." (If Jefferson expressed similar thinking in private conversation with Lewis, it would provide an insight into why Lewis later seemed to privilege his exploration of the upper Missouri over every other facet of the country he traversed.)[11]

Jefferson's more specific guidance to Michaux underscores these points. The French botanist was to pursue "the largest streams," by which Jefferson meant tributaries, "as shall lead by the shortest way, & the lowest latitudes to the Pacific ocean." At an unspecified juncture, but one that does not seem to have been necessarily adjacent to the headwaters of the Missouri, he was told: "You may get by the shortest & most convenient route to some principal river of the Pacific ocean ... & pursue it's course to the ocean." The "latest maps" indicated "a river called Oregan interlocked with the Missouri for a considerable distance, & entered the Pacfic ocean, not far Southward of Nootka sound." Jefferson cautioned that the APS did not warrant the trustworthiness of such depictions, "leaving to yourself to verify it," but the larger point is the secretary of state's awareness of Nootka Sound's centrality to the maritime fur trade that had emerged on the Northwest Coast in Cook's wake. Like his adoption of Ledyard's scheme, this shows Jefferson's openness to a mercantilist outlook, which his own countrymen, led by Gray, were then pursuing.[12]

Prefiguring an allowance later granted to Lewis, Michaux was advised that if, upon reaching the Pacific Slope, he conveniently found a "settlement of Europeans," he was to provide them with a duplicate version of his journal that could be dispatched to Philadelphia "by sea." He was then to return "by the same, or such other route, as you think likely to fulfill ... the objects of your mission," which he was reminded, for the third time, was "that route which shall form the shortest & most convenient communication between the higher parts of the Missouri [with] the Pacific ocean."[13] Note that Jefferson avoided specifying the headwaters of the Missouri as the jumping-off point to the Great River of the West.

There was with Michaux one last foreshadowing of a future event, Lewis's problematic encounter with Spanish officials in St. Louis in December 1803.

Jefferson instructed Michaux to access the Missouri from a point above St. Louis ("the Spanish settlements") to "avoid the risk of being stopped." After Michaux reached Kentucky, Jefferson noted in the Lewis eulogy, "he was overtaken by an order from the minister of France then at Philadelphia to relinquish the expedition."[14]

Here Jefferson sanitized what actually occurred, a function of his long-standing and well-known affinity for France and its revolutionary fervour. By this time, France had entered a bloodier and more complicated phase, including war with Great Britain, whose ally was Bourbon Spain. The Washington administration was doing its best to maintain neutrality, a stance that was severely tested by the same French minister, the famed Citizen Edmond Genêt, who was attempting to mobilize American public opinion in favour of joining his country in belligerency. Genêt had conducted a backcountry journey from his port of entry in Charleston to Philadelphia, during which he engaged frontier leaders, including George Rogers Clark, with a view to fomenting an attack on Spanish territory in Louisiana. In 1784, Spain closed New Orleans to American shipping from the interior, reopening the right of deposit in 1788 but subject to stiff duties. At the same time, France was interested in regaining its former colony, lost in the peace treaty ending the Seven Years' War in 1763, an interest Napoleon later picked up on. In the moment, and in a manner typical of such figures, the flamboyant Genêt overplayed his hand in trying to force the United States into war not only with Britain but also with Spain. Among the consequences, the APS/Michaux expedition came to naught.

The fifth (in Jefferson's mind the third) and final attempt at western exploration was prompted by the December 1801 publication of Alexander Mackenzie's *Voyages from Montreal.* It was the most popular account of North American exploration to that time. In it, Mackenzie laid the foundation for a British counterclaim to Gray's discovery of the Columbia River, founded on what turned out to be his mistaken belief that he had discovered its headwaters. This misidentification would not become evident until 1808, two years after Lewis and Clark had reached the Pacific Ocean and returned to the United States. When he was in the field, Mackenzie referred to this river by its Indigenous name, "Tacoutch Tesse." Only later would Mackenzie learn of the Gray/Vancouver/Broughton discourse, including the name "Columbia." In Mackenzie's concluding paragraphs, he pointed out that this river was the long-awaited corridor linking the Pacific Slope with waters flowing into the Atlantic. By dominating this new Northwest Passage, British imperial interests could encircle the Americans and close off their extension to the Pacific.

An account of Mackenzie's exploits was long-awaited. As early as 1795, news of his reaching the Pacific by an overland route was in the American press. The *Kentucky Gazette,* recycling a notice from a New York newspaper, published in its February 28 issue the news that Mackenzie's accomplishment would "in the course of time, be of the utmost consequence to this country, as it opens a direction communication with China, and may doubtless lead to further discoveries."[15] This excerpt is all the more remarkable because of its overt reference to the Asian market for furs from the Northwest Coast that both British and American traders were conveying to Canton.

Mackenzie suffered from a case of writer's block that foreshadowed Lewis's difficulties, but when his account finally appeared in print, it created a sensation on both sides of the Atlantic. In January 1802, the APS's Caspar Wistar wrote Jefferson: "Have you seen McKenzie's account of his journeys across the Continent & to the Northern Ocean."[16] Given his history and influence, Jefferson would have initiated exploration of the Missouri at some point in his presidency, but Mackenzie's mercantilist vision was an obvious threat to American pretensions on the Pacific. In most accounts, the Lewis and Clark Expedition begins here, but there was another sequence of events that informed Jefferson's decision to respond with a western expedition.

Over the same winter of 1801–02 that Mackenzie's book was causing a literary stir, rumours of a diplomatic transaction in Europe drifted into the consciousness of the Jefferson administration. In October 1800, Napoleon cajoled Charles IV of Spain into a preliminary and nominally secret arrangement, known as the Treaty of San Ildefonso, by which France's former Louisiana colony, like Quebec lost in the treaty ending the Seven Years' War, might be restored. The timeline was unknown, seemingly at France's option. So long as Spain controlled Louisiana, the United States could wait for that declining empire to drop the territory into its lap, or could simply acquire it gradually through cellular colonialism. This new development was cause for concern because France, conversely, was the greatest military power in Europe, run by a mercurial dictator. In April 1802, Jefferson wrote Robert Livingston, the American minister in Paris, confiding that the day France took possession of Louisiana, the United States would be circumscribed geographically and it would have to marry the British fleet.

Concern became alarm in October 1802 when Spanish authorities in New Orleans, annoyed by American smugglers avoiding duties imposed back in the Genêt era, suspended all rights of deposit. This was authorized in Madrid and served a further purpose of sending a message to Paris that Spain was resolved to actively govern its colony. As a practical matter, the suspension meant that American boatmen could not convey agricultural produce from the Ohio River Valley and the east bank of the Mississippi to ocean-going ships destined for Atlantic port cities or overseas. The political ramifications were felt immediately. There was talk of secession or a "filibuster" (an unauthorized civilian attack) by Kentucky and Tennessee frontiersmen upon New Orleans.

Jefferson, concerned about the diplomatic situation and hair-trigger westerners, settled on a multi-faceted course of action to avoid a potentially disastrous war with Spain or France. He swallowed hard and began cozying up to the British chargé in Washington. To placate his base in the West, he took military measures in the Mississippi River valley in advance of the day when France might reacquire Louisiana. These included authorization of an 80,000-man militia and, though never thought of as originating in this context, the Lewis and Clark Expedition.

Jefferson's first diplomatic overture relative to the latter occurred in late November 1802, the month after the rights of deposit were suspended. On December 2, Carlos Martínez de Yrujo, head of the Spanish legation in Washington, dispatched a letter to Pedro Cevallos, minister for foreign affairs in Madrid, positing that the president aimed "to send travelers to explore the course of the Missouri River, and for them to penetrate as far as the Southern Ocean." This was Spain's preferred name for the Pacific, which it had historically claimed as entirely its own. Jefferson attempted to minimize the scale of this endeavour. De Yrujo characterized it as "a small caravan." Jefferson explained to the Spanish ambassador that he would have to sell this idea to Congress as a contribution "to the progress of commerce; but that in reality it would have no other view than the advancement of the geography." De Yrujo explained this constitutional nicety. A "mercantile" purpose was the only way to procure "the necessary funds; it not being possible to appropriate funds for a society, or a purely literary expedition." The ambassador told Jefferson that this venture "could not fail to give umbrage to our Government." Jefferson questioned why it should be bothersome. The only goal was to "unite the discoveries that these men would make with those the celebrated Makensi made in 1793" – this with a view toward establishing "a continual communication, or little interrupted, by water as far as the South Sea."[17] Jefferson's dropping of Mackenzie's name is indisputable evidence linking the publication of *Voyages from Montreal* with the origin of the Lewis and Clark Expedition.

De Yrujo recognized this talk of a "communication" as code for the Northwest Passage, proving also the durability of Alexander Dalrymple's framing of the concept. He told Jefferson that this quest "was already a determined point," citing the "fruitless attempts" of a long line of maritime explorers such as Cook, Martínez, Vancouver, Quadra, and Malaspina. These men had collectively surveyed "in the most careful and scrupulous manner all that coast, from the south of the famous strait of Juan de Fucca, situated at 45° latitude, up to … which the English call Cook River situated at 60° north latitude." As he related to Cevallos, de Yrujo stated to Jefferson that "all these examinations and attempts evidently prove that there does not exist this passage of the Northwest, sought with so much anxiety by the most famous navigators of all the nations in the last two centuries." He told Jefferson that John Meares "in these later times had inclined to believe in the existence of this passage" but "his error had been later demonstrated" – by Vancouver, though de Yrujo did not name the British explorer.[18]

Like Jefferson, de Yrujo had read Mackenzie, and he stated that the Scotsman had shown "that there does not exist such a communication by water" because he had encountered "a considerable cordillera of mountains which runs north and south parallel to the ocean coast between 50° and 60°." This necessitated "traveling by land a very great distance, and the practical Indians of the country could not give him news of any considerable river whatsoever which from those mountains … empties into the Pacific Ocean." In some ways de Yrujo demonstrated a more realistic understanding of western geography than the American polymath, though he mischaracterized Mackenzie's understanding of the "Tacoutch Tesse." Mackenzie knew it reached the sea, but only by way of turbulent canyon waters he dared not venture himself. De Yrujo believed that Jefferson was attempting "to perpetuate the fame of his administration … by discovering or attempting at least to discover the way by which the Americans may some day extend their population and their influence up to the coasts of the South Sea."[19]

This was an astute observation. The ambassador detected that this American version of Mackenzie's treks, following the other cited "fruitless attempts" at discovering a Northwest Passage, was intended to cloak a mercantile opening into Spanish territory, if not a military reconnaissance – a reasonable deduction given the larger diplomatic context. In other words, de Yrujo discerned that the commercial qualification for passing constitutional muster was in fact Jefferson's actual purpose, not a way of dodging Congressional objection. Even so, de Yrujo thought the project's prospects before Congress doubtful, though he promised to keep the foreign minister apprised.

January 1803 was Jefferson's month of decision on his two-track approach to avoid catastrophe in the West. On the 12th, James Monroe was confirmed as a special envoy to France, where he was to join Robert Livingston with instructions to offer Napoleon $10 million for New Orleans and West Florida. If that failed, the American diplomats were to open negotiations for an alliance with Great Britain. Even before Monroe reached Paris in mid-April, French foreign minister Talleyrand unexpectedly asked Livingston what the United States would offer for all of Louisiana. After two weeks of negotiation, a deal was reached on the basis of an offer of $15 million. When the news of this settlement arrived in the Ohio and Mississippi river valleys in summer of 1803, there was unrestrained joy.

By that point, Congress had already authorized the Lewis and Clark Expedition. On January 18, one week and a day after Monroe's confirmation, Jefferson tendered a secret message to Congress that kick-started the western expedition. Secrecy was the watchword for the entire enterprise. Even before Meriwether Lewis, Jefferson's personal secretary, delivered the proposal to Congress, the president was toying with a formula for a cipher so his exploratory agent could communicate with him confidentially. During the Enlightenment, every nation tried to shield their discovery ventures from the prying eyes of other powers. Captain Cook had a second set of secret orders for the exploration he was to conduct in search of *Terra Australis Incognita* after the public purpose of his first voyage, observing the transit of Venus, was completed. Cook never took a cipher system to sea, but the great navigator might have been on Jefferson's mind as he developed the cryptic formula. The code word he used was "antipodes," defined as the places on the earth's surface diametrically opposite each other, and a term Cook made famous on his second voyage by reaching the antipode of Greenwich southeast of New Zealand.

The trans-Mississippi Indian trade was the ostensible context of Jefferson's proposal to Congress. He warranted that the tribes along the settler frontier were uneasy "at the constant diminution of the territory they occupy." Worse, from an American policy perspective, they were increasingly resistant to "any overture for the purchases of the smallest portions of their land." To counter that pattern "and to provide an extension of territory" driven by "the rapid increase of our numbers ... two measures are deemed expedient." The first was active avowal of a cultural suppression strategy that was the underlying dynamic for centuries: convince the Indians to "abandon hunting" in favour of stock raising, agriculture, and "domestic manufacture." Such an economy required much less land than the "extensive forests necessary in the hunting life." The second requirement was an expansion of the trading post system, which would "place within their reach those things which will contribute more to their domestic comfort than the possession of extensive, but uncultivated wilds." American land hunger was rarely expressed more directly.[20]

In furtherance of this long-standing continental objective, Jefferson reminded Congress that it had previously authorized "trading houses" for establishing "commerce" with the western tribes, in part to counter foreign influence among them. This was an important constitutional reminder. As he explained retrospectively in the Lewis eulogy, the act establishing such posts was about to expire and "some modifications of it were recommended" in his January 1803 message, principally "an extension of it's views to the Indians on the Missouri." In other words, this formerly internal tactic was to become an aspect of foreign policy, involving both the sovereign Indian nations outside the American boundary that needed to be seduced into selling their land, plus any European country (Spain and/or France) claiming international title to Louisiana. This nexus was complicated by the president's acknowledgment that the Missouri's peltries were going "to the trade of another nation carried on in a high latitude," a discreet reference to Great Britain. Drawing on Mackenzie's book, Jefferson alluded to the fact that the networks of the Hudson's Bay and North West Companies were connected by "an infinite number of portages and lakes" that were "shut up by ice through a long season." Lest Congress be concerned about taking on that competition, he posited that the commerce on Mackenzie's "line could bear no competition with that of the Missouri, traversing a moderate climate."[21] Jefferson, like many a promoter of the Northwest Passage, was not averse to seeding his proposal with an optimistic forecast of climate, distances, and ease of travel.

The problem, as Jefferson stated in his message to Congress, was that the "river Missouri, & the Indians inhabiting it, are not as well known as is rendered desireable by their connection with the Missisipi, & consequently with us." He referenced uncited accounts suggesting that "a continued navigation" to the Missouri's headwaters was possible where a single portage allowed access to the "Western ocean." In this vision, which borrowed heavily from Mackenzie's geographical and commercial hypotheses, the Missouri corridor could funnel trade to the Atlantic seaboard through "a choice of channels," including (1) the Great Lakes and Hudson River, (2) via the watersheds of the Illinois and Wabash Rivers in the middle states, or (3) through the Tennessee and Savannah Rivers further south. The only way to unravel these interconnected issues, Jefferson wrote in the prefatory section of Biddle's edition of the Lewis and Clark Expedition's journals was "sending an exploring party to trace the Missouri to it's source, to cross the highlands and follow the best water communication which offered itself from thence to the Pacific ocean."[22]

Having borrowed from Mackenzie to describe western geography and trade patterns, Jefferson did the same to frame the human element necessary to implement the mission. The situation required an "intelligent officer with ten or twelve chosen men, fit for the enteprize and willing to undertake it, taken from our posts." This mirrored the size and a comparable point of origin for the parties Mackenzie took on his voyages to Arctic and Pacific tidewater. It was sufficiently scaled, Jefferson told Congress, to "explore the whole line, even to the Western ocean." Alluding to the larger policy context, including a constitutional basis for such a venture, Jefferson said this expedition would conduct "conferences with the natives on the subject of commercial intercourse" and "get admission among them for our traders." Jefferson lowballed projected expenses with a request for an appropriation of $2,500, arguing that the soldiers who engaged in the effort were going to be paid for some duty anyway.[23]

Near the end of his message, Jefferson appealed to Congress's sense of national pride:

> While other civilized nations have encountered great expense to enlarge the boundaries of knowledge, by undertaking voyages of discovery, & for other literary purposes, in various parts and directions, our nation seems to owe to the same object, as well as its own interest, to explore this, the

only line of easy communication across the continent, and so directly traversing our own part of it.

There is a lot to unpack here, including Jefferson's sophisticated understanding that unless a voyage of discovery was described in print, it was as though it had not actually taken place. Cook's influence seems to have loomed over this conceptualization, as well as the evolved framework for the Northwest Passage developed by Dalrymple and Mackenzie. In case Congress missed the president's nuances, he argued that the "interests of commerce place the principal object within the constitutional powers and care of Congress, and that it should incidentally advance the geographical knowledge of our own continent can not but be an additional gratification."[24]

Ignoring the views a hesitant Spanish ambassador conveyed to him the previous month, Jefferson told Congress: "The nation claiming the territory, regarding this as a literary pursuit which it is in the habit of permitting within it's dominions, would not be disposed to view it with jealousy, even if the expiring state of it's interest there did not render it a matter of indifference." True, Spain suffered British trading interests to dominate the upper Missouri, but the notion that it would approve this American incursion because of its "literary" quality was a lie. This statement graphically demonstrates how Jefferson contorted his message(s) to reflect what he thought his several audiences wanted or needed to hear. Thus, his Congressional missive concluded with the request for an appropriation "for the purpose of extending the external commerce of the U.S."[25]

Ambassador de Yrujo had left his meeting with Jefferson in December thinking he had talked the president out of the western expedition, but three days after the (secret) message was delivered to Congress, he knew about it. He confessed to the foreign minister, Cevallos: "I thought that ... he would desist in his attempt, and in this way I hinted this to Your Excellency, but later I have learned that he has communicated his design to the Senate, which has already taken a step towards the execution." The ambassador held out hope that some opposition might arise from some subset of the "European nations" that might take offence, and consequently "the project will not proceed."[26]

This was wishful thinking. Napoleon, confounded by the rebellion in Hispaniola led by Toussaint L'Ouverture, was losing interest in the Western Hemisphere and Jefferson was a determined man. During the winter/spring of 1802–03, the president unleashed a torrent of meetings and correspondence laying the groundwork for Lewis's venture. On February 24, he wrote to Bernard Lacépède, a French naturalist working on a multi-volume encyclopedia, in response to a recent inquiry about the sources of the Missouri and Mississippi Rivers. In his letter, physically carried to Paris by Monroe, Jefferson said: "It happens that we are now actually sending off a small party to explore the Missouri to it's source, and whatever other river, heading nearest with that, runs into the Western ocean; to enlarge our knolege of the geography of our continent, by adding information of that interesting line of communication across it." There is no mention of trading posts or external commerce here, nor in his correspondence that same week with the members of the APS to whom he would send Lewis for scientific training. Jefferson's note to Benjamin Rush, a physician and APS member, is representative: "I wish to mention to you in confidence that I have obtained authority from Congress to undertake the long desired object of exploring the Missouri & whatever river, heading with that, leads into the Western ocean."[27]

Jefferson's letters to Caspar Wistar and Robert Patterson were similarly peppered with cautionary expressions as to the "confidential" nature of Lewis's mission and the "secret" act by which Congress authorized it. But loose lips, probably Federalists, immediately shared the news with the diplomatic corps. Jefferson himself, on February 28, took receipt of Lewis's British passport, which was secured for exploration of "the western parts of the North American continent," and again limited to "scientific and literary objects" alone. The forthcoming French passport carried the same devious rationale. Louis André Pichon, head of the French legation, advised Talleyrand on March 4 that he had provided Lewis with a passport at Jefferson's personal request: "For a long time Mr. Jefferson has been concerned with the means of exploring the sources of the Missouri beyond which he supposes must be found [near] those of the river Origan which flows into the Pacific Ocean, and of which only the mouth is known." Much of this had been suspected since the 1780s, when John Ledyard was attending Jefferson's Independence Day parties in Paris, though Pichon was obviously confused when he added that the Great River of the West had been renamed "Colombia by the explorer McKenzie."[28]

Pichon had been clued into the underlying constitutional dynamic, telling Talleyrand that Jefferson was able to secure underwriting only because of the "scruples" Congress had "over the right of the general Government to do anything which might tend toward the encouragement of the Sciences." It was approved only for the purpose of "encouraging foreign trade." Pichon related asking Jefferson whether he had secured a passport for Lewis from the Spanish minister. Jefferson replied that "he ought to grant one." This was another example of wishful thinking on Jefferson's part, if not outright subterfuge. Pichon ended by explaining to Talleyrand that the only reason he granted a passport was to countenance the prospect that after reaching the Pacific, the Americans might return by sea. That was "the only detail of the expedition in which a French passport might be of some use." Fortunately for Pichon and Jefferson, Livingston and Monroe would soon take care of Lewis (and Clark's) Spanish and French problems.[29]

The British ambassador, Edward Thornton, could not understand why all the mystery and diplomatic manipulation was necessary. On March 9, he wrote Foreign Secretary Lord Hawkesbury that Congress had authorized an expedition "in secret sessions ... with closed doors." He related that the western mission was for "the encouragement and extension of the external commerce of the United States," but professed that "on the face of it," this did not "present any circumstances requiring concealment." He allowed that America's "able jurists" had determined that the United States Constitution "does not permit the general government to offer bounties for the promotion of discoveries or for the advancement of science." Therefore, under the "ostensible object of extending the external commerce of the country," Congress proceeded "with a complete understanding of the real nature of the plan in contemplation." According to Thornton, Jefferson emphasized this point in a conversation between them: the "real object ... is exclusively scientific. He assured me that it was in no shape his wish to encourage commerce with the distant or indeed with any Indian tribes."[30] Once again, Jefferson was telling his audience what he believed they needed to hear to assuage their concerns.

In his elaboration of the scheme, Thornton struck the same themes as Pichon. This suggests that in laying out his plan, Jefferson used much the same language to describe his intentions to both individuals. According to Thornton: "The President has for some years past had it in view to set on foot an expedition entirely of a scientific nature for exploring the Western Continent of America by the route of the Great River Missouri, and for tracing the proximity of the sources of this river to the streams, which fall on the other side into the Pacific Ocean." Thornton's "Great River Missouri" was a garbling of the traditional name of the Columbia – the Great River of the West. The larger point, Thornton added, was that Jefferson believed this linkage provided "the most natural and direct water-communication between the two Oceans, and he is ambitious in his character of a man of letters and of science, of distinguishing his Presidency by a discovery, now the only one left to his enterprize – the Northern Communication having been so ably explored and ascertained by Sir Alexander Mackenzie's journeys." It is not clear from this construction whether Jefferson had introduced the Mackenzie factor into the conversation or whether Thornton had deduced it, but the description was a crystallized version of the latest image of the Northwest Passage. Thornton closed by acknowledging the larger diplomatic context, namely, that the "apprehended occupation of Louisiana by the French seems to have accelerated the determination of the President, as he thinks it certain that on their arrival they will instantly set on foot enterprizes of a similar nature."[31] Historians have long recognized the Mackenzie connection relative to the origins of the Lewis and Clark Expedition, but it is clear that the Treaty of San Ildefonso was far more important to the timing of the venture than generally credited.

Having secured Congressional funding and (except for Spain) the concurrence of the key foreign powers, Jefferson set his administration into motion to plan the expedition. Nicholas King, the War Department's principal cartographer, collated geographical information from several sources, principally Aaron Arrowsmith, the great British map-maker, to create a chart for Lewis's use in the field (see Figure 8.1). A draft of Lewis's instructions was floated to the cabinet. Their responses showed varying levels of engagement. Curiously, given the diplomatic context, the least expressive was Secretary of State James Madison, who was having to navigate the American ship of state between France and Britain in their war with each other. Madison's concern centred on the mission and its relationship to "commerce." He had not yet heard from Livingston and Monroe, and a mercantile excursion into contested terrain would only make his job more difficult.[32]

On the other hand, Secretary of the Treasury Albert Gallatin was very much engaged in launching the expedition. He, not Secretary of War Henry Dearborn, commissioned and scoped King's map and reported to Jefferson that its projection included "the coast of the Pacific from Cook & Vancouver" as well as Mackenzie's track, all borrowed from Arrowsmith. He was also more expressive about the diplomatic consequences than Madison. After reviewing the draft of Lewis's instructions, Gallatin related to Jefferson that

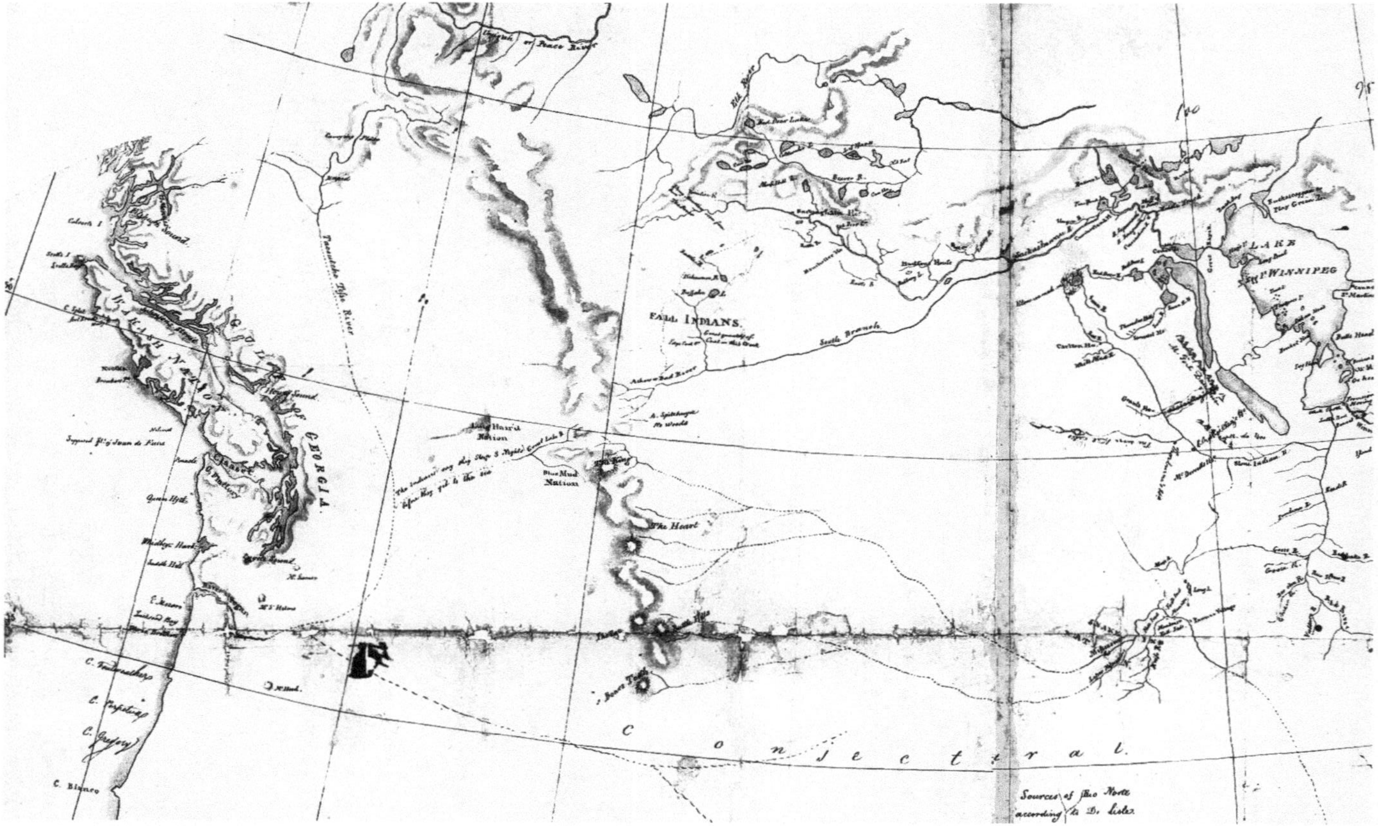

Peace River
Elk River
LAKE WINNIPEG
FALL INDIANS
South Branch
Taoutche Tesse River
Long Hair'd Nation
Blue Mud Nation
The Heart
GULF OF GEORGIA
C. Blanco
Conjectural
Sources of Rio Norte according to D. Lisle

Figure 8.1 Manuscript map prepared by Nicholas King in 1803 for Lewis and Clark's use in the field. Based on the published cartography found in the accounts of George Vancouver (1798) and Alexander Mackenzie (1801), it projected a "conjectural" southern branch of the Columbia River. Drawing on decades' worth of geographical speculation and a strong measure of wishful thinking, King posited a southern fork conveniently located opposite the western headwaters of the Missouri. | Library of Congress, Geography and Maps Division

"the present aspect of affairs may, ere long, render it necessary that we should, by taking immediate possession, prevent G.B. from doing the same." Accordingly, he argued that an assessment of "the posts, establishments & force kept by Spain in upper Louisiana, and also of the most proper station to occupy, for the purpose of preventing effectually the occupying of any part of the Missouri country by G.B., seems important." Not only was it vital to keep upper Louisiana from falling into the hands of a stronger power but Gallatin believed that the forthcoming expedition might also shed light on British operations emanating from the Hudson Bay watershed. The specific goal was discernment of "the mode in which a small but sufficient force could best be conveyed to the most proper point from whence to prevent any [British] attempt from Lake Winnipec."[33]

This was bold thinking but Gallatin thought the diplomatic stakes were high. As related to Jefferson, the "Missouri country" was "of vast importance to the United States, it being perhaps the only tract of country, and certainly the *first* which lying out of the boundaries of the Union will be settled by the people of the U. States." For that reason, he believed that the "precise extent" of the Missouri, including "all the waters emptying into that river," ought to be "ascertained," especially "that particular branch which may be followed for the purpose of examining the communications with the Pacific Ocean." Gallatin was here alluding to the blank spot on the Arrowsmith and King maps, in contrast to the country west and north of Hudson Bay that had been "tolerably ascertained by Mackenzie & others," including the unnamed Peter Pond. In a line of text overlooked by many students of the expedition, Gallatin averred that the Columbia River was "the principal object of this voyage to explore."[34]

Before April 1803 was out, Jefferson's diplomatic offensive began to bear fruit. Spain reinstituted the right of deposit. Lewis was in training up in Philadelphia, but the president apprised his successor as personal secretary, one Lewis Harvie, that the Louisiana retrocession "is probably not yet finally settled between those powers," referring to Spain and France. In fact, Livingston and Monroe were just then putting the finishing touches on the Louisiana Purchase agreement. In the absence of such knowledge (which would not arrive in America until late June and early July), Jefferson was still proceeding cautiously. He wrote Lewis on April 27: "The idea that you are going to explore the Missisipi has been generally given out: it satisfies public curiosity, and masks sufficiently the real destination."[35] This gambit could only have been intended to keep western frontiersmen and Spanish colonial officials in St. Louis in the dark. The diplomatic corps in Washington well knew what was happening, and soon their governments across the Atlantic would too, as well as Jefferson's friends in the Republic of Letters in both Philadelphia and Europe.

That spring, from the holdings of the APS or one of its members, Lewis forwarded to the capital "some sketches taken from Vancouver's survey of the Western Coast of North America." This was to help King complete the map he was working on at the direction of the War Department. They were drawn, Lewis stated, "in a haisty manner," thereby diminishing their potential reliability. Subsequent journal entries from when his expedition was at the mouth of the Columbia suggest that he also kept a copy of these tracings for his use in the field. Lewis was forced to do this because, as was explained to Jefferson, "the maps attached to Vancouver's voyage cannot be procured seperately from that work, which is both too costly, and too weighty, for me either to purchase or carry." Lewis's naïveté about the resources King had at his disposal is remarkable, but at least he was sensible enough to know the impracticability of carrying several large quarto volumes into the wilderness. At the same time and for the same reason, Jefferson was on the lookout for a more economical octavo edition of *Voyages from Montreal.* He bought one from a bookseller in New York that, he happily apprised Lewis, contained "the same maps which are in the 4to edition" that was also "too large & cumbersome" for Lewis to take west.[36]

Lewis was back in Washington by late June 1803, at which time two developments redirected the course of Jefferson's western expedition. On the 19th, the day before his final instructions were issued, Lewis, with Jefferson's concurrence, extended an invitation to William Clark, George Rogers Clark's youngest brother. Lewis covered many of the common elements treated in this chapter. Congress had appropriated funds "for extending the external commerce of the United States." This provided government sanction "to exploreing the interior of the continent of North America, or that part of it bordering on the Missourie & Columbia Rivers." Drawing on his forthcoming orders, Lewis informed Clark that the mission was to proceed up the Missouri "as far as it's navigation is practicable ... and proceed to it's source, and if practicable pass over to the waters of the Columbia or Origan River and by descending it reach the Western Ocean."[37]

This letter, and the larger body of correspondence between the two men, has been much studied for insights into their relationship with each other. Our principal interest is the mercantile outlook it displays, which, as noted earlier, was the constant subtext of Jefferson's several attempts at launching an expedition linking the Pacific with the Atlantic. Lewis apprised Clark that the mouth of the Columbia "lies about one hundred and forty miles South of Nootka-Sound, at which place there is a considerable European Tradeing establishment, and from which it will be *easy* to obtain a *passage* to the United States by way of the East-Indies in some of the trading vessels that visit Nootka Sound annually, provided it should be thought more expedient to do so." This text was assuredly intended to convey to Clark a sense of confidence and manageable risk, but its underlying sentiment was the geography of hope that lay at the heart of most rationalizations about the practicability of a Northwest Passage and the effort required to discover it.[38]

The other major news concerned one of the most important developments in the history of the United States. Word finally reached the capital that Livingston and Monroe had far exceeded expectations, acquiring by way of a prospective treaty not merely New Orleans and West Florida but the entirety of Louisiana. Lewis told Clark that "the whole of that immense country wartered [sic] by the Mississippi and it's tributary streams, Missourie inclusive, will be the property of the U. States in less than 12 Months from this date." As a consequence, it was important to commence "an early friendly and intimate acquaintance with the tribes that inhabit that country." Not quite as an afterthought but nonetheless following an adumbration of mercantile potential and tribal objectives, Lewis advised Clark that "the other objects of this mission are scientific, and of course not less interesting to the U. States than to the world generally," referring to geography, ethnology, botany, and zoology. For Jefferson, the Louisiana Purchase relieved him of the diplomatic anxiety he had been feeling about the expedition since he first determined on it. He related in Lewis's eulogy that this news "increased infinitely the interest we felt in the expedition, & lessened the apprehensions of interruption from other powers."[39]

Lewis's final instructions, a draft of which he was long familiar with, were formally issued on June 20. The key paragraph, reflecting a refinement of thought on Jefferson's part for twenty years running, stated: "The object of your mission is to explore the Missouri river, & such principal stream of it, as, by it's course and communication with the waters of the Pacific ocean, whether the Columbia, Oregan, Colorado or any other river may offer the most direct & practicable water communication across this continent for the purposes of commerce." This statement drew on the semantic influences of Dalrymple and Mackenzie and their respective visions for an evolution of the Northwest Passage. If the internal trade and consolidation of sovereignty questions with the Indian tribes were the only objectives, there was no point in crossing the Continental Divide. Jefferson targeted the southern branches of the Missouri for possible linkages to other southward-flowing rivers, including the Colorado and the Rio Grande by name, because, as the instructions read: "The Northern waters of the Missouri are less to be enquired after, becau[s]e they have been ascertained to a considerable degree, & are still in a course of ascertainment by English traders, and travellers." This was an allusion to the Saskatchewan River drainage flowing into Hudson Bay via Lake Winnipeg, and the Athabasca/Mackenzie basin that drained to Arctic, in contradistinction to the Missouri's temperate range, which Jefferson had earlier established as a geographical desideratum.[40]

Jefferson's instructions then reverted to the mercantilist subtext. Lewis was advised: "Should you reach the Pacific ocean inform yourself of the circumstances which may decide whether the furs of those parts may not be collected as advantageously at the head of the Missouri (convenient as is supposed to the waters of the Colorado & Oregan or Columbia) as at Nootka sound, or any other point of that coast." With no American settlers, or even trading posts, on the Pacific Slope at that time, let alone clear international title to what became known as the Oregon Country, this coded language was clearly aimed at laying a foundation for commercial opportunity. The tactical aim laid out for Lewis was to determine whether the Northwest Coast fur trade could be redirected "through the Missouri & U.S. more beneficially than by the circumnavigation now practised."[41] In short, Jefferson was scoping a new, river-based American Northwest Passage shortcut from the Pacific to the Atlantic. In the event, Lewis paid very little attention to the maritime fur trade. He never encountered any traders on the Pacific coast – not that it would have mattered in the long run.

Despite what Lewis would write once he returned to St. Louis, the continental interior's geography was not conducive to Jefferson's vision.

Further evidence of Jefferson's gradual multi-decade shift of focus from continental infilling by American settlers to an international outlook is reflected in his backup plan for the expedition's return. Lewis was authorized to send "two of your trusty people back by sea" with copies of his journal. Jefferson instructed him: "Should you be of opinion that the return of your party by the way they went will be eminently dangerous, then ship the whole, & return by sea, by the way either of cape Horn, or the cape of good Hope." To expedite that option, the president provided Lewis with "open letters of credit" to be presented to any American official "in any part of the world," or the representative of a foreign government or commercial agent "of any nation with which we have intercourse."[42] He referenced by name the American consuls in the Dutch East Indies, Île de France in the Indian Ocean, and Cape Town.

What Jefferson had done was delineate the route of the East India Company and other mercantile entities. Later on, he effectively admitted that the expedition's goals had shifted, telling Robert Patterson in November 1805 that when Lewis's mission was first "under contemplation," its "principal object" was "obtaining a correct map of the Missouri."[43] Lewis, however, never seemed to fully catch on to this transition, which is why his journaling emphasized the previously unexplored extent of the Missouri from the Mandan villages to the river's headwaters.

If Jefferson's instructions to Lewis were not as explicit regarding the value of a Northwest Passage as those from the era of Cook or Dalrymple, that changed after the Louisiana Purchase. Just before news reached Washington about the diplomatic success in Paris, Jefferson received another letter from the French naturalist Bernard Lacépède. Writing on May 3, and perhaps with knowledge of what Livingston and Monroe had accomplished the previous month, Lacépède responded to the note Jefferson wrote him the previous February. He was delighted that Jefferson was "going to have the sources of the Missouri explored." After a brief discussion of western geography focusing on what Vancouver's voyage seemed to have disclosed about the Columbia River, he launched into a full-throated version of the mercantile vision that Dalrymple and Mackenzie had touted in their publications of 1789 and 1801, respectively: "If your nation could establish an easy communication route by river, canal and *short* portages, between New Yorck, for example, and the town which would be built at the mouth of the

Columbia, what a route that would be for trade from Europe, from Asia, and from America, whose northern products would arrive at this route by the Great Lakes and the upper Mississippi."[44] The "northern products" were peltries.

Not content with just the commercial sphere, Lacépède then exclaimed: "What greater means to civilization than these new communication routes!" As he explained to Jefferson, American progress along these lines was all part of the "movement of enlightenment" that had proceeded from east to west since ancient times. He added: "The inhabitants of the United States, if they do not reject their destiny, will one day halt and reverse this movement." Lacépède was invoking a cultural trope about the fated succession of power coined by Bishop George Berkeley in the 1720s: "Westward the course of empire takes its way." This theme became particularly popular in American landscape painting and narrative mythmaking in the middle of the nineteenth century. It took such forms as Hudson River School progenitor Thomas Cole's *The Course of Empire* pentalogy (1833–36), Emmanuel Leutze's mural in the Capitol, *Westward the Course of Empire Takes Its Way* (1861), and the imperialist concept of Manifest Destiny.[45]

We know that Jefferson found Lacépède's thinking conducive to his own emerging world view because he immediately shared its contents with Lewis, who had left the capital in early July for Pittsburgh, where he would commence his river journey westward. The Louisiana Purchase Treaty arrived in Washington on July 14 (Bastille Day), and the very next day the president quoted "Mr. La Cepede at Paris" thus: "If your nation can establish an easy communication by rivers, canals, & short portages between N. York for example & ... the mouth of the Columbia, what a route for the commerce of Europe, Asia, & America." In 1786, in the first flush of national independence, Jefferson had written fondly about Americans peopling the continent, for "wherever we reach ... the inhabitants become uneasy, as too much compressed, and go off in great numbers to search for vacant country." By the summer of 1803, he was anticipating Walt Whitman's "Passage to India." Lewis himself later lamely adopted this same trope to warrant what he imagined as the success of his expedition.[46]

Though American sovereignty over the Missouri River Basin itself seemed fated when Lewis headed west, in fact the US Senate had yet to approve the treaty effectuating the Louisiana Purchase. This would not be accomplished until October 20, 1803, on a vote of 24–7 that overcame Federalist opposition. Napoleon had implemented the Treaty of San Ildefonso by conveying Tuscany to the Spanish crown while pledging never to convey Louisiana to a third party, a promise he broke flagrantly when it suited his ends. Indeed, France itself did not take legal possession of the territory until November 30. On December 20, the tricolour was lowered in New Orleans and the American flag raised in one of the great moments in American diplomatic history. Lewis later witnessed a similar ceremony in St. Louis.

This chronology explains the guarded nature of Jefferson's external correspondence for the balance of 1803. On July 3, he dispatched a letter to Henri Peyroux, whom he had met in Philadelphia in 1791 but recently learned was the head of a trading firm in St. Louis. With Lewis heading Peyroux's way, Jefferson invoked their "former acquaintance" based on the interests of "science and business." He renewed it on this occasion to assess Peyroux's knowledge of the "geography of the Missouri and the most convenient water communication from the head of that to the Pacific ocean." The president advised Peyroux: "Since coming to the administration of the U.S. I have taken the earliest opportunity in my power to have that communication explored, and Capt Lewis with a party of twelve or fifteen men is authorised to do it." (The exploratory party would grow, with forty-five men heading up the Missouri in 1804, and the permanent party that reached the Pacific in 1805 numbering thirty-three.) Since the Senate had not yet acted, Jefferson told the Frenchman that Lewis's journey was "merely literary, to inform us of the geography & natural history of the country." This was the fiction that was occasionally useful to the president. Peyroux was advised that Lewis had a passport from the French minister in Washington, even if subsequent events, referring to France's cession of Louisiana that spring, outran its procurement. Nevertheless, Jefferson pleaded that "for an object as innocent & useful," he hoped Peyroux might use his patronage to help Lewis "going & returning."[47]

The diplomatic community was not fooled by Jefferson's attempts at literary legerdemain. On March 30, 1804, mere weeks before Lewis and Clark jumped off from St. Louis, Marqués de Casa Calvo, the Spanish commissioner for the retrocession of Louisiana, told Foreign Minister Cevallos that the intent of the American expedition up "the Missouri River towards the direction of the South Sea" was the establishment of "a port within five years." Casa Calvo advised Madrid that the Americans were "directing towards that district all their views and voyages, not losing from sight the forming from the interior of their states a chain of establishments, making themselves masters of our rich possessions, which they desire."[48]

Casa Calvo was implying that with Louisiana now in hand, the United States, via the Lewis and Clark Expedition, was casting covetous eyes on the Spanish realm from California to Nootka. He was prescient, anticipating the subsequent Freeman-Custis and Pike Expeditions into the central plains plus the Astorians' settlement at the mouth of the Columbia in 1811, and, for that matter, the Adams-Onís Treaty of 1819 by which Spain conveyed its title to the Oregon Country to the United States. Spanish colonial officials never reconciled themselves to France's liberality with the United States, sending out several teams from their base in New Mexico to intercept Lewis and Clark, but failing in their attempts.

The Expedition, Briefly

I have written extensively elsewhere about the course of events surrounding Lewis and Clark's western adventure, including the serial influences of Cook, Vancouver, and especially Mackenzie.[49] Space here allows only brief mention of my major themes, all of which contest the mythology that took shape during the expedition's bicentennial observance two decades ago. First, when Lewis and Clark reached the actual discovery zone, they had a far more nuanced understanding of the complex geography of the mountain west than Jefferson and his correspondents initially brought to their attention, especially ideas about a single range of mountains and the supposedly short portage between watersheds.

The breakthrough in comprehension came in St. Louis with their cognizance of the 1795–97 expedition of James Mackay and John Evans to the upper Missouri. These men were British expatriates enlisted in a last-ditch effort by Spanish colonial officials to reimpose sovereignty in the face of British incursions into present-day North Dakota. Only Evans made it to the Mandan villages, but in getting that far, he gathered Indigenous insights into the tributary system further west and the transverse layering of mountain ranges in the vicinity of the river's great falls. Lewis and Clark were able to clarify and expand upon Evans's insights when they reached the Mandan and Hidatsa themselves, including such information as the Missouri's three forks forming its headwaters and the interface with streams that flowed west. Even before reaching the Great Falls of the Missouri, Lewis and Clark could see that several ranges of mountains would need to be crossed. There was no great geographical disappointment for Lewis at Lemhi Pass on the Continental Divide as legend has it. That is a historiographic invention; a rigorous crossing had long been anticipated.

Once the expedition reached the Rocky Mountains, Mackenzie's *Voyages from Montreal* was regularly consulted as an advisory text. This we know from the more candid notes to that effect by the enlisted men, as opposed to the captains, who were chary to admit their reliance on the Scotsman. Indeed, through a close examination of text, it is easy to see that Lewis and Clark borrowed from Mackenzie for stylistic enhancements, and at times engaged in outright plagiarism. Their legendary scrawls on trees when they reached Pacific tidewater followed Mackenzie's scripting. They also drew upon the literature of Cook for occasional literary inspiration, and Vancouver's charts of the lower Columbia were of great practical value.

Notwithstanding the advance notice and logistical insights secured from other explorers and Indigenous sources, Lewis and Clark did some significant geographical problem solving in the Rockies. Contrary to conventional understanding, Clark was the primary detective. He discovered, in the years after Lewis's death, when he was working with Nicholas Biddle to prepare the expedition's journals for publication, that his co-commander had tried to expropriate insights he had gleaned in the field. This reflected a larger pattern of behaviour. First, as every student of the expedition knows, Lewis was one of the least consistent journal keepers in the history of Enlightenment-era exploration. If he had been in the Royal Navy of Cook and Vancouver, he would have been court-martialled for such inconsistency in journal keeping. No wonder, then, that he had to steal Clark's material and make it his own. Second, this correlated with Lewis's tendency to jump ahead of the expedition in quest of glory while Clark was left behind to manage the expedition. This happened at the forks of the Yellowstone, the Great Falls, at the Continental Divide, and upon reaching the Pacific Ocean. On the other hand, Lewis was such a fragile explorer that when he realized that Mackenzie had (seemingly) discovered the main fork of the Columbia, and that the Clearwater/Snake system he was on was a mere tributary, he stopped writing altogether.

When the expedition returned to St. Louis in late September 1806, the first thing Lewis did was write a letter to Jefferson reporting on his findings. In the second sentence, he declared: "In obedience to your orders we have penitrated the Continent of North America to the Pacific Ocean, and sufficiently explored the interior of the country to *affirm with confidence* that we have discovered the most practicable rout which dose exist across the continent by means of the navigable branches of the Missouri and Columbia Rivers." Explorers, by their nature and because of their psychological investment, were not prone to trumpet failure, and there is no better example than this. Lewis, to his credit, did offer a clarification – nonetheless in conflict with his major point – that the stretch of trail from the Great Falls and "a navigable part" of the Clearwater, across the Bitterroot Mountains, was "the most formidable part of the tract proposed across the Continent." Two hundred miles was good road but another 140 traversed "tremendious mountains which for 60 mls. are covered with eternal snows." The route was passable only from late June to late September.[50]

With that weak foundation, Lewis expanded upon his initial finding to address the mercantilist objectives that were over time ever more important to Jefferson, as reflected in his correspondence of over twenty years' duration, starting with George Rogers Clark and running through Ledyard, Michaux, and ultimately Lewis himself. In a prolix passage evoking the central texts defining the mission of Cook's final voyage and Dalrymple's adaptation, Lewis asserted:

> We view this passage across the Continent as affording immence advantages to the fur trade, but fear that the advantages which it offers as a communication for the productions of the Eeast [sic] Indies to the United States and thence to Europe will never be found equal on an extensive scale to that by way of the Cape of Good hope; still we believe that many articles not bulky brittle nor of a very perishable nature may be conveyed to the United States by this rout with more facility and at less expence than by that at present practiced.[51]

The imperishable product Lewis referred to was probably tea, and possibly spices.

The classic version of the Northwest Passage as a shortcut to Asia dating to Cook was long gone, but the challenge Mackenzie issued could still be met and a passage to India achieved. Lewis pointed out that the furs of the Missouri River Basin, plus "the immence country watered by the Columbia, may be conveyed to the to the mouth of the Columbia by the 1st of August in each year and from thence be shiped to, and arrive in Canton earlier than the furs at present shiped from Montreal annually arrive in London." He even suggested that if the North West Company's furs aggregated in the

Athabasca district and upper Saskatchewan were redirected to his route via Lake Winnipeg, then "the productions [of] nine-tenths of the most valuable fur country of America could be conveyed by the rout proposed to the East Indies."[52]

This statement was followed by a lengthy paragraph reminiscent of James King's fur trade prospectus in the Admiralty account of Cook's third voyage, including a recitation of schedules, seasonal factors (ice), and trading posts. Reaching a mercantile fervour that was largely out of character for Lewis but of extreme interest to his patron, he repeated: "By this means the furs not only of the Missouri but those also of the Columbia may be shiped to the East indies ... and the comodities of the East indies arrive at St. Louis or the mouth of the Ohio by the last of September in each year." Invoking the classic mercantilist formula, Lewis stated that with financial assistance from the government, "I am fully convinced that we shal shortly derive the benifits of a most lucrative trade from this source, and that in the course of ten or twelve years a tour across the Continent by the rout mentioned will be undertaken by individuals with as little concern as a voyage across the Atlantic is at present."[53] In some measure, this was a prescient glance ahead to the foundation of Astoria at the mouth of the Columbia in 1811, in half the time Lewis forecast, though the gruelling nature of that venture's overland party was conducted with much more than "little concern."

Lewis's letter to Jefferson was intended as a private communication to the head of state. Nevertheless, it was newsworthy that his and Clark's expedition had safely returned. Accordingly, Lewis drafted a letter, nominally from Clark, that they could expect his relatives on the Ohio to "leak" to the public via local newspapers. From there, the news would spread to the wider world of readers, as indeed happened. This public relations document replicated some of the contents found in the letter to Jefferson, describing the route taken, certain topographic particulars, and fur trade prospects. It reached a similar crescendo by contending that a "practicable rout ... dose exist across the continent by way of the Missouri and Columbia rivers. In this we were completely successfull and have therefore no hesitation in declaring that such as nature has permited it we have discovered the best rout which dose exist across the Continent of North America in that direction."[54]

Jefferson's message to Congress, delivered on December 2, 1806, was also predictably upbeat, but more measured in its enthusiasm. He reported that Lewis and Clark "had all the success which could have been expected," which was defined by tracing "the Missouri nearly to it's source," then across "the high mountains dividing it from the Columbia." The latter carried them "to the Pacific ocean." Only one man died, the president reported; contact was made with many Indian nations, and his explorers "informed themselves of the trade which may be carried on with them, the best channels & positions for it." More of a realist than Lewis, and having only to defend the expenditure to Congress and not seek the approbation of a patron, Jefferson added demurely: "The desideratum therefore of the interior of our continent along this important channel of communication with the Pacific, is now obtained."[55]

No passage to India there. As Jefferson's muted message indicates, in its grandest state of idealization the Lewis and Clark Expedition failed. There were, of course, many additions to geographical comprehension and natural history but, as Henry Nash Smith pointed out, the expedition's importance "lay on the level of imagination: it was drama, it was the enactment of a myth that embodied the future." Explaining the expedition's long-standing attraction for a variety of publics ranging from government officials in the nineteenth century to re-enactors in the twenty-first, Smith states that Lewis and Clark "gave tangible substance to what had been merely an idea, and established the image of a highway across the continent so firmly in the minds of Americans that repeated failures could not shake it."[56]

Exploration as Literature

The history of Lewis and Clark's journals serves as a fitting coda to the American response to the British search for the Northwest Passage. At the end of his first two voyages, Cook largely succeeded in stymying the publication of unauthorized versions of expeditionary journals; his successors at the conclusion of the third had far less success. John Ledyard was attracted to the fame and wealth such publications were thought to bring, and it should not have surprised Lewis that, in the wake of his expedition, the same thing happened. Even before he could publish his own prospectus, Lewis posted a public notice in the *National Intelligencer* for March 18, 1807, warning his future readers (as he supposed) against "spurious publications now preparing for the press, on the subject of my late tour to the Pacific Ocean." Probably provoked by these reports, he issued a small pamphlet less than two weeks later touting his forthcoming multi-volume report, highlighted by his itinerary along "the most direct and practicable rout across the continent of North America, from the confluence of the Missouri and Mississippi rivers to the discharge of the Columbia into the Pacific Ocean."[57]

Lewis's attempt to quash illicit accounts, combined with the publication of his prospectus, caught the attention of David McKeehan in Pittsburgh, the publisher of expedition member Sergeant Patrick Gass's journal. In an open letter addressed to Lewis, published in the Pittsburgh *Gazette* on April 14 (and running to nine printed pages in Donald Jackson's archival compilation), McKeehan levelled a broadside severely critical of Lewis's literary pretensions and leadership of the expedition. This was unprecedented in the history of travel literature. Of particular note, given our international theme, was McKeehan's piercing accusation that Lewis had inflated the "hazardous nature of the enterprize." In his key point, he observed that Mackenzie's party, "one fourth part of the number under your command, with means which will not bear a comparison with those furnished you, and without the *authority,* the *flags,* or *medals* of his government, crossed the Rocky mountains several degrees north of your rout, and for the *first time* penetrated to the Pacific Ocean. You had the advantage of the information contained in his journal, and could in some degree estimate and guard against the dangers and difficulties you were to meet."[58]

Though little recognized in the literature of Lewis and Clark, this charge was accurate. Lewis and Clark regularly relied on Mackenzie's *Voyages,* but never with attribution. But McKeehan dug deeper, mocking what he regarded as Lewis's excessive compensation that had already preceded "the fortune to be made" from the publication of his book. In effect, he accused Lewis of double-dipping, asserting that his journal, "taken during a military expedition ... ought to be considered the property of the public and delivered up to the government." This was an astute invocation of the legal principle known as replevin, and as a practical matter, given that Lewis himself was never able to surmount the challenges of getting his journal into print, it would have been a good idea. Jefferson has largely escaped the blame for failing to give Lewis the kind of support that the Admiralty gave to its voyagers, notably Cook.[59]

Less fairly, but indicative of what actually made travel literature popular, McKeehan ridiculed Lewis's scientific airs, asking: "Of what consequence is it to the generality of readers to know how a plant has been classed, to what order, genus or species it has been confined by Botanists?" No one seemed to miss that in Mackenzie's account, he implied. McKeehan questioned the value of another of Lewis's "great points of investigation," namely, "the cause of muddiness of the Missouri." Citing Sergeant Gass's journal, he pretended to resolve the question right there and then. Gass, after observing some bluffs on the upper Missouri, stated: "The earth of which they are composed dissolves like sugar; every rain washes down great quantities of it, and the rapidity of the stream keeps it mixing and afloat in the water until it reaches the mouth of the Mississippi." The teasing continued: "Now who can relish this homespun account without a spice of 'scientific research!'" McKeehan summed up this aspect of his criticism by charging that Lewis was "not a man of science ... not a man of letters ... not qualified for scientific research." Presciently, he taunted: "I shall wait with some impatience for your voluminous work."[60]

Incisively, but with a characteristically heavy dose of sarcasm, McKeehan delved into the sensitive matters of Lewis's relationship with Clark, the larger complement of men, and Indigenous nations. He was sympathetic to Clark's plight, having to cope with Lewis's ego, and other "*companions,*" who, in Lewis's published comments, were degraded "to mere *Followers.*" He chastised Lewis for his "unhappy affair with the Indians on the plains of Maria's river," referring to the fatal fracas with Blackfeet Indians in northern Montana during the return journey. There was also the embarrassing incident shortly thereafter "affecting one of your own posteriors," a reference to Lewis's getting shot in the buttocks by Pierre Cruzatte during the same leg home. McKeehan charged that Lewis was lucky that Gass's journal had not fallen in the hands of a less scrupulous editor, "who might have insinuated that your wound was not accidental, but that it was the consequences of design." In others words, McKeehan intimated that Lewis had either been shot intentionally (or not) by one of his men, the ultimate insult to a commanding officer, or the episode was orchestrated for dramatic effect. Either way, the scene was worthy of becoming one of "the plates embellishing your second volume!" He even lampooned Lewis's love life, alluding to "some favorite widow Wadman" (a figure from Laurence Sterne's 1759 novel, *The Life and Opinions of Tristram Shandy, Gentleman*), "who might have been languishing during his absence."[61]

In the literature of the expedition, it is often observed that Meriwether Lewis never wrote for publication. This opinion is based on prospective publisher John Conrad's comment to Jefferson, written the month after Lewis's death, that "Govr. Lewis never furnished us with a line of the M.S." Jefferson must have passed this observation on to eventual editor Nicholas Biddle, who in turn wrote a bookseller in July 1810 that "not a word was prepared for the press by Captain Lewis." This appeared in a longer letter complaining about how "exceedingly troublesome" the work was turning out to be. Lewis might not have delivered a workable manuscript to Conrad, but anyone who has seen the clear pages and polished text of his journal for the Mandan-to-Rockies segment, and for the return leg from Fort Clatsop to the Yellowstone River, including the literary convention of applying carry-over words as subtext from one page to the next, must conclude that these journals, at least, were intended as the proto-manuscript. Lewis never delivered it because he never composed the connecting narrative, work that Biddle had to complete. The reason for Lewis's nonfeasance has been much discussed by students of the expedition, but not enough weight has been given to the psychological damage McKeehan inflicted upon him, which more than suggested that he was not an Enlightenment voyager like Cook or a terrestrial explorer like Mackenzie. Lewis's prospectus, published just a few weeks before McKeehan levelled his charges, was the high-water mark of his attempt to publish an account of his expedition.[62]

This juxtaposition is not mere speculation because we know that Mackenzie was on Lewis's mind in the spring of 1807. A few weeks after McKeehan's diatribe appeared in print, Lewis posed for a portrait by Charles Willson Peale in Philadelphia. Following this encounter, Peale wrote on May 5 to one John Hawkins, apprising him that Lewis "is the first Person who has crossed to the Pacific Ocean." Much like Gass could have been the only source of McKeehan's insights into how much the Lewis and Clark Expedition relied on Mackenzie, only Lewis could have been the source of Peale's ensuing addendum. In it, he clarified the main point by telling Hawkins that Mackenzie "only went to a River which ran into the South Sea," whereas "Lewis and Clark have actually visited the sea shore." On this technical point Lewis was correct, because Mackenzie only reached tidewater, and never secured a view of the ocean that was otherwise screened by the coastal littoral.[63]

We end with Biddle's own prospectus for the journals, which finally appeared in 1814 under the title *History of the Expedition under the Command of Captains Lewis and Clark*. Prospective readers were informed that this account would define the "sources of the Columbia and Missouri rivers, which had eluded all former research." This linkage formed "a line of intercourse – the future path of civilization" that connected "the Atlantic and Pacific oceans."[64] The passage to India had been sketched in outline; a future generation would have to complete the canvas.

9

Following in the Footsteps of Lewis and Clark

Political considerations were prominent in the backdrop to the Vancouver and the Lewis and Clark quests for the western gateway to the Northwest Passage, but they were paramount during what turned out to be the final disposition of the matter in the last half of the nineteenth century. A technological innovation also figured markedly.

A prototype of the steam engine was invented by Thomas Newcomen in 1712. As refined by James Watt in 1776, the year James Cook commenced his final voyage, it became the driving force behind the Industrial Revolution, which transformed manufacturing in eighteenth-century Great Britain. The early applications of the technology were, for the most part, stationary – fixed in place within an ever-growing number of factories. There were intermittent efforts to apply the technology to boats, but when the key breakthrough came that enabled the consistent powering of transport on water (and eventually across land as well), the result was world shattering, literally and figuratively. The key innovation was Robert Fulton's development of the first commercially successful steamboat, the *Clermont,* on the Hudson River in 1807. This was the year after Meriwether Lewis and William Clark returned from the Pacific; it was also on the cusp of the construction of the Erie Canal, a corridor that, not coincidentally, was authorized by the state of New York two years later. The canal era was short-lived, however. In less than two decades, steam engine technology was harnessed to power the pulling of cars on a fixed bed of rails. The first railroad was built in Great Britain in 1825, followed shortly thereafter by the United States in 1827.

The American railroads began extending their lines along the Atlantic Seaboard in the 1830s, gradually crossing the Appalachians into the Ohio and Mississippi valleys, linking with lines that were forming there separately. By the 1840s, it was clear that railroads would become an increasingly important network for the transport of goods, services, and people, and even Henry David Thoreau was enthusiastic about the technology. But the Oregon Trail migration starting in the early 1840s and the California Gold Rush at the end of the decade changed the equation spatially and temporally.

The influx of American settlers into the rich Willamette Valley of Oregon created a sense of urgency in American policy circles. Two explorers, Charles Wilkes and John C. Frémont, visited the region in 1841 and 1843, respectively. However, unlike their more prominent predecessors in the work of discovery, such as Cook, Vancouver, Mackenzie, and Lewis and Clark, they gave little thought to the Northwest Passage, in theory or practice. Instead, their missions were more instrumentally focused on the incorporation of Oregon into the domain of the United States. The Democratic Party's candidate in 1844, James K. Polk, ran successfully on a national expansionist platform that, in part, called for the absorption of the entire Oregon Country, up to latitude 54°40'N, even at the risk of war with Great Britain. Once the United States announced that it had no interest in an extension of the joint occupancy of Oregon, a convention that dated to 1818, the British hoped to divide the territory at the Columbia River. The Americans, in the wake of Wilkes's survey of the Columbia River, including the loss of one of his ships while crossing the bar, combined with the attractiveness of the many deepwater and protected harbours in Puget Sound, insisted on an extension of the 49th parallel from the Rocky Mountains to the Salish Sea as the border. This proved to be the line of compromise adopted by treaty in 1846.

Frémont's first two expeditions scouted the Oregon Trail for prospective overland emigrants. He extended the second in 1843 to include an extra-territorial survey of California (at the time Mexican territory). He undertook a third expedition in 1845 that was less a matter of exploration and more a nationalist filibustering venture. It culminated with him in a prominent role in the creation of the Bear Flag Republic, the precursor to the present-day state of California." The ensuing cession of land to the United States by Mexico in the Treaty of Guadalupe Hidalgo, which ended the Mexican-American War in 1848, immediately compounded by the Gold Rush, resulted in California's admission to the Union in 1850. Thus, within a decade, American citizens created two colonies significantly detached from the rest of the nation as then constituted. The railroad, that wondrous compressor of time and space, was immediately discerned as the obvious and fortuitous linking device. The same industrial dynamic would be replicated a decade later as the solution to the political isolation of British Columbia from eastern Canada.

One of the early figures speeding this development on the American side was Thomas Hart Benton, senator from Missouri. Benton was an advocate of American expansion to Oregon as early as 1825. The year before, he had visited Thomas Jefferson. The "Sage of Monticello," as we saw in the previous chapter, had been discreet in the way he framed the nature of the Lewis and Clark Expedition. Benton, in his autobiography, published four years after he left the Senate in 1850, wrote that Jefferson "was the first to propose the North American road to India, and the introduction of Asiatic trade on that road." He further maintained that all he had done over his thirty-year senatorial career was "nothing but the fruit of the seed planted in my mind by the philosophical hand of Mr. Jefferson." Benton's daughter Jessie, not coincidentally Frémont's spouse, was even more expansive. Writing as her husband's amanuensis in 1887, she drew on her deceased father's comment to state that Jefferson told Congress that the expedition jointly commanded by Lewis and Clark would "*open overland commercial relations with Asia*; and enlarge the boundaries of geographical science."[1] The former aspect might have been implied in Jefferson's message to Congress and his instructions to Lewis, but only the last part was explicitly stated. In any event, the Benton family began the process of recasting the Northwest Passage from its classic mercantilist roots to a concept that included a continental expansion quotient.

In a fashion, Benton was succeeded by Asa Whitney, a New York merchant who had travelled to Asia after the Panic of 1837, returning in 1844 with a fortune that enabled him to become what would today be called a public intellectual. His particular claim to fame is that he is generally conceded by historians to be the first well-known proponent of a Pacific railway. In 1845, contemporaneous with the widespread expansionist fervour subsumed by the term "Manifest Destiny," Whitney tendered a proposal to Congress for a gigantic land grant that he could translate into the monetary resources necessary to build a railroad from Lake Michigan to the Pacific. The cabinet of President James K. Polk also discussed the matter.

Whitney's plan was audacious and entirely impractical at the time, not least because the United States did not secure uncontested title to the Oregon

Country from Great Britain until the following year. Still, Whitney's vision was highly influential, if only because it seemed to follow the path first pointed out by Jefferson and then Benton. It is important to note, in light of subsequent events, that Whitney's was to be a *northern* railroad. The Oregon Trail, notwithstanding its utility as a means of conveying thousands of migrants to the Pacific Slope, was as inconsequential in the matter of sponsoring commerce to Asia as Lewis and Clark were. Writing for a commercial magazine in November 1848 – that is, before news of the discovery of gold in California reordered perceptions – Whitney heaped scorn on the idea "of an overland communication with [ox]-teams through a wilderness and desert of more than two thousand miles in extent."[2] The next year, he solidified his thinking in a pamphlet that evoked the legacy of Alexander Dalrymple, titled *A Project for a Railroad to the Pacific.*

Whitney was dismissive, however, only of the overlanders' means of travel to the valleys of western Oregon and gold fields of northern California, not what their physical presence portended in terms of a railroad's potential. Like Benton, he accelerated the transition of the Northwest Passage away from its classic saltwater (Cook and Vancouver) and revisionist river (Mackenzie) origins to include commercial development within the interior of the continent. To the end of his career, Benton never fully envisioned a railroad running all the way to the Pacific. At best he thought of it as a part-rail, part-trail combination – a hybrid between the Oregon Trail and the Northern Pacific. Benton never quite understood, as even Thoreau did, not to mention Whitney, that a railroad was the only way to enhance trade with Asia and settle the Trans-Mississippi West. He was perhaps circumscribed by St. Louis's normative self-image as the junction of river systems draining the continent.

Not that Benton was alone in this limited vision. One of Whitney's early critics was Stephen A. Douglas from neighbouring Illinois. He too saw the promise of this new technology but thought Whitney's plan was too bold, perhaps even contrived, since it granted too much power to a single holder of the land patent. At mid-century, Douglas thought the more natural way to build a transcontinental line was through the gradual accretion of segments slowly moving westward. Incremental interior development, and the commerce derived thereby, would underwrite the next section. This way, the individual farmer or merchant, not the industrial visionary, would dictate the rate and range of progress. Though never quite explained this way, Douglas was advocating a system of cellular growth much the way Canadian fur traders like Peter Pond and Alexander Mackenzie expanded their network. In the end, however, Whitney's "one giant leap" (to borrow a metaphor from the race to the moon) won out over Douglas's more conservative approach. The generative power of the railroad was an irresistible option.

Whitney's grand vision of a northern transcontinental railroad, the latest incarnation of the Northwest Passage, was so bold that it quickly prompted competing schemes for other potential east-west corridors. Given the vast distances between the centres of population in the Mississippi Valley and the Pacific Slope, it was commonly thought that even with the necessary financial involvement of the federal government, only one transcontinental rail line was achievable. As every student of American history knows, the 1850s were a notoriously fractious period. Accordingly, there was no political or commercial consensus about where between St. Paul and New Orleans a line might jump off to the west; or, correspondingly, where between San Diego and Puget Sound it might terminate. Since Congress was chronically deadlocked on sectional issues, it naively looked to *science* to settle the matter of how to link California and the Oregon Country with the rest of the nation.

The central personage in this ostensibly unifying effort for the United States turned out to be one of the most divisive figures in American history. The new secretary of war in the administration of President Franklin Pierce that took office in March 1853 was Jefferson Davis, formerly senator from Mississippi. Two days before Pierce was inaugurated, Congress passed the Pacific Railroad Survey act. This was a policy response to the series of railroad conventions (a hybrid phenomenon that was half political caucus and half commercial gathering) that occurred in many interior cities throughout the late 1840s and early 1850s. The bill directed the secretary of war to commission a survey of all practicable routes for a railroad across the muscular chest of America. The surveys were to be conducted by the Army's Corps of Topographical Engineers (CTE), predecessor to the modern Corps of Engineers, whose findings were to be reported within ten months.

Davis, not surprisingly, had a history as an avid promoter of any development that would benefit the southern section of the country. The head of the CTE, Major William H. Emory, was his best friend and a southern sympathizer (though he would later serve in the Union Army). This was a fortuitous alignment for regional self-interest. As historian William Goetzmann phrased it, "because of his close identification with the Southern cause Davis was prepared, all things being equal, to give his own section the benefit of the doubt."[3] When it became clear that a southern route was disadvantaged by the southern boundary of the Mexican Cession, Davis masterminded the Gadsden Purchase, securing the more topographically conducive tract running from Las Cruces, New Mexico, to Yuma, Arizona. That accord was negotiated on December 30, 1853, just before the New Year and the War Department's deadline for a recommendation on the transcontinental line.

The surveys were intended as a high-level reconnaissance, only slightly more detailed than what Lewis and Clark, and other army explorers like Frémont, had done earlier. To use an architectural analogy, these surveys were envisioned as a conceptual image, not a construction drawing (or blueprint) that a contractor could build from. The investigations were intended to highlight the relative merits of the several corridors, not to delineate a prospective right-of-way. These efforts entailed a broad assessment of elevations, mountain grades, climate, and exploitable resources that could underwrite the costs of construction. Each party was to have a team of professional scientists, in the manner popularized by Captain Cook, to complement the army engineers' work with assessments of natural history within each corridor explored. The Pacific Railroad Surveys were the greatest investment the American federal government had ever made in theoretical science, especially biology and geology.

Exploring parties were formed for four east-west corridors, each reflecting both rival sections and urban centres both within and between sections that had the most political support in Congress. Denominated by latitudinal ranges, they were (running from south to north): (1) the 32nd parallel (from El Paso through the Gila River basin in southern Arizona to San Diego); (2) the 35th parallel (from Fort Smith, Arkansas, through Albuquerque and northern Arizona into the Los Angeles basin via Tehachapi Pass); (3) the 38th–39th parallel corridor, leaving Independence, Missouri, and reaching Salt Lake City via Cochetopa Pass in Colorado, eventually linking to the emigrant route across Nevada into northern California; and (4) the 47th–49th parallel corridor, running from Lake Superior and the upper Mississippi to Puget Sound.

The northern effort was the first formal government survey of that corridor since Lewis and Clark. Davis, in an economy move, appointed Isaac I. Stevens, already named by Pierce as the first governor of the Washington Territory, to command the northern survey on April 8, 1853. The memory of Lewis and Clark's encounter with rivers and mountains was not far from the minds of the CTE's planners. As Stevens wrote in his memoir: "My instructions required me to examine carefully the passes of the several mountain ranges, the geography and meteorology of the whole intermediate region, the character, as avenues of trade and transportation, of the Missouri and Columbia rivers, the rains and snows of the route, especially in the mountain passes, and, in short, to collect every species of information bearing on the question of railroad practicability."[4] This was the Lewis and Clark agenda, modified to incorporate modern technology.

Scoping the magnitude of his challenge, Stevens, alone among the four corridor commanders, subdivided his expedition by creating a stand-alone western detachment. He appointed Captain George B. McClellan, later of Civil War fame, the leader of a unit that headed east from Fort Vancouver on the lower Columbia to assess the prospect for passes through the Cascade Mountains from its eastern flanks. Lewis and Clark had transected this range by skirting it, running the Columbia narrows and rapids known collectively as The Dalles. Therefore, Stevens wrote, the Cascades were the only part of his corridor that "was comparatively new and unexplored."[5] After searching the mountains for any passes they might yield, McClellan was directed to proceed eastward to meet Stevens and the main party at some point west of the Rockies. In the event, McClellan and Stevens met near the present-day city of Spokane, Washington.

In effect, there was actually a third division of Stevens's command, what might be thought of as a central unit, created from elements of the main and western commands. Lieutenant Rufus Saxton (later a brigadier general for the Union and Medal of Honor recipient) was to detach from McClellan's group to establish a supply depot in the Bitterroot Valley in order to reprovision Stevens and the main party. If he beat Stevens to that point, the meeting was to occur at Fort Benton, the head of navigation on the Missouri River. Another Stevens adjutant, Lieutenant A.J. Donelson, headed a fourth detachment. Donelson was directed to confer with the Hudson's Bay Company's Sir George Simpson in Montreal over a wide range of topics: western topography, the identification of guides that could be procured from the Red River settlements south of Winnipeg, and the prospect of resupply from the firm's stores in case of need. Thus informed, Donelson was to proceed to St. Louis. From there, he was to assess the current state of navigability of the Missouri to Fort Benton before joining Stevens for the final assault on the Rocky Mountains.

Stevens was nothing but thoughtful in his approach to his mission. He toyed with the idea of securing a steamboat in Pittsburgh (where Lewis had jumped off) for himself and the main party, aiming for Fort Benton. That way, he could "throw it upon the mountains at once" while a smaller party handled the more prosaic duties crossing the plains west of St. Paul. He ended up rejecting this idea because "it introduced an element of uncertainty" in his operation. He could not be sure that any craft he secured on the Ohio would make it that far up the Missouri, nor was he confident it could hold a sufficient number of horses and cattle "required for the examination of the passes."[6]

Stevens's corridor, already underprivileged because of its length, complexity, and nordicity was, in the end, further disadvantaged by his gubernatorial appointment because he was thought certain to exaggerate the northern line's potential. Indeed, the proponents of the other routes were quick to criticize his report as a special pleading grounded in an obvious conflict of interest. His biographer, Kent Richards, asserts that Stevens's reports did "overstate the advantage of the northern route," but no more so than those of any other survey commander. This was to be expected, Richards believes, because each leader had "the natural tendency" to cast his route in "the best light possible."[7]

A native New Englander and West Point graduate, Stevens made his mark during the CTE's coastal survey. He had been a booster of Franklin Pierce during the election campaign, and was rewarded first with the territorial appointment and now the railroad survey, a military effort that he was to command as a civilian. This unusual arrangement would precipitate some dissension. Stevens, recognizing his special challenges and reflecting his personal and professional ambitions, was the first to deploy his party in the field. He also took the most expansive view of what the survey was to entail in terms of gathering ethnological, geological, botanical, zoological, and especially meteorological data. He drew on the nation's scientific establishment in Washington, DC, including the nascent Smithsonian Institution, for staffing and scoping of the endeavour. One of his best hires was Dr. John Evans, who had previously conducted a survey of the Oregon Territory (before it was split to create the Washington Territory) for the Department of the Interior. Evans was appointed the expedition's geologist but Stevens valued him especially because he "had passed over portions of the route" beforehand and had personal insights about the Lewis and Clark trail from Travelers Rest in the Bitterroot Valley to the Columbia.[8]

Stevens replicated this strength-through-experience strategy by choosing John Mix Stanley as the expedition's artist (and photographer – Stanley brought along a daguerreotype camera and experimented with its use from time to time, although he left no images to posterity). A portrait artist inspired by the success of George Catlin, Stanley first visited the western frontier in the late 1830s and took his gallery on tour thereafter. He then travelled to the Pacific Northwest by steamer from Panama in 1847. Like Catlin, Stanley painted his Indian portraits under the common conceit that the continent's Indigenous People were disappearing. When Stevens approached him, Stanley had just deposited his collection with the Smithsonian (most of which was later lost due to a fire). He had travelled over much of the old Oregon Country. Indeed, Stanley barely missed the fracas at Whitman Mission in November 1847 that had resulted in the death of Marcus and Narcissa Whitman and others. He ended up being one of Stevens's most frequent companions, much in the manner of the artists who accompanied Cook ashore at his many destinations.

The Stevens team was a strong one. Kent Richards asserts that "Stevens was determined that the northern railroad survey would be the greatest scientific, topographic expedition since Lewis and Clark and would provide a fitting opening chapter to the new era about to dawn in the northwest quarter of the nation." Stevens's "excess of zeal," Richards adds, was not reflected by "inflating his route's desirability, but rather in his exceeding somewhat the scope of his instructions by attempting to produce a report on an entire region." Stevens believed that his team's many treatises would enable federal officials and his own territorial associates to develop an orderly transportation plan, encompassing roads and rivers as well as the railway that would spur the region's growth.[9]

Stevens reached St. Louis on May 15, 1853, accompanied by Stanley. There he oversaw the outfitting of an American Fur Company boat for Donelson's use. Destined initially for Fort Union, the company's post at the Missouri's strategic confluence with the Yellowstone, Donelson was instructed to survey the river as best "permitted by his circumstances as a passenger on the steamboat, and to prepare a report on the capacity of the Upper Missouri for steamboat navigation, giving particular attention to obstructions in the river, and the cost and means of removing them."[10] Unique to the four corridors under investigation, the legacy of an earlier image of the Northwest Passage, the idea of connecting rivers dating to Mackenzie and Lewis, lingered in the background of Stevens's effort in the north.

Stevens left St. Louis for St. Paul on May 23. On arrival, he bought "strong, well broken mules" and horses right at the wharves to replace and complement the young but "unbroken" animals his quartermaster had secured in St. Louis.[11] After setting up camp near Fort Snelling, the army post overlooking St. Paul at the head of navigation on the Mississippi, Stevens began inspecting sites for a railroad bridge across the river. This was essential for a connection to the head of Lake Superior, whose incorporation in the original scoping of the endeavour by Congress was another legacy of the Mackenzian imagination linking the Atlantic with the Pacific. The maturation of Chicago at the mid-continent nexus of the national railroad network after the Civil War eventually made the Great Lakes an outmoded means of transport for most cargos and human travel, but that was not yet clear when Stevens set out.

After a quick, small-party reconnaissance of the terrain from Lake Superior to St. Paul, Stevens dispatched several advance contingents to the west "in order to infuse hope into the whole party and avail myself of the present high spirit of the camp." This early period of good feelings was enhanced by the amusement derived from breaking mules for the trail. "Much hilarity was produced," Stevens reported, "by the efforts of different persons, and each fall occasioned a laugh."[12] Camp broke on June 6, 1853.

Within a week, Stevens conducted the time-tested exercise in naval and military history by causing "to be read the camp regulations which I had prepared for the government of the party." He also addressed the gathering, informing "them that every man would be expected to look to the safety of his comrades; that all alike, whether soldier or civilian, would be expected to stand guard, and in case of difficulties to meet them promptly." In his memoir, Stevens confessed that he "exaggerated the difficulties which lay before us," a litany that included physical obstacles like rivers and ravines, mountains and snow, but human as well. In an echo of Lewis's fracas, he referred to "the Blackfoot Indians, a tribe proverbially treacherous and warlike." Employing the common leadership trope of expeditionary leaders of several nations during the preceding century if not longer, though one usually expressed in retrospect, Stevens stated that "every man would be expected to follow wherever he might be led; that no one would be sacrificed, nor would any one be subjected to any risk, which I would not freely incur; and that whoever was not willing to co-operate with us had better at once retire."[13] This threat might have been more productively rendered back in St. Paul, when the option was more, as it were, practicable. That it was delivered by a civilian in such an authoritative manner is all the more remarkable.

Stevens's immediate goal was to reach the Missouri at its confluence with the Yellowstone, where he could, in effect, intersect with the Lewis and Clark trail. To get there, he had to cross the prairie of western Minnesota and the Dakota Territory, as then configured. This terrain, interspersed with small sloughs filled with rainwater and some minor creeks (the largest of which of which were bridged, not forded), offered "no special difficulties in the way of railroad practicability." The party, with its heavy gear (including two rubber boats) and provisions, usually made eighteen to twenty miles a day. Pike Lake near Minnesota's western border, which was reached on June 23, was the end of the shakedown phase. A few minor officers and other men were dismissed. "This," Stevens wrote, "I consider as the real starting point of the expedition."[14]

A little more than a fortnight later, the expedition had its most stunning encounter with the sublimities of the northern West. Near Lake Jessie, in present-day south-central North Dakota, the group ascended a hill and took in an expansive view. Stevens is worth quoting at length:

> For a great distance ahead every square mile seemed to have a herd of buffalo upon it. Their number was variously estimated by the members of the party – some as high as half a million. I do not think it any exaggeration to set it down at 200,000. I had heard of the myriads of these animals inhabiting these plains, but I could not realize the truth of these accounts till to-day, when they surpassed anything I could have imagined from the accounts which I had received. The reader will form a better idea of this scene from the accompanying sketch, taken by Mr. Stanley on the ground, than from any description.[15]

The trope of ineffability – the supposed inability of an explorer to describe a scene – was common in travel literature. Even Cook resorted to it, though it became more common in the Romantic period that followed his Enlightenment-era voyages. Often, the artist-illustrators who accompanied these many explorers failed to live up to the grandeur being experienced. Not so here. Stanley's sketch, even after its translation to the dreamy style of lithographic depiction, is one of the most famous images in the Western art history oeuvre (Figure 9.1). It is the best-known scene in Stevens's account, still much sought after by collectors. Before leaving his description of this tableau, Stevens added another element common to travel literature – the antics of the faithful canine companion. Zach, a dog owned by one

Figure 9.1 John Mix Stanley, *Herd of Bison, near Lake Jessie,* 1860. The American *Pacific Railroad Reports,* published before photographic imagery was ubiquitous, conveyed a sense of western landscapes to a wide audience through the use of two- and three-colour lithography. The image reproduced here, from volume 12, Isaac I. Stevens's memoir of his 1853–55 survey, is one of the best. | Washington State Historical Society, 2005.0.21.9

of the soldier/packers, dashed into the middle of the herd, and followed them "a considerable distance" before being recalled.[16]

In an interesting convergence of exploration history, Stevens's next trail station, Lake Jessie, had been named by Frémont in 1839 after the daughter of Missouri senator Thomas Hart Benton, Jessie Ann Benton, who was to later marry Frémont. Theirs was one of the West's great romances. French emigré geographer Joseph Nicholas Nicollet, as the first head of the CTE, assisted by Frémont, reached this area before turning back to the Mississippi. His multi-year effort to map the north central plains was the essential cartographic foundation for the eastern segment of Stevens's corridor. Lake Jessie, Stevens noted, was "the last point on the trail at which our work will connect with Nicollet's survey."[17]

The Stevens party reached the American Fur Company's Fort Union on August 1. Located on the Missouri 2¾ miles above the Yellowstone confluence, this was an essential spot of western geography since Lewis spied it in 1805. When the river had come into view the day before, Stevens stated, "the whole party gave three cheers as its beautiful bluff banks, dotted with timber, came into view." It had taken fifty-five days of travel to cover 715.5 miles, measured precisely by an odometer wagon. With considerable exaggeration, indeed in contradiction to what he had earlier averred with regard to Nicollet, Stevens asserted that he had travelled "over a country almost unknown, without roads." Stanley took his daguerreotype out of storage to take portraits of the Indigenous inhabitants. Stevens stated that "the Indians were greatly pleased" with the imagery that resulted.[18] If only they had been preserved to the present!

The good times lasted until, Stevens reported, his men became "exceedingly alarmed as to the difficulties to be expected from snow in the westward journey." The voyageurs employed by the fur company "thought it a good practical joke to spread bugbear stories about the immense snows to be expected early in the season, and many of the men got to believe that they would find snow knee-deep before they reached Fort Benton, and that it would be twenty feet deep in the passes of the Rocky Mountains in October." This was doubly damaging. Having his men spooked was bad enough, but by snow was the worst thing possible. These rumours struck at the heart of his line's prospects. To assuage these concerns, Stevens was forced to resort to "some books of travel in that country." The only work he cited in his memoir was "De Smet's Oregon Missions."[19] The Jesuit missionary had first reached the continental interior as a supernumerary on one of Nicollet's expeditions before heading further west.

Getting to Fort Union was the easy part. Stevens departed St. Paul with the knowledge that his survey would rise or fall, like Lewis and Clark's mission, on finding a practicable route linking the Missouri River Basin with the upper Columbia. Not surprisingly, in the course of his expedition from this point westward, Stevens actively relied upon Lewis and Clark, and subtly tried to burnish his own report by continued references to them. His expedition left Fort Union on August 10, 1853. Across the rolling prairie of what is now southeastern Montana, there was no discernible change in topography or mood. A comet (Klinkerfues) blazed across the southeastern sky, offering distraction if not enjoyment on the night of the 24th, and the next night the aurora borealis presented to Stevens "a beautiful appearance" in the northwestern sky. The party was upbeat, probably in anticipation of exploration glory. Donelson, the executive officer, started to pitch to Stevens an idea about how to approach the Rockies, still over the horizon. Stevens was "much grateful at such marks of interest in the expedition."[20]

Travelling westward inland from the north bank of the Missouri, the party reached the first significant landmark west of Fort Union when Lewis and Clark's Marias River came into view on September 1. Its confluence with the Missouri, site of the captains' famed decision point, presented Stevens with "a beautiful view."[21] The appearance of the Marias had presented a geographical problem of the first magnitude for the co-captains, and there was an echo of their difficulty in sorting out river systems during the Stevens expedition as well. The day before, Stevens sent a small detachment consisting of Lieutenant Cuvier Grover, engineer Frederick Lander, and Stanley on a side excursion overland toward Bear Paw Mountain to the north, south of present-day Havre, Montana. On their return, they got lost in their confusion between the Teton and Missouri Rivers (both flow directly west to east), having actually travelled beyond Fort Benton. They arrived just as Stevens was about to send out a search party.

Fort Benton, located twelve miles above the Marias and approximately triple that distance below the Great Falls, was the centre of the American Fur Company's trade in buffalo hides with the various subtribes of the Blackfeet Nation. Once again, Stanley took out his daguerreotype. According to Stevens, the stunning, seemingly magical qualities of the technology ("likenesses produced by the direct action of the sun") imbued the practitioner with the qualities of "a great medicine man." By association, Stevens found himself "much occupied" with the same Indians and was able to secure valuable "information about the country."[22]

The pivotal point of the expedition – establishing the practicability of a northern railroad – was now at hand. The prospect of extensive snows in the mountains, directly encountered and graphically described by Lewis and Clark on their return from the Pacific coast, had already been imprinted on the national imagination. From the captains and others such as Jesuit missionary Fr. Pierre-Jean De Smet, Stevens knew of several passes crossing the Continental Divide leading to the watershed of the Bitterroot/Clark Fork River. In order to finesse the Bitterroot Mountains annex to the Rockies, Stevens would have to choose between a pass, likely requiring a tunnel, or following the Clark Fork – a longer route by seventy miles but a relatively level one. This predicament presaged the companion problem of the Cascade Mountains barrier closer to the Pacific coast. The options there would be similar to that presented by the Bitterroots: cross the mountains head-on (with all the tunnelling and bridge building it entailed), or follow the level-water route of the Columbia to Fort Vancouver and then

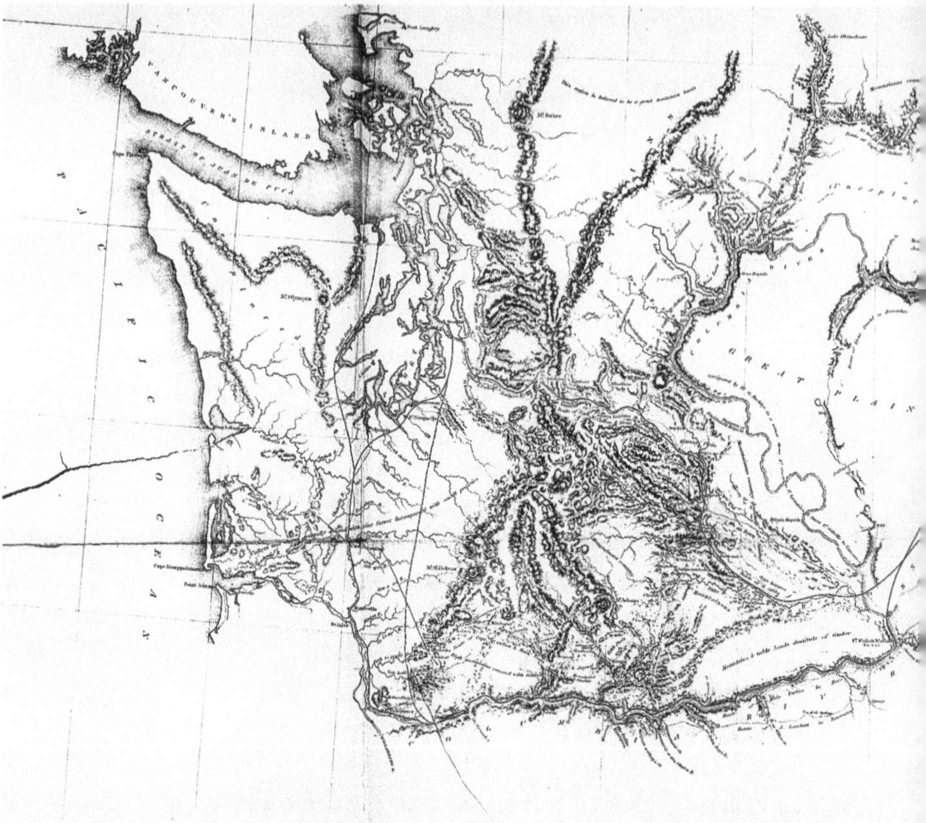

Figure 9.2 Isaac I. Stevens, *Rocky Mountains to Puget Sound,* 1860. Stevens's railroad survey resulted in a significant improvement in geographical comprehension of the northern Rockies and Cascade Mountains in Washington, and became a major contributory element to the comprehensive chart of the American West prepared by G.K. Warren. Among other elements, Stevens's chart showed possible alignments for a railroad across the "Great Plain of the Columbia." | Washington State Historical Society, 1999.0.13

north to the Puget Sound lowlands (see Figure 9.2). In the event, Stevens determined that even the latter route would require a tunnel through the bluffs known as Cape Horn, downstream from the present-day Bonneville Dam.

In all, Stevens would identify five passes in the Rocky Mountains – Marias, Lewis and Clark, and Cadotte's in the main chain, plus Lolo and Hell Gate in the Bitterroot annex. Four of the five drew on Lewis and Clark place names or their expeditionary experience, the only exception being Cadotte's Pass (called Cadotte Pass today). In his initial assessment,

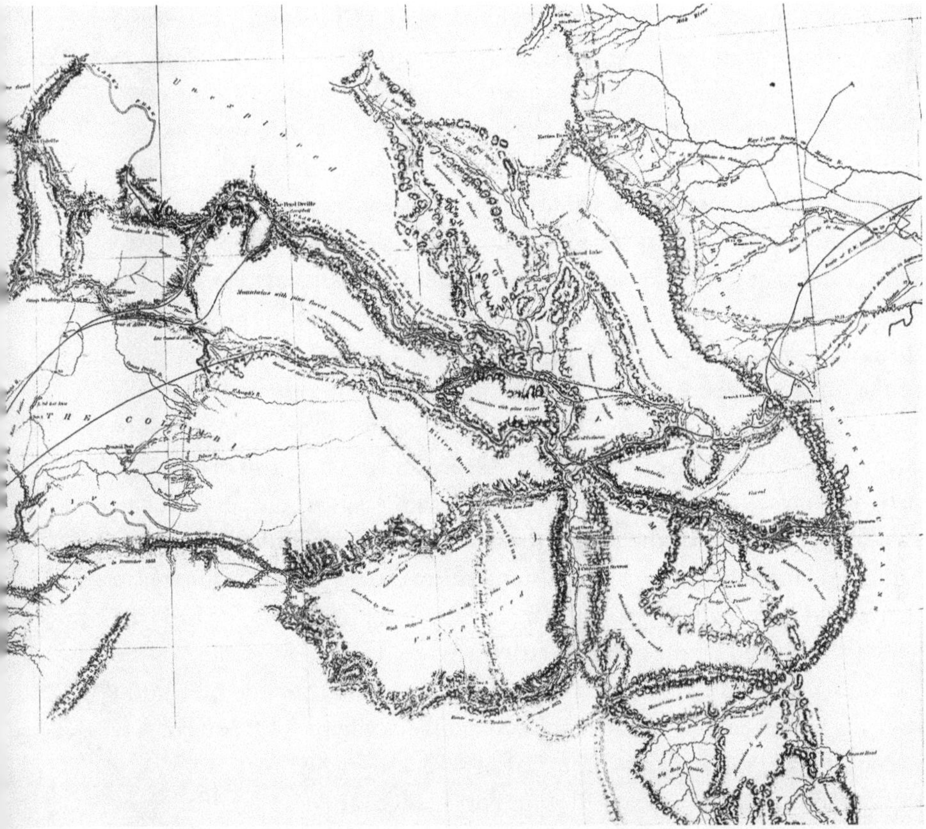

Stevens concluded "that there must be a good and practicable pass leading from some branch of the Marias." At Fort Benton, he reflected: "I was sanguine that we should find there [the Marias valley] the best solution of the question of railroad practicability of the Rocky mountain range."[23] In this regard, Stevens was eventually proven correct, as the Great Northern Railway later adopted that route for its line, but Cadotte's Pass ended up being his favourite. From the east, it was accessed via the upper Dearborn River (the second major tributary to the Missouri above the Great Falls, also named by Lewis and Clark for their boss, Jefferson's secretary of war).

Though not explored by Lewis and Clark, Cadotte's Pass (named after an early settler/trader operating out of Fort Benton) came to be discerned as the closest approximation to the short portage crossing the Continental Divide that Jefferson and other theoretical geographers had long speculated about. This route emerged on the west side of the divide into the watersheds of the Little Blackfoot and Blackfoot Rivers, whose combined flow transects Hell Gate near present-day Missoula, Montana. Eastbound, it approximated Lewis's shortcut to the Great Falls after he and Clark separated at Travelers Rest on their homeward journey. Writing after the fact in his geographical memoir, using language reminiscent of Alexander Mackenzie, Stevens said of the Cadotte's Pass corridor: "Nature seemed to have intended it for the great highway across the continent, and it appeared to offer but little obstruction to the passage of a railroad."[24] Generally speaking, this became the path of the Northern Pacific Railway though the line used adjoining Mullan Pass, not Cadotte's, for the tunnel that transected the Continental Divide.

Stevens approached the Rocky Mountain challenge with a thoroughness that would have made Cook and Vancouver proud. Like them, he had able assistants. Pressed by the administrative need to set the government of the Washington Territory in motion before the fall was over, Stevens left behind over the winter several competent lieutenants (of whom John Mullan was the most accomplished) in command of several detached parties for sub-explorations. These units ranged from Fort Hall (near Pocatello, Idaho) on the Emigrant Trail at latitude 43°N to slightly beyond the 49th parallel into Canada. He also used part of his own treaty tour in 1855 (holding also a *third* commission, as superintendent of Indian affairs for the territory) into eastern Washington (which in its foundational form ran to the Continental Divide) to supplement the original forays.

Rushing ahead of the expedition with a small party (including Stanley) to meet his timeline for establishing the territorial government, Stevens passed north of the Great Falls of the Missouri (without mention) that had so dominated Lewis and Clark's narrative. Coursing up the Sun River (Lewis's "Medicine River"), he entered the Rocky Mountains, crossing over to the headwaters of the Dearborn River via Cadotte's Pass on September 24, 1853. With a flourish reminiscent of Lewis's when he reached the Continental Divide nearly five decades earlier, Stevens wrote:

> It was with great gratification that we now left the plains of the Missouri to enter upon the country watered by the Columbia; and it was the more

especially gratifying to me as, looking to my future duties in the Territory, I felt that I could welcome to my future home and the scene of my future labors the gentlemen of the party, which I did very cordially and heartily.[25]

Descending into the Blackfoot River Valley west of the divide, Stevens noted a geological phenomenon that had also caught Lewis's attention – the "Prairie of the Knobs," an allusion to the knolls that dot this region. After reaching the Bitterroot Valley through Hell Gate, Stevens proceeded south on the Bitterroot River's west bank, where he came to the "Traveller's Rest creek of Lewis and Clark," south of Missoula. His goal was Fort Owen (later to become Mullan's winter headquarters, renamed Cantonment Stevens) and the nearby Jesuit Mission. For his geographical assessment, Stevens invited his readers "to study carefully the narrative of Lewis and Clark; for to us it was a matter of the greatest gratification, with their narrative in hand, to pass through this valley and realize the fidelity and the graphic character of their descriptions."[26]

After crossing the Bitterroot Mountains to the north of Lewis and Clark's more treacherous Lolo Pass route, Stevens descended into the Spokane River Valley. He met McClellan on October 18 at the Hudson's Bay Company's Fort Colville. After a raucous party, Stevens learned the dismaying news that McClellan had failed to run a topographic line to the crest of any Cascade Mountains pass. The governor did not have time to investigate the matter himself because of his need to reach Olympia, which he would designate as the territorial capital. (From there he would direct McClellan to try from the Puget Sound side, and when that failed he commissioned an alternative assessment.) His intention was to reach the Columbia by way of the Snake River, which he accessed where the Palouse River merged with it. At the confluence, he met "a chief from a band but a few miles distant from our camp." This individual, recorded by the governor as "Wi-ti-my-hoy-she," was wearing "a medal of Thomas Jefferson, dated 1801, given to his grandfather, as he alleges, by Lewis and Clark."[27]

Following a brief stop at Vancouver Barracks on the lower Columbia, Stevens arrived in Olympia on November 26. After setting up the machinery of territorial government, he toured the inland sea Vancouver had first charted, to "learn something of the general character of the sound and its harbors." Travelling by sail, he called at Steilacoom, Seattle, the mouth of the Skagit River, Bellingham Bay, Port Townsend, "and several other points on the western shore of the sound." He was immediately "impressed with the important advantage of Seattle," though "Port Townsend, as described by Vancouver, is also an admirable harbor."[28]

Returning to the territorial capital, Stevens pushed himself and his team of military and civilian engineers into the hard work of preparing the reports that by statute were due to Congress in early February 1854. The principal document was Stevens's preliminary report to Davis, dated December 5. This was his first update to the secretary of war since writing him from Fort Benton. It was a pivotal piece of correspondence because the key mountain passes (or most of them) had now been transected. Stevens got right to the heart of the matter in his first paragraph:

> The three great western ranges have been crossed in the fall; no snow whatsoever seen, except on the topmost peaks till the last days of October, and then only for a few inches, passing away in two or three days. I can pronounce, with the greatest confidence, the route to be eminently practicable; and in so doing, I feel repaid for the great exertions I have made, and the great anxiety which at times has almost overcome me, in preparing for and conducting the exploration.

He added that once the many other contributions his expedition made toward geographical comprehension, including the northern corridor's natural resources, were seen by the general public, they would prove the worth of his endeavour "even should other railroad reports be finally determined on." The journey had also proved advantageous toward "management of the Indian tribes, restraining their predatory habits, and disposing them to rely upon the general government to guaranty to them peace and protect them in their rights."[29] Subsequent events would put this naive remark in stark relief.

With these last two embellishments, Stevens was attempting to burnish his case for a greater allotment of resources. His Fort Benton letter had warned Davis that the expedition was proving more costly than projected. In the event, Davis rejected the request for an additional $150,000, telling Stevens curtly: "I have no means of meeting any further demands for the expenses of your party." Davis also reminded the governor that his instructions directed him to close up operations by early February 1854, when his report was to be forwarded to Congress by the War Department. "All arangements looking to the extension of your operations beyond the time indicated in those instructions," Davis advised, "are without authority; and you are directed to close your work in the manner therein prescribed."[30] Stevens would work around this prohibition in the short term by using resources available from his concurrent authorities as territorial governor and Indian commissioner. Congress eventually responded to his plea and those of other commanders by appropriating an additional $190,000 to address deficiencies in the original surveys and to publish their results.

Notwithstanding his quick determination of railroad practicability, when Stevens wrote Davis, he was evading the fact that a topographic line had not actually been carried through the Cascade Mountains of western Washington. After reaching Olympia, he immediately directed McClellan to accomplish that objective by conducting a small-party survey from the Puget Sound side of the mountains. While Stevens toured Puget Sound for harbours that would serve as a suitable terminus, McClellan was to head inland. His scope extended as far north as Bellingham Bay, but the key target was Snoqualmie Pass east of present-day Seattle. Characteristically, McClellan was easily foiled by the terrain and heavy snowfall near Snoqualmie Falls and turned back. This was a mortifying circumstance, as his discouraging report would inevitably become part of the stream of information flowing to the desk of Jefferson Davis.

Fortunately, Stevens had a backup plan. Frustrated with an army officer whose performance perfectly prefigured his deportment as a Union general during the Civil War, Stevens had previously arranged for a civilian engineer left behind with Mullan's Bitterroot Valley command, Abiel Tinkham, to run a topographic line westward up the Yakima Valley and across the Cascades to Puget Sound (see Figure 9.3). Stevens had learned at Fort Walla Walla in the fall of 1853 that pioneer parties were taking wagons up that corridor and over the mountains. He reasoned that one of his surveyors could do likewise. By letter, Tinkham was detached from Mullan's command and in January 1854, he crossed Snoqualmie Pass. There, as the governor recorded in his memoir, he found "but six feet of snow for a short distance." This was a significant finding because now Stevens had "the actual depth of the snow as determined by admeasurement." Here, Stevens was taking a page out of John C. Frémont's playbook. The great pathfinder had argued in the late 1840's that the only way to prove the practicability of a railroad through the Rockies was to attempt a traverse in the winter. The Tinkham initiative was also a not-too-subtle shot at McClellan, whose reconnaissance had forecast "the probable depth of snow in that pass" in places at twenty feet deep or more.[31]

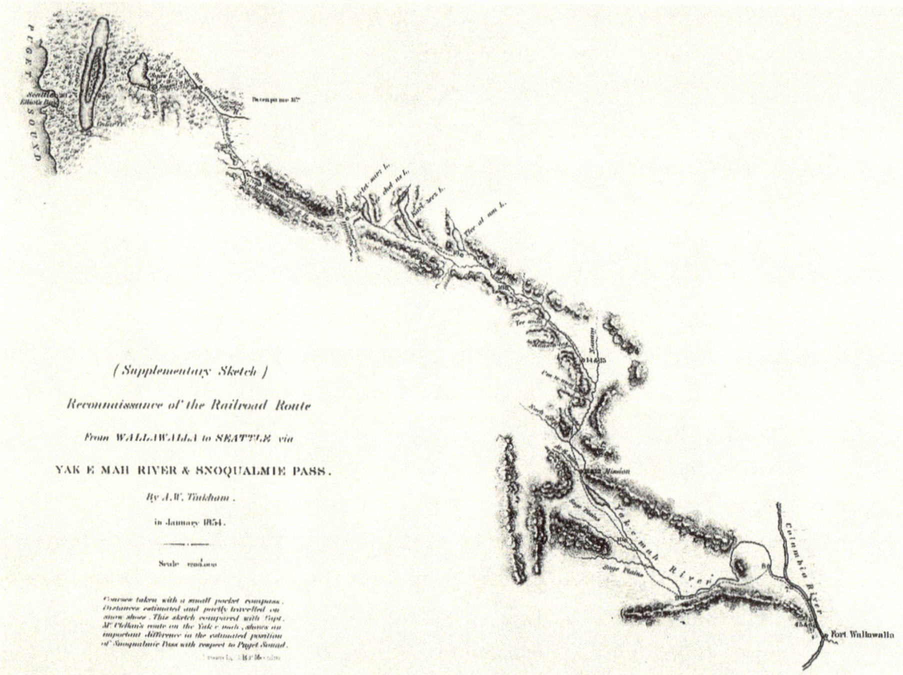

Figure 9.3 Isaac I. Stevens, *Reconnaissance of the Railroad Route from Walla Walla to Seattle,* 1860. This chart is an inset within Stevens's larger work showing the geography of his route from the northern Rockies westward. It shows the survey line Abiel Tinkham ran through Snoqualmie Pass in the Washington Cascades in January 1854. | Washington State Historical Society, 1990.0.13

As Stevens described it, Tinkham first ran a line following Lewis and Clark's trail from the Bitterroot Valley to the Clearwater River, which he confirmed was "steep and laborious, wholly impracticable for wagons. It was rendered more difficult by the snow, which covered the summit to the depth of three feet". On the downhill slope, "the trail led over the summits of high and very rugged hills covered at their tops with deep snow." It took Tinkham fifteen days to reach the Nez Perce villages on the Clearwater, several days longer than it took Lewis and Clark, though he was travelling in late fall/early winter, they in late summer. As an aside, Stevens pondered the question of why the Flathead and Nez Perce Indians at Travelers Rest "impressed" Lewis and Clark "with the idea that the route they pursued was altogether the most practicable of any route across these mountains, and yet it is unquestionably the worst route of the whole." He concluded, not improbably, that it was "in order to keep white people out."[32]

After reaching the Columbia River in early January 1854, Tinkham's charge began in earnest, as he followed the Yakima River into the heights of the Cascade Mountains and its headwaters near Snoqualmie Pass. Stevens later reported that this route "afforded fair facilities both in ascent and descent for a wagon and railroad" through step gradients and a "short tunnel, or with a longer tunnel and easier grades." In the event, this became the route by which the Northern Pacific's rails first reached Puget Sound. Turning to an even more important question, Stevens wrote:

> Wishing to know the real difficulty to be apprehended from the passage of these mountains in the winter season by railroad trains, he [Tinkham] paid particular attention to the measurement and examination of the snows on the route. From Lake Kitchelus to the summit, some five miles, and where occurs the deepest snow, the average measurement was about six feet, but frequently running as high as seven feet.[33]

In his account, Stevens took pains to point out that the snow Tinkham encountered was "very light, dry," which would "present little obstruction to removal in comparison with the compact, drifted snow of the Atlantic States, and would cause very little detention to the passage of trains." Better yet, on the downslope toward Puget Sound, "the snow rapidly disappears; fourteen miles from the summit there were but eight inches of snow, and thence it gradually faded away as approach was made to the shores of the sound."[34] Tinkham reached Seattle on January 26, and met Stevens in Olympia on February 1.

With Tinkham's reconnaissance in hand, Stevens felt he had the key piece of information to sell his corridor to the War Department, overriding McClellan's pessimistic forecast, based as it was on third-hand reports from Indian informants he encountered in the Cascade foothills. Nonetheless, he must have grimaced when he read McClellan's report/rebuttal, dated February 8, 1854, which duty required him to forward to the War Department. McClellan employed two arguments to discount Tinkham's lower estimates. Although he himself two months earlier had encountered a snowfall that was so severe it dictated abandoning an attempt from the west, he asserted that Tinkham had crossed the mountains during an unusually dry season, compounded by the fact that his transect occurred before the greater part of the Cascades' annual snowfall descended. (In fact, the most intense snowfall in the Washington Cascades occurs from mid-November to mid-February.) McClellan believed that a route down the Columbia River was preferable. In two cutting lines of text that must have pleased Davis as much as it annoyed Stevens, he contended: "The question is, after all, reduced to a choice between the shorter line, high grades, a very long tunnel, and almost certain difficulty from the snow, in one case; and the longer line, low grades, little or no tunnelling, and no trouble from the snow, in the other. I prefer the latter."[35]

Stevens left Olympia by Pacific steamer on March 26, 1845, bound for Washington, DC, via Panama. He was able to tender his preliminary findings to the CTE and Davis on June 20. In addition to expeditionary correspondence, he attached forty-two separate sub-reports detailing the various mountain surveys, including McClellan's damning one, plus other side explorations and documents describing river navigability, climatology, and Indian affairs. This collation formed the entirety of volume 1 of the *Pacific Railroad Reports,* published in 1855. Stevens had succeeded in having the first word.

For the purposes of this book, and probably in the governor's mind at the time as well, the most important addendum in Stevens's compendium was his report titled "Resources and Geographical Importance of Puget Sound and Its Relation to the Trade of Asia." As related to Davis in a letter dated January 31, 1854, this work was founded in a tour of the sound concurrent with his dispatching of McClellan to the Cascade's western foothills in December 1853. Stevens described leaving Olympia to "acquaint myself with the harbors of the sound with reference to the termination of the railroad line." He told Davis he "was agreeably impressed with Elliott's bay, on which are the flourishing towns of Seattle and Alki, and I agree entirely with Captain McClellan, in the opinion that it is the best harbor of the sound, and unless the approach to it from the pass should, on a more minute examination, prove less favorable than to some other point, and which is hardly to be expected, that it is the proper terminus of the railroad."[36] Two ships from San Francisco were already in the bay when he toured it.

Knowing that his superiors had full access to Charles Wilkes's insights regarding the inland waters of western Washington, Stevens began his report with a quick introductory allusion to Puget Sound's "many capacious harbors and roadsteads, accessible, commodious, and entirely land-locked." This was a body of water "particularly adapted to steam navigation." He considered Steilacoom, Seattle, and Bellingham Bay as the best prospects for a terminus, but then quickly reconfirmed what he had told Davis in correspondence: "Seattle combines the greatest number of advantages."[37]

Stevens then turned to the heart of his argument, which he hoped would overcome any hesitation about his northern corridor. He stipulated that the "question of the highest importance in connexion with the proposed railroad, is the effect which it will have in securing for this country the control of the Asiatic trade." This, he posited, was "the great commercial prize in ancient and modern times." Expanding and amplifying on themes popularized by Asa Whitney in the previous decade, he argued that whichever nation controlled commerce to the east "swayed the world." Persia and Rome, through Venice and Portugal in early modern Europe, and then later Amsterdam and London "each in its turn attained commercial supremacy, as it became the dispenser of Eastern luxuries to the Western world." The British were currently dominant with their trade through Canton, but, noting the "comparative nearness" of the Pacific Slope of America to Shanghai, a port closer to the silk- and tea-producing region of China, Stevens concluded: "Nature has clearly indicated the northern pathway for the commerce from the future mart of Asiatic trade to this country and Europe." That phrasing was right out of Mackenzie, but by adding the Great Lakes, which could carry waterborne commerce "half-way across the continent," the vision was purely Dalrymplian.[38]

Turning to his project, Stevens noted that the northern railroad corridor communicated "on a direct line with the northern lake trade – the most wonderful internal commerce the world has ever known," and it also intersected with the Mississippi River valley. Unstated but implied in this visualization was a transportation network that could tie the country together, an appealing notion to a nation divided by sectional strife. Not only that, the northern route was also, "above all ... the shortest distance between the centres of European and Asiatic population." Computing the distances between New York and Shanghai, via Puget Sound by way of either Cape Horn or the Cape of Good Hope, thousands of miles would be shaved off sailing times. Stevens realized that his concept was predicated on the ability of steamships to run from the Pacific coast to China, but even here his line was conducive to international trade because the coal fields already identified in western Washington would "furnish the necessary combustible." In his mind, it seemed destined that "when an uninterrupted line of steam communication is established, a portion of the European trade, and nearly all the travel to Asia, must take its course across our continent, and on the northern road, as the shortest route." This 5,000-mile shortcut was "equal to a voyage across the Atlantic."[39] Not only was the northern steam corridor faster for passenger travel and mail but, by avoiding the tropical latitudes, some cargoes, such as tea, could be imported with less damage.

Stevens then addressed the nexus between commerce and national politics that he must have considered his clinching argument: a shorter route to China would "facilitate the exportation of goods manufactured from the great American staple" – cotton. At the time, this trade was dominated by Great Britain, and, Stevens pointed out, "hitherto we have had no advantage of distance in our competition for this trade." With the maturation of the American manufacture of cotton goods, "millions of people in China" and elsewhere in Asia could be supplied. Reaching a visionary crescendo, Stevens argued: "When the proposed railroad is completed," the "manufacturing skill and enterprise of the North" combined with "the resources of the South" were "adequate to meet the future demands of an unparalleled trade."[40]

After tendering his reports to Davis, Stevens returned to Olympia, which he reached in December 1854. He immediately began laying the groundwork for his treaty tour. He used that expedition to supplement the work from 1853, reaching as far east as Fort Benton in October 1855. After his controversial term as governor, Stevens was elected the territory's delegate to Congress, and arrived in the nation's capital in the fall of 1857. There, with access to materials prepared by his expedition and from external sources, he began to write "the final report contemplated in the first volume." In May 1860, at the conclusion of that effort, he stated that the intervening years had been spent "industriously occupied in preparing the report." He considered the delay fortuitous, as it increased his "opportunities of information and judgment."[41]

By this Stevens meant his ability to digest and interpret the data from the mountain reconnaissance and consideration of the strategic importance of Puget Sound. Tinkham seems to have been most valuable with regard to the former. With the possible exception of Mullan, he was Stevens's most helpful aide. In the spring of 1854, while Stevens was on his way to Washington, DC, with the original field reports, Tinkham returned to Cantonment Stevens in the Bitterroot Valley with the mission of running a topographic line across the Continental Divide to Fort Benton. This assessment resulted in the identification of Prickly Pear Creek (named by Lewis and Clark) as an alternative to the Dearborn/Cadotte's Pass route. This is now the course of Montana Rail Link, a descendant of the original Northern Pacific. Nevertheless, the practicability of Snoqualmie Pass was the larger issue, politically as well as from a meteorological perspective. As Stevens framed it in his final narrative: "When the time comes for building the railroad over the northern route, the engineers will then [determine] if they are not satisfied that the snow is not an obstruction. They will make the instrumental survey of the Snoqualmoo Pass to the sound, because it is a question of the saving of some 130 miles over the route by the Columbia and the Cowlitz" rivers.[42]

A Return to Maritime History

Deep into his final report Stevens pursued a narrative digression on relevant geographical nomenclature. In 1798, Vancouver's published account defined "Puget's Sound" as the many-fingered inlet south of the narrows near present-day Tacoma, satisfying the classic meaning of the term "sound." (Another good example is Prince William Sound in Alaska.) But by the time the Washington Territory was created sixty years later, Stevens noted, the "whole system" – Puget Sound proper, plus Admiralty Inlet and Hood Canal, also named by Vancouver – were all "popularly known as Puget Sound."[43]

Stevens fully understood Vancouver's original hydrological typology. We know this because he appended to his geographical memoir a report prepared by Johann Georg Kohl, the least appreciated figure in the history of Stevens's work as an explorer. (The standard treatment on Stevens's career, *Young Man in a Hurry* by Kent Richards, does not even mention him.) Kohl was a German travel writer, historian, and geographer. After immigrating to the United States in 1854, he prepared maps and reports for the United States Coast Survey, founded in 1836 (and after 1878 subsumed into the Coast and Geodetic Survey, now the National Geodetic Survey). Stevens met Kohl when he became a member of Congress in 1857. Titled "Hydrography of the Coasts and Navigable Waters of Washington Territory," Kohl's report was excerpted from a larger work that he compiled for the Coast Survey office, *Notes of Discovery on the Western Coast of the United States,* complete with maps, a taxonomy of place names, and bibliography. His analysis is still in some respects a serviceable history of the early exploration of the inland waters that now collectively go by the name of the Salish Sea. Not familiar with the evolving usage that Stevens's reports routinely reflected, Kohl drew heavily on Vancouver's typology, in which Admiralty Inlet, not Puget Sound, was to be understood as the principal body of water southeast of the Strait of Juan de Fuca. In his (and Vancouver's) construction, Admiralty Inlet was not merely the strait between Whidbey Island and the

Olympic Peninsula, the common understanding in modern times, but the entire basin as far south as Tumwater, Washington.

The collateral question, which, given this book's theme, supersedes what might be considered toponymical antiquarianism, is why Stevens went to such lengths as including a forty-six-page hydrographical report in his railroad report. The answer was most forthrightly stated by Kohl. He provided a strategic thought fully in the mode of Alexander Mackenzie but converted into the more modern vision of Asa Whitney: an "immense system of commercial connexions ... foreshadowed on our railroad maps, where one of the principal projected lines which commenced at Chicago and Lake Superior has its western terminus at Puget Sound and De Fuca strait."[44]

To Stevens the matter of where the northern railroad crossed the Rockies was much less important than where it ended. Having the railroad transect the Cascades via Snoqualmie Pass meant that it would terminate in his new jurisdiction, not the lower Columbia Valley, where Portland, Oregon, dominated the political economy. In this respect, the Kohl report was intended to empirically demonstrate the superiority of Puget Sound over the Columbia River and, for that matter, every other port on the West Coast of the United States. This was the same logic behind the original conception of the several railroad surveys; that is, having "science" determine policy. There was no analysis comparable to Kohl's in the surveys of the other corridors and their western termini.

Kohl pointed out that the classic version of Admiralty Inlet ran "more than a whole degree of latitude" on a north-south line with "branches run out from it in all points of the compass," filling a basin 70 nautical miles in length and "30 miles in breadth from east to west." He likened the inlet to a tree "of which the body is recognizable, which is called Admiralty Inlet proper, and the side branches have their particular names." To his eye, there were five major branches: (1) the entrance, as far as the tip of Kitsap Peninsula; (2) Hood Canal, the great western branch; (3) the main stem as far as the narrows at Point Defiance (feeding the present-day ports of Seattle and Tacoma); (4) Puget Sound, the "southern cluster" of inlets connected to the larger entity at the narrows (echoing George Vancouver); and (5) Possession Sound (also named by Vancouver), which held the San Juans and the other island groups. In the report's key line, Kohl observed that in its entirety there was "nothing like this" north of San Diego, although he conceded that north of the de Fuca strait there were "many other complicated channel systems very similar in their principal features to this." All of the harbours in Stevens's Washington Territory were "comparatively narrow and long," and deep. Quoting Wilkes, Kohl averred that "in many places a ship's side will strike the shore before the keel will touch the ground."[45]

Summing up the whole, Kohl asserted: "Nothing can exceed the beauty and safety of these waters for navigation. Not a shoal exists within them; not a hidden rock ... no such strong flaws of the wind as in other narrow waters; for instance, as in those of Magellan's strait. And there are in this region so many excellent and most secure ports that the commercial marine of the Pacific ocean may be here easily accommodated." The adjoining terrain was described as a "broad valley" that was "upon the whole, of moderate elevation, and presents a pretty level depression." The shorelines were generally not like the fjords found further to the north in Canada. The mountains here did not "come down to the water's edge." Instead, Puget Sound was bounded, for the most part, by wooded hills, and in some places prairie.[46]

Kohl recorded the particular attributes of the several waterways. He quickly dismissed Hood Canal as "less attractive and diversified" than neighbouring Admiralty Inlet, principally because of its resemblance to the fjord-like inlets further up the British Columbia coast. He noted the "extraordinary depth" of the water at its northern end, near present-day Bangor, Washington, home to the US Navy's submarine base. On the other hand, several paragraphs were devoted to the qualities of "Duwamish Bay." Vancouver had not named this roadstead, and Wilkes had called it Elliott Bay, after one of his officers. Kohl noted: "On the northwestern shore of the bay a little town or settlement has sprung up called Seattle." It had the other advantage of being "one of the principal and most useful harbors of the whole inlet," dozens of fathoms deep. Further south was another "deep and spacious inlet," Commencement Bay, another Wilkes place name, derived from the starting point for his survey, and the heart of present-day Tacoma. Kohl noted its "square form" fronting "a valley which seems to come down and to open here from the Cascade mountain range."[47] Indeed, there is no more spectacular eastward view toward Mount Rainier than what can be secured from these waters.

Across from Duwamish and Commencement Bays to the west, Kohl referenced Port Orchard and Dyes Inlet indenting the Kitsap Peninsula, now home to American naval base at Bremerton. To the south, through the narrows where "the tide runs with great velocity, causing many eddies and whirlpools, through which the vessels are carried with extraordinary rapidity," Kohl came to "Puget Sound proper." This was constituted by "a most variegated compound" of inlets, "interlinked among each other by passages and channels, and connected with Admiralty Inlet only by the above named Narrows." From a commercial point of view, this body of water contained only one noteworthy "bay or harbor," that where "the flourishing town Steilacoom has sprung up." Kohl clearly misjudged Steilacoom's potential, probably a function of writing from afar and trusting too much what Stevens had said about it after his initial survey of the sound in December 1853. The only other "Puget Sound" destination mentioned was Olympia, the territorial capital. Located on Budd Inlet, which constituted "the most southern waters of Admiralty Inlet," this too was a "promising town" that had "sprung up."[48]

Vancouver had once mocked the idea that the waters accessed by the de Fuca strait off the Pacific formed an inland "Mediterranean." This idea had gained currency after Cook had merely skirted the Northwest Coast, and speculative geographers, always eager to fill a vacuum in geographical knowledge, imagined the existence of a western analogue to Hudson Bay. In the age of sail, Admiralty Inlet, Hood Canal, and Puget Sound were all disappointments, at least in terms of serving as a shortcut to the Atlantic. But the emergence of railroad technology by the middle of the nineteenth century changed that equation entirely. As Kohl phrased it for Stevens's memoir:

> It is very probable that De Fuca strait, from its remarkable geographical position, will, in the course of time, prove to be the most important entrance from the Pacific ocean on the whole northwest coast of America. It will surpass, with respect to commercial and naval advantages, all its rivals to the north and south. It is the outlet of that great and fertile valley into which Admiralty Inlet branches, filling it with innumerable excellent harbors, and the greatest facilities for navigation. And also of that still greater northern valley which is partly occupied by the Gulf of Georgia.

With this, Kohl had rather diminished San Francisco Bay and San Diego, but his real target was "the entrance to the Columbia river, with its sand banks and breakers," which "will at all times be inconvenient, compared with the broad, deep, safe navigation of De Fuca strait; and when these countries" (the low-lying terrain of western Washington and the Fraser River Delta) "become densely settled, and railroads come their aid, a great part of the trade of the Columbia river valley may possibly be drawn towards Puget Sound."[49] This was prophetic. Portland, Oregon, was the dominant commercial centre of the region until the transcontinental railroads transferred that honour to the ports of Puget Sound, namely, Seattle and Tacoma.

Stevens made separate note of an additional problem attendant upon a route following the course of the Columbia, namely, the "high floods to which the river is subject," which were "serious obstacles" to construction. So, too, were the dense forests of southwestern Washington. To Stevens's thinking, all this led to the advisability of establishing the terminus "higher up the sound" where the prospective line came down from Snoqualmie Pass. Seattle drew particular mention due to its "admirable harbor for a great railroad depot."[50]

Stevens's narrative goal with the Kohl report's extensive admiration of Admiralty Inlet/Puget Sound, conjoined with the corresponding derogation of other waterways, was to cement Whitney's vision of a transcontinental railroad's "immense system of commercial connexion" that would not merely terminate in the Washington Territory but reach across the Pacific Ocean. This presciently posited Puget Sound as the gateway to Asia – long the dream of America's commercial expansionists. The Great Circle Route puts ports in China only 5,000 miles away, those in Japan or Korea even less. Kent Richards asserts: "To Stevens the only logical connection from Europe and North America to East Asia lay on the northern route across the Atlantic, the Great Lakes, and Puget Sound."[51]

Stevens went to these lengths because from the time of his arrival in Olympia in November 1853, he was forced into a rearguard defence of what might be called his "mountains-to-sound" corridor. William Goetzmann, assessing Stevens's work in the field in 1853 and 1855, plus the exemplary work of his subordinates at all times of year but especially during the Rocky Mountain winter of 1853–54, asserts that "the Stevens expedition was a remarkable accomplishment because it covered so much unknown territory in such a short time with reasonable efficiency." But his effort was not without discord. Stevens had resigned his army commission to become the civilian head of Washington's territorial government, yet he was simultaneously the commander of a military unit conducting the survey. This only complicated matters more because, in line with the long tradition of expeditionary leaders tangling with their scientific complement (Cook and Vancouver come to mind), both the army men *and* the naturalists chafed under Stevens's command. One of the latter, George Suckley, wrote his family after the expedition reached the Pacific Slope: "There was the greatest amount of disaffection in the command throughout amounting almost to open mutiny – arrests, shooting, kidnapping, and violence were openly breached and everything in the way of short rations, lawlessness, and sour looks were to be seen."[52]

Writing at the conclusion of the expedition from Fort Steilacoom, near present-day Tacoma, Suckley asserted less melodramatically: "The general feeling here is in favor of a railroad route through the South Pass having a Y-shaped bifurcation, the one branch to San Francisco, the other to Puget Sound." This is precisely how the rail connections to California (via the Central Pacific and Union Pacific Railroads) and the Northwest (via the Oregon Short Line Railroad) were eventually effectuated, as in earlier times during the heyday of the emigrant trails, and afterwards with the construction of the Interstate Highway System (in the form of I-80S and I-80N). For Suckley, Stevens's northern corridor, "to the mind of all who went over it ... seems *impractically expensive.*" He likened its prospective miles-long tunnels, steep grades, high bridges, and complicated radii to a road over the Himalaya Mountains. Suckley reflected contemporary opinion and helped seed an axiomatic understanding about Stevens in the region's historiography when he opined: "The Governor is a very ambitious man and knows very well that his political fortunes are wrapped up in the success of the railroad making its Pacific terminus in his own territory."[53]

Suckley's critique took an operational form. Washington's inaugural territorial assembly, convening on February 27, 1854, turned to this issue in one of its first steps. On March 4, it authorized the exploration of a substitute corridor that would link Puget Sound, via the Columbia River, "with a railroad to California."[54] The legislators authorized Frederick West Lander, one of Stevens's civilian engineers, to survey, at his own expense, the northern leg of this Y, and to make recommendations to the War Department and Congress. Lander tendered his preliminary findings to Jefferson Davis on November 23. The secretary, who needed little encouragement to disfavour anything associated with Stevens's northern corridor, printed a revised version of Lander's lengthy memorandum in the second volume of the *Pacific Railroad Reports.* This, of course, followed the initial volume that was completely dominated by Stevens's largely enthusiastic narrative on his corridor's prospects. Not coincidentally, Volume 2, printed in the summer of 1856, was more largely dedicated to conformance of the documentation associated with Captain John Pope's 32nd parallel corridor, running west up the Red River Valley through El Paso, the southernmost route favoured by Davis and other sympathizers.

Lander defined his project as a reconnaissance, not a survey like Stevens's, arguing that "several lines might occupy relative positions on a single route." In his introductory paragraphs, he cited "the patriotic Whitney," but quickly made it clear that he was not the enthusiast of trade with Asia as a rationale for a Pacific railroad that Whitney was. More specifically, Lander took exception to the way Stevens extolled those prospects in Volume 1 of the *Pacific Railroad Reports.* Lander asserted that the federal government was "not particularly interested in the question as to whether" a transcontinental line would carry "the commerce of the Pacific seas." No doubt reflecting the attitude of the one man who was his principal audience, Secretary Davis, he argued: "Government is interested in the solution of the problem only so far as the results of the experiment tend towards the extension of a speedily consummated effectual means of overland mail and military transportation." Without citing Stevens by name, Lander suggested that translating the governor's logic into action was an exercise of "doubtful constitutional powers." The only legitimate rationale for federal railway support was "the cheapest and earliest consummation of the military defence of the Pacific possessions."[55] This was an echo of the arguments surrounding the appropriation for what became the Lewis and Clark Expedition.

Moving from the philosophical to the practical, Lander pointed to Frémont's explorations in the mid-continent emigrant trail corridor as the compromise solution to the national problem, thereby underlining Congress's mistake in not commissioning a formal survey along that route. Lander no doubt saw himself as the corrective element in this regard, though he deviated from Frémont's corridor by suggesting that the Platte River route (westward from present-day Omaha) was the best point of access to the interior of the continent, not the towns of western Missouri (near Kansas City) that had primarily served emigrant traffic. A Platte River road would be "a surface graded and ballasted by the act of nature." The river emptied a basin that was extraordinarily level for 500 miles "on the grand approach to ... South Pass of the Rocky mountains," thereby avoiding the undulations Stevens encountered west of St. Paul and the Ozark Plateau obstacle immediately west of St. Louis. The mouth of the Platte, where it emptied into the wide Missouri, not only facilitated logistical support for the shipment of rails and construction equipment but was "advantageously situated" for connection to eastern railways extending to Lake Michigan (by which Lander meant Chicago) or southward to St. Louis and beyond. This central line would become "to Kansas, Nebraska, and Utah, what the great rivers of our country have been in their natural advantages to the country east of the Mississippi."[56]

After a long digression into the relative merits of private versus federal capital construction, Lander turned to an extended analysis of the St. Paul–to–Puget Sound corridor. Describing the nature of conversation within the Washington Territorial Assembly, he stated that "the extreme difficulties" of Stevens's route "gradually led to the necessity of seeking" an alternative, "eventually to the exploration of the present line." Reading between the lines of Lander's text, it seems clear that territorial legislative leadership, reflecting the doubts of some of the governor's major associates, such as Suckley and McClellan, came to fear that, notwithstanding Stevens's enthusiasm for his northern route, the nation's politics and topography suggested that he was leading them into a dead end. Scary talk that the prospects for a northern corridor were, in Lander's phrasing, "destroyed" even before the expedition analyzed the Cascade Mountains caused them to hedge their bets, and Lander was their man. As he put it (in part quoting the actual territorial resolution), "information was needed ... prior to any final action of Congress on the Pacific railroad question." Otherwise, "*should* a decision occur ... previous to such information being offered, the difficulties existing upon the extreme northern route would evidently defeat that project, [and were] a railroad to California be placed too far south to favor a branch connexion," it would have "a fatal effect upon the interests of the important northwestern Territories," meaning Oregon and Washington.[57]

In Lander's view, the citizens of the Washington Territory had to "surrender the line of direction" (the Stevens corridor) and "deflect" to the Columbia River route if they wanted a rail connection to Puget Sound. Calling upon a geographical trope first trotted out by Alexander Mackenzie, Lander said that the Columbia was "the only natural pass to the ocean from the great interior of the American continent."[58] A level-water railroad would avoid the snow and tunnelling of the Cascades, and the stream itself mirrored the capability of the Missouri to the mouth of the Platte – a means of conveying supplies and equipment to build the western segment. By using these rivers in that fashion, one could start construction at either end of the line and meet in the middle of the country. This scheme would have anticipated the post–Civil War scenario by which by which the Central Pacific and Union Pacific Railroads were eventually linked.

This efficiency, Lander argued, more than countervailed the supposedly "superior and distinctive feature of an extreme northern route to the Pacific." He was referring to "the apparent short distance between the navigable waters of Lake Superior and Puget Sound" created by the curvature of the earth. Besides, the Great Lakes froze in winter, negating any natural advantage that might otherwise exist. A northern line's traffic was also potentially susceptible to diversion to Canadian railroads once they too were built, or worse, since it would run "so near an exposed frontier," to military entanglements.[59]

This line of thinking, which Suckley had described, spurred the Washington Territorial Assembly to action. As Lander put it (without a nod to modesty): "It therefore became necessary for some professional party to attempt" an examination of the southern alternative to what he routinely referred to as the *extreme* northern route. His effort did not require a "ponderous" expedition, complete with odometer wagon or a litany of astronomical calculations. None of that was needed "to obtain the features of the country regarding facilities for construction."[60]

Lander left Olympia on March 18, 1854, two weeks after the territorial resolution and one week before Stevens did the same. Though he was assisted by military detachments and fur trade posts along his route, Lander was accompanied by just three men. Only one of them, J.F. Moffet, a meteorologist in Stevens's eastern division, made it all the way to St. Louis with Lander.

The heart of Lander's report was his outline of how the "northern fork or branch of any central railroad to California" would connect to Puget Sound. It was not until page 31 of a 45-page printed document that Lander mentioned Stevens by name. Even in this first of two instances, it was to verify the governor's confirmation of McClellan's determination that Seattle was the best choice for a terminus because it had the best harbour, was close to the de Fuca strait, and had coal deposits only fourteen miles away. Lander did not minimize the challenges presented by railroad construction in the Columbia River gorge, but he reminded Davis and other readers that its associated difficulties could only be compared to building a railway through the Sierra Nevada or Cascades. The gorge, Lander professed, would be "the most severe section" of a line connecting to a Platte River railroad, "but by no means the worst section upon the extreme northern route." Lander proposed transiting the Blue Mountains via "side-hill location," thereby avoiding cuts, switchbacks, tunnels, and trestles. Once onto the Snake River plain, only a comparatively minor height of land separated his route from the Great Salt Lake basin, where it could meet the main trunk heading to California.[61] Imagining a flow of prospective traffic across his line back toward Puget Sound, Lander added gratuitously that a railroad emerging from the Blue Mountains onto the Columbia Plain and up the Yakima River to Snoqualmie Pass at its head was actually more linear in nature "than a continuation of the northern route" to the same terminus.

In Lander's scheme, the gateway to the twin termini in northern California and Puget Sound, as it had been to that date by emigrant wagons, was South Pass in southwestern Wyoming. He accurately characterized this famed locale as "but an extended plain, slightly broken towards the south into an undulating country." This terrain was a far cry from the passes Lewis and Clark and Stevens had identified further north. To the east of South Pass was the Sweetwater River valley, home to many of the famous landmarks on the Oregon/California/Mormon trail. The Sweetwater was tributary to the Platte, whereupon, Lander maintained, "all difficulties of location cease, and a broad bottom land, falling at scarcely perceptible inclination to the very banks of the Missouri, and overlapping a substratum of clear gravel or sand." This offered "every facility for cheap construction." The only conceivable objection to this route was that for 200 miles west of the Missouri, the land was destitute of usable timber; only cottonwood was available near the river bottom. And unlike the Sweetwater, which was rimmed by timbered mountains, the Platte Valley's comparative scarcity could not be remediated by recourse to adjoining terrain. Iowa was also "scantily timbered."[62] Lander had put his finger on a reason why this corridor had been neglected by railroad planners in the Corps of Topographical Engineers (CTE).

Summing up, Lander argued that his proposal was in effect a two-railroads-for-the-price-of-one bargain: "By adopting a line to California located so far north as to admit of the connexion of the branch-road to Puget Sound, the amount required to construct the spur to Puget Sound becomes a clear profit." This reasoning, Lander continued, was "obvious if the northern, the line of extreme cost, were to have been adopted to Puget Sound, and a distinct route built to California." Making his best case for the "northern branch reaching Puget Sound, through the Grand Level Pass of Columbia valley," Lander asserted that "the impracticable nature of the western mountains shuts the northern route from the Pacific terminus." Besides which, the governor's proposal "passe[d] along exposed frontier for a distance of fifteen hundred miles, and in direct vicinity of a great navigable river," the Missouri. The southern line, on the other hand, would help form "the main trunk of a more important road through central American territory which is entirely undeveloped." Presented with a choice of his own making, Lander wrote: "I claim, then ... for the southern."[63]

Lander's report appeared in print nearly half a decade ahead of Stevens's full explication of the practicability of the northern corridor, with the inevitable effect of making the latter anticlimactic. Stevens, sensing that, inadvertently or not, Lander was only serving the southern interests, threw all his explanatory power into his conclusive narrative that completed the *Pacific Railroad Reports* series. He realized that the whole northern corridor had to work across a spectrum of evaluative frameworks, ranging from the agricultural to the meteorological.

Stevens expended the most effort defending his route against the expected charges that the corridor was both cold and snowy. He did this by going on the offensive, emphasizing its other advantages. Though often criticized, and deservedly so, for his governance of the Washington Territory, the 1854–55 treaty tour, and the Indian wars that broke out on his watch, Stevens has never gotten his due as an observant, practical geographer. As a function of his treaty tour exposure to the Palouse country east of the Columbia's Wallula Gap toward the Nez Perce villages, he determined that the terrain there was "not the barren desert it has been represented to be." He asserted that "grapes would be a very profitable crop." Wine is now an ever more prominent agricultural crop in that district, but Stevens discerned other opportunities that also proved out. He wrote: "The whole country is exceedingly well adapted to grazing, and especially to wool growing," an enterprise that would come to dominate that region. Linking to his predecessors, Stevens referred readers "anxious to understand this country" to the journals of Lewis and Clark. He considered it "a work of supererogation to do anything more than simply quote their narrative. They speak of the soil of the country on this day's route as fertile."[64]

Broadening his perspective in a section titled "General Conclusions," Stevens admitted that "the agricultural capacities of the country" along the entire route was "inferior to many parts of the United States in its adaptation for great variety of crops." On the other hand, it compared "favorably with the best portions of the empire of Russia for the cultivation of the great staples of agriculture, and west of the Rocky mountains far surpasses them, approaching the most productive countries of Europe in character." Stevens maintained, correctly, that contrasted with the other routes under consideration, his was the one with a climate "superior for agriculture." Even the semi-arid Columbia Plain, ranging between the Bitterroot Mountains and the Cascades, he again correctly asserted, had a soil that was conducive to agrarian pursuits. These were not merely climatological observations but prospectively demographic ones as well, envisioning as they did an influx of agrarian settlers who would both travel upon and ship produce along the northern line. Even the mountainous country, which was not cultivatable but contained "patches of arable land," was good for stock raising and "the nuclei for mail and railroad stations."[65]

Stevens was anxious about the allegation "that the weather is so cold on this route that it will be impracticable to work men in the construction of the road for a large portion of the year, and that it would be impracticable to run [rail]cars for many days in the winter." To assuage these concerns, he presented climate data gathered over the course of 1853–54 and compared them to existing railroads in Canada, the eastern United States, and even "the great Russian line between St. Petersburg and Moscow." His data showed that the coldest part of the northern route was between the Great Bend of the Missouri and the Mississippi (Minnesota and eastern North Dakota), where it was indeed cold but "not more severe than at Quebec or Moscow." Fort Benton, at the foot of the Rockies, was warmer than Montreal and St. Paul. There were more cold days (defined as below 0°F) on the "Grand Trunk line and the railroad lines in Minnesota than in the Rocky mountains." John Mullan provided data indicating that Cantonment Stevens, in the Bitterroot Valley, had more days above 32°F in winter than Montreal, Albany, or Boston.[66]

Stevens had been accused by associates of underestimating the snowfall on his route in his preliminary report in 1854, but in his geographical memoir published six years later, he denied that charge. The mild winters in the Pacific Northwest, he asserted, prevented a great accumulation of snow, principally as a function of the constant rains that fall instead. This was particularly true for the Cascades, but even for the Rockies, Stevens concluded, "evidence favors the belief that there is not a greater depth of snow in the highest six miles of the pass than on the line of the Portland and Montreal railroad, and that it passes off about as early, while the rest of the route is never liable to be impeded by snow." The greatest depth of snow one of his lieutenants found near the Bitterroots was two and a half feet deep. In conclusion, Stevens pointed to "the accounts of all who have traversed the Rocky mountains, during almost every winter month the snow there met with would not present the slightest impediment to the constant passage of railroad trains." As for the cold, "the whole of these mountains between Fort Benton and the Bitter Root valley, a distance in a straight line of 190 miles, have a milder winter climate than Wisconsin or Iowa, or any part of Nebraska east of the 100th meridian."[67]

Ultimately, Stevens's narrative effort came to naught, though not his fieldwork and practical insights. In the end (early on, actually), Davis recommended the southern line, stating that "the most practicable and economical" route "was that of the thirty-second parallel."[68] For him an ice-free southern line was the only sensible solution and he was zealous in his demonstration of the northern route's limiting features. Of great strategic value to Davis was McClellan's estimate of a twenty-foot snowfall in the Cascades. Stevens's overly optimistic cost estimate and the proximity of his corridor to the always potentially hostile British Empire just above the 49th parallel, a particular point of emphasis by Lander, were other deterrents. Stevens's route was the only corridor for which the War Department adjusted the field estimate of costs upward, in this case from $117 million to $140 million, a price that conveniently doubled the cost of the least expensive corridor. The 32nd parallel route came in at $68 million.

But if the secretary of war gladly seized upon the negative portions of Stevens's survey (and especially the statements of the governor's former associates), he casually overlooked those of his favourite corridor. Snow and ice disqualified the northern route, but the lack of timber and water along the southern line were viewed as trifling matters. CTE captain A.A. Humphries, in his composite "Report on the Pacific Railroad Explorations and Surveys," issued in November 1855, following his preliminary assessment in February of that year, did his best to sell the southern corridor. He stated: "On the routes of the 35th and the 32nd parallels, the structures of desert areas has been minutely studied with special reference to the practicability of obtaining water by wells or by boring, and generally with the most satisfactory results."[69]

Davis congenially copied this finding in his report to Congress, noting "an examination into the practicability of procuring water along certain parts" of the 32nd parallel where it had been perceived to be deficient.[70] But the perception of Davis as a sectional partisan was as obvious, if not more so, than the one affecting Stevens. The political circumstances made it look as if the other assessments were a ritual performance prerequisite to proceeding with what had been his preferred route all along.

In the end, the Pacific Railroad Surveys, intended as a conclusive answer to a transportation problem, only deepened the confusion. The lack of resolution doomed prospects for a federally sponsored railroad before the Civil War. Born of sectional differences building up to 1853, the Kansas-Nebraska Act of 1854 only widened the national split during the post-survey evaluative phase. Arguably, Davis's biggest mistake was not surveying the 41st parallel that Lander investigated for the Washington Territory after the fact. This route, overlying the Emigrant Trail preliminarily surveyed the previous decade by Frémont and scoured into the earth by Mormons, gold rushers, and agrarians headed for Oregon, was, in retrospect, an obvious compromise. Originating in the border state of Missouri, and terminating in the bifurcated Y reaching both California and the Northwest as recommended by Suckley and Lander, it ended up being close to the corridor chosen after the Civil War for the first transcontinental railroad as built by the Central Pacific/Union Pacific combine.

Exploration as Science

The *Pacific Railroad Reports* containing the many field reports created by the survey teams began appearing in print in 1855, one year late, in three octavo volumes. They included summaries of all the surveys except the north-south corridor investigated from northern California through central Oregon. These volumes contained no scientific data, no illustrations, and only a few hastily drawn maps. Later that year, the CTE transitioned to a quarto format, the one most students of western history are familiar with, starting with Stevens's initial assessment of the northern corridor and followed by the southernmost. In its entirety, the quarto edition comprised thirteen massive tomes, of the sort the British Admiralty would have been proud to issue in an earlier era. The last (Volume 12, by Stevens, in two parts) was published in 1860, virtually on the eve of the Civil War. The several topographic assessments of the other corridors, including those running north and south through California and up to the Columbia River, filled out most of Volumes 3 to 7, issued in 1856 and 1857. None of them addressed the hydrographic qualities of the ports in San Diego, San Pedro (Los Angeles), or San Francisco Bay. Unlike Stevens, those surveyors had no interest in waters reaching the Far East. Thoughts of how to find water in the Mojave, Sonoran, and Llano Estacado deserts south of the 40th parallel that could be fired to power steam engines dominated their thinking instead. Every artesian spring was mapped assiduously.

Volumes 8 to 10, devoted almost entirely to the natural history of the various corridors, were published in 1857–58. The cartography in Volume 11, which appeared in 1859, was the most revolutionary since William Clark's pioneering efforts. G.K. Warren's compilation of charts showing the evolution of geographical comprehension of the American West since 1803, complemented by a series of engravings showing the topography of the mountain west, is still stunning to the eye of the modern observer. His comprehensive map of the West, drawing on all the surveys, reduced every previous attempt at large-scale cartographic projection of the region to a mere sketch. In the same fashion that Henry Roberts's master chart tracking Cook's three expeditions became the first modern map of the world, Warren's map, to which Stevens and Mullan contributed mightily, is the first modern map of the western half of the country.

The *Pacific Railroad Reports* were the last distant echo of the golden era of scientific travel literature first popularized by Cook in the 1770s, even if the empirical logic of that earlier time was now overlain by a cultural commitment to the picturesque in artistic representation of the land and the romantic in literature. This was typified by the reports' colour lithography, especially in Stevens's report, the best illustrated of any. These images drew on the tradition of the crisp engravings etched onto copper during Cook's time, but reflected the gauzy delineation of reality that the grease pen upon limestone surfaces yielded in the supposedly more modern lithographic style. In that sense, these views were in equal measure half scientific but also half dreamscapes. They prefigured the voluminous promotional literature and imagery touting settlement and tourism in the American West that was issued later in the nineteenth century and into the twentieth by the same railroad whose general route Stevens first mapped out.

Davis may have rejected Stevens's blandishments but the governor's efforts bore fruit shortly thereafter, if posthumously. Stevens left the world of politics to join the Union Army once the war broke out. He was killed at the second battle of Bull Run near Chantilly, Virginia, in September 1862. Coincident to his death, Stevens's survey, touting the advantages of a northern railroad, helped convince Congress (now unburdened of Southern influences) to charter the Northern Pacific Railroad in 1864. This occurred a mere two years after the central route to California had been authorized.

Over the course of the three decades following the Civil War, two other railroad lines were built in the corridor Stevens studied: the Great Northern Railway and the Milwaukee Road. At times, one or the other followed Stevens's projected route, though each also deviated from it. But perhaps his most important contribution was to identify the commercial possibilities that could be drawn from the region covered by his survey if it was served with a suitable transportation infrastructure. No previous report on the Northern West was as wide-ranging, meticulous, and exact. The region's timber resources were the most obvious opportunity, but even from an agrarian perspective, Stevens provided a corrective for Lewis and Clark, who had emphasized the land's aridity in their course west of the Missouri's big bend in present-day North Dakota. It is no exaggeration to state that Stevens rescued the reputation of the northern plains, the Washington Territory's intermountain valleys (including the Bitterroot at that time but also the eastern section of the Great Columbia Plain, especially the Palouse and Walla Walla districts), which have proved constantly productive for various crops and grazing.

10

The Canadian Dream

The evolution of the Northwest Passage via railroad technology was replicated north of the 49th parallel. Though there were similarities in these national stories, the Canadian version had a distinct feature: fear of the United States annexing some (or all) of its territory. The concern was justified. Benedict Arnold led Continental Army troops to the gates of Quebec in 1775, in the first months of the Revolutionary War. That campaign presaged even greater American ambitions manifested in the War of 1812. British-American diplomatic relations took a conciliatory turn with the settlement of the Maine and Oregon boundary disputes in the 1840s, which ironically exacerbated concerns among many Canadians that the mother country, in a pinch, would not come to the defence of its colonial holdings. These fears peaked in the 1860s. William Seward from upstate New York and the American secretary of state from 1861 to 1869, continued the long American tradition – from before, during, and after the era of Manifest Destiny – of casting covetous glances northward. The purchase of Alaska from Russia in 1867 reflected his ambitions. Canada was able to avert the comparable fate of being subsumed into the republic to its south through two interrelated strategies: forming its own country through a confederation of the various British colonies, which became provinces, and building a railroad that would tie them together.

The earliest proponent of a Canadian transcontinental railroad was Thomas Dalton, editor of the Toronto *Patriot.* Writing in 1834, and possessing the same limited vision that hindered Thomas Hart Benton south of the border, Dalton touted an all-steam river/canal/rail corridor from Toronto to the Pacific. With the settlement of the Oregon boundary dispute in 1846, British engineer Robert Carmichael Smyth could, in 1848, write more credibly about an "Atlantic and Pacific Railway." In 1851, another promoter, Toronto business magnate Allan Macdonell, actually applied for a charter to build such a line from the Legislative Assembly of the Province of Canada, then consisting of the former provinces of Lower Canada (present-day Quebec) and Upper Canada (present-day Ontario). Though the Crown's title to British Columbia was now secure, the land in between was still the property of the Hudson's Bay Company.

The Canadian equivalent of Asa Whitney was Joseph Howe, head of the Railway Ministry of Nova Scotia colony and later a parliamentary leader. His visionary 1851 address, when there were only 200 miles of rail north of the border (compared with 10,000 miles in the United States), is worth reproducing at length.

> Think you that we shall stop even at the Western bounds of Canada [at Lake Superior], or even at the shores of the Pacific? Vancouver Island with its vast coal treasures, lies beyond. The beautiful islands of the Pacific [Hawai'i], and the growing commerce of the ocean, are beyond. Populous China and the rich East are beyond, and the sails of our children's children will reflect as familiarly the sunbeams of the South [Sea] as they now brave the angry tempests of the North [Atlantic]. The Maritime Provinces which I now address are but the Atlantic frontage of this boundless and prolific region – the wharves upon which its business will be transacted ... I believe that many in this room will live to hear the whistle of the steam engine in the passes of the Rocky Mountains, and to take the journey from Halifax to the Pacific in five or six days.[1]

Alexander Dalrymple would have been pleased.

The individual who began the process of translating Howe's vision into action was Sandford Fleming, best known for his tireless advocacy of a universal prime meridian and the invention of standard time. Fleming was born in Scotland and after his schooling apprenticed himself to a land surveyor. He immigrated to Canada in 1845 at the age of eighteen. The transformative event in his life occurred in 1863, when he returned to North America on a steamship that took one-sixth the time of his first voyage by sail. He was inspired to write an essay that was published in the *New York*

Herald on July 5, 1863, just as the dead at Gettysburg were being buried. Under the headline "New Canadian Passage Proposed," Fleming expanded upon Howe's sentiments from twelve years earlier.

He began by commenting on the abbreviated nature of his recent Atlantic crossing, which formerly took five to seven weeks, depending on the weather. In the spirit of Robert Fulton and other champions of industrial technology, he declared that "iron and steam" were destroying "the terrors of an ocean voyage." Shorter times and greater comforts could be expected. Indeed, Fleming considered these improvements so much of a given that he moved to his next point, the need to link Atlantic Canada's maritime regions "with the railway system of the interior." He imagined travelling from London to Chicago in eight days![2]

Returning to Nova Scotia, Fleming became the head engineer and surveyor charged with establishing the feasibility of the long-delayed Intercolonial Railway from Halifax to Quebec. The prospective road's very name presaged the other event that looms large in the Canadian analogue to transcontinental railroad planning in the United States – the Charlottetown Conference, at Prince Edward Island's colonial capital, which convened on September 1, 1864. Post Gettysburg, the historic dominance of the behemoth to the south was now guaranteed. To fend off Seward and other advocates of American expansion in the face of an unlikely defence by Great Britain, Canadian colonial officials decided to seize control of their destiny and create their own country.

After negotiating terms with the mother country, the Dominion of Canada was instituted on July 1, 1867. It was formed by the previously separate colonies of "Canada" (defined above), New Brunswick, and Nova Scotia. Prince Edward Island, the ostensible birthplace of Confederation, did not join until 1873. In 1869, the Hudson's Bay Company relinquished its quasi-governmental control over the Northwest Territories – essentially all of Canada west of Lake Superior to the crest of the Rocky Mountains. British Columbia joined Confederation on July 20, 1871. This was the Canadian analogue of the American "Year of Decision" (1846), when Oregon and the Mexican Cession, including California, were appended to the United States. As the Reverend George Grant, a Presbyterian minister whose congregation in Halifax included Fleming, put it, "the whole mainland of British America became one political State under the aegis of the Empire."[3]

Like the US Constitutional Convention with its pivotal large state/small state compromise, the admission of British Columbia incorporated a grand bargain, and one central to our theme – John A. Macdonald's rash promise to complete a railroad to the Pacific Slope within ten years. Macdonald's bold move might be thought of as the early Canadian equivalent to John F. Kennedy's decision to launch the United States on a course to the moon within a decade. Macdonald, a Tory and Canada's first prime minister, tied his political fate to a project that has been referred to as the National Dream (a term popularized by historian Pierre Berton), plausibly premised on the belief that Confederation could endure only if the country ran coast to coast. Otherwise, the thinking went, Seward and his successors would chip away at the Dominion, in effect reversing Alexander Mackenzie's encirclement strategy that was so alarming to Thomas Jefferson. Without a railroad to the west, Canada itself was truly more a dream than a nation.

Only two years earlier, a US Senate committee report postulated that "a Northern Pacific railroad seals the destiny of the British possessions west of the ninety-first meridian." (Thunder Bay, Ontario, is approximately 2 degrees of longitude east of this line.) This same policy document asserted that the Northern Pacific would inevitably Americanize western Canada "in interests and feelings," such that it would be "severed from the new Dominion, and the question of their annexation will be but a question of time." British Columbia was always considered the most vulnerable candidate. As recently as the Manifest Destiny heyday of President James K. Polk, this province's far western exposure and its northern boundary at latitude 54°40′ drew the active attention of American policy-makers. Subsequently, thousands of former California gold miners migrated into British Columbia's interior in the late 1850s, making British and Canadian diplomats very nervous. With Confederation, as the Reverend Grant put it, "by uniting together, the British Provinces had declared that their destiny was – not to ripen and drop, one by one, into the arms of the Republic."[4]

Grant succeeded to the Whitney/Howe/Fleming mode of thought. He believed that the addition of British Columbia enabled Canada to start "dreaming magnificent dreams of a future when it shall be the Greater Britain, and the highway across which the fabrics and products of Asia shall be carried, to the Eastern as well as the Western sides of the Atlantic." A transcontinental railroad would do for the west what the Intercolonial was intended to accomplish for the Maritime provinces. Pierre Berton writes: "It would be a rare example of a nation created through the construction of a railway."[5]

To realize the National Dream, and in a seeming replay of Isaac Stevens's railroad planning strategy, on the same day British Columbia joined

Confederation, preliminary survey parties departed for the Rockies from the provincial capital in Victoria, and corresponding units headed westward from the upper Ottawa River. The reports of these expeditions were laid before Parliament in April 1872. Here Fleming rejoins our story. To help the Macdonald government make sense of the early reconnaissance, Fleming was appointed to the position of chief engineer of the Canadian Pacific Railway (CPR), at its formation a wholly owned enterprise of the federal government. Fleming may have worked in the visionary idiom of Whitney and Howe, but by 1872 he was also an experienced railroad man. He had a surveyor's sense of the toil and treasure it would take to build a railroad to the Pacific Slope. Accordingly, he wanted a clear and comprehensive view of the entire proposed route as discerned through his own eyes. That year, he conducted his first of many surveys into the vast expanses of the Canadian West. He left Toronto on July 16, having appointed Grant (also an educator associated with Dalhousie College in Halifax) as the recording secretary of his expedition. They reached Victoria the following October and published a report before year's end.

Fur traders dating to Peter Pond and Mackenzie had traversed the rivers and lakes trending in a northwesterly direction toward the Arctic from the western end of Lake Superior. Simon Fraser and David Thompson had established the north-south courses of the Fraser and Columbia Rivers in 1808 and 1812, respectively. Their findings were confirmed by the numerous fur trade brigades that followed in their wake. But comparatively little was known of the British Columbia interior other than where those channels ran. Even the prairie west of Winnipeg, over which any railroad linking to British Columbia would have to run, was not fully comprehended by CPR planners. Grant, summing up the geopolitical context of Fleming's expedition, stated that "our North-west, in some parts of it, was less an unknown land to the people of the States along the boundary line than to the people of the Dominion; and, in other parts, it was unknown to the whole world."[6]

In a similar vein, historian Clark Blaise argues, "Canada's knowledge of its own far west in 1872 ... was hardly greater than President Thomas Jefferson's had been in 1803 when he appointed Lewis and Clark to survey the vast new lands of the Louisiana Purchase."[7] This is a catchy analogy but a misleading one. As noted, Fraser and Thompson had effectively paralleled the American captains' corresponding understanding of the Missouri-Columbia nexus less than a decade later. Furthermore, there was a partial analogue to Stevens in the form of Robert Palliser's late-1850s survey of the

Saskatchewan River Basin that emptied into Lake Winnipeg. A mid-Victorian adventurer, Palliser, like many of his generation, was drawn to the prospect of a trendy excursion into continental interiors that replaced the more genteel Grand Tour of Europe of earlier times.

Young British noblemen like Palliser were elsewhere venturing into the heart of Africa looking for the headwaters of the Nile, to the Arctic searching for Sir John Franklin (discussed below), and to Australia. Palliser's self-conceived expedition was avowedly inspired by the American railroad surveys, though his leadership of it more closely resembled the model exemplified by Prince Maximilian's tour of the upper Missouri with Karl Bodmer in 1833. Palliser secured backing from the Royal Geographical Society, which, following a model dating to the Royal Society's sponsorship of Captain Cook's voyages, was translated into government support. He put together a team of scientists, including specialists in astronomy, meteorology, geology, and botany. His charge from the Colonial Office was to determine whether "the country presents such facilities for the construction of a railway as would at some period, though possibly a remote one, encourage ... the belief that such an undertaking between the Atlantic and Pacific Oceans will ever be accomplished."[8] Palliser's British North American Exploring Expedition (1857–60) published several reports between 1859 and 1863, and issued a comprehensive map in 1865.

Echoing Stevens's treatment of the Columbia Plain, Palliser wrote convincingly of the agricultural promise of the Saskatchewan River Basin. He especially liked what he called the "fertile belt" of the north fork. For him, its fecundity, and corresponding prospect for agrarian settlement, dictated that a western railroad must take that route. Palliser and his naturalists misjudged the corresponding capability of the South Saskatchewan (its propinquity to the US border may have flavoured their judgment), but his report imprinted the northern route upon the national imagination.

Unlike Stevens, Palliser never made it all the way to Pacific tidewater. His parties investigated six passes in the Rocky Mountain chain, reaching as far as the upper extent of the Columbia River. (One of them, Kicking Horse Pass, west of Fort Calgary, through which the CPR would be built, was named after an incident in which one of the expedition's members was knocked senseless.) The limit of Palliser's scope might have mattered less prior to Confederation, but by 1871 the western gap was retarding the National Dream. Worse was what Pierre Berton describes as Palliser's "bluntly negative" opinion on the practicability of running a line across the Rockies to the Pacific exclusively through Canadian territory. Palliser, per Grant, blamed this on "the unfortunate choice of an astronomical boundary line" that not only isolated Britain's "American possessions" from eastern Canada but "also almost debarred them from any eligible access from the Pacific Coast on the west."[9]

Grant (as Fleming's amanuensis), much like Palliser, believed that the Stevens expedition was the one to model a Canadian effort after. He was envious of the *Pacific Railroad Reports,* "thirteen large quarto volumes, richly embellished, stored with valuable information concerning the country." Yes, the reports pointed out difficulties related to the practicability of each of the prospective railroad corridors, but, Grant noted, the United States "always knew what it wanted," whether it was conducive boundary lines that secured prizes like Puget Sound, or other national objectives. He had a point. The American railroad surveys capped off a half-century of official government expeditions, including those of Lewis and Clark, Frémont, and Wilkes. Combined with travel accounts produced by missionaries like Father De Smet and fur trade explorers like the Astorians in the north or Jedediah Smith further south, "the whole country had thus been gradually examined from every possible point of view." Summing up the importance of Fleming's prospective effort, Grant said that the challenge for Canada was doing "in one or two years, what had been done in the United States in fifty."[10]

This rhetorical flourish was also slightly misleading. The operative timeline did not run from Thomas Jefferson to Jefferson Davis; rather, it covered the twenty years from Asa Whitney to Promontory Point. Nevertheless, considering Grant's vantage point – he started his journal documenting Fleming's expedition in the trough of Palliser's despond – juxtaposed with the celebrated meet-up of the Central Pacific and Union Pacific in 1869, his sense of urgency was understandable. "Man was thus triumphing over all the obstacles of nature in the Territory of the United States," while Canada was just entering the field north of the border.[11]

Fleming was a naturalist with competence in geology and meteorology, and could hold his own as a scout with an Indigenous conductor at his side, but he had to be a surveyor and railroad right-of-way engineer as well. His Canadian line had its own idiosyncratic challenges, just like American planners such as Stevens had faced. Chief among them were the muskeg swamps north of Lake Superior. This 300-mile pockmarked stretch of the Canadian

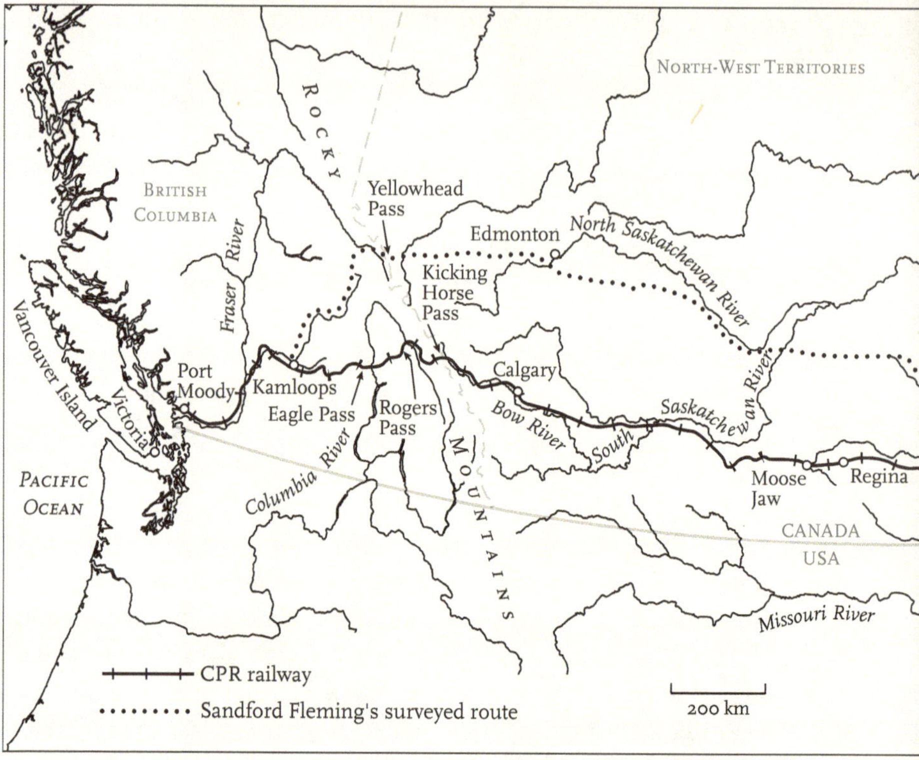

Figure 10.1 The evolution of the CPR's alignment. This map shows the divergence between the original corridor surveyed by Sandford Fleming and the one William Cornelius Van Horne completed. | Based on Pierre Berton, *The Last Spike*, 16–17; redrawn by Eric Leinberger

Shield would devour more financial resources per mile than the Central Pacific did in the Sierra Nevada. Of course, the CPR route would have to transect the Rockies at some point, and twelve different options were surveyed. As was the case south of the border, they were geographically disparate. The early favourites were the northern route through Yellow Head Pass, and Kicking Horse Pass to the south.

Fleming needed little convincing to recommend a North Saskatchewan corridor through Fort Edmonton toward Yellow Head Pass, at the head of the north fork, as the gateway to British Columbia. As someone who had actually run a railroad, he was predisposed to a national route that represented the least topographic resistance. Yellow Head Pass was far more amenable to construction than Kicking Horse Pass west of Fort Calgary, whose western descent into the upper Columbia River valley was truly

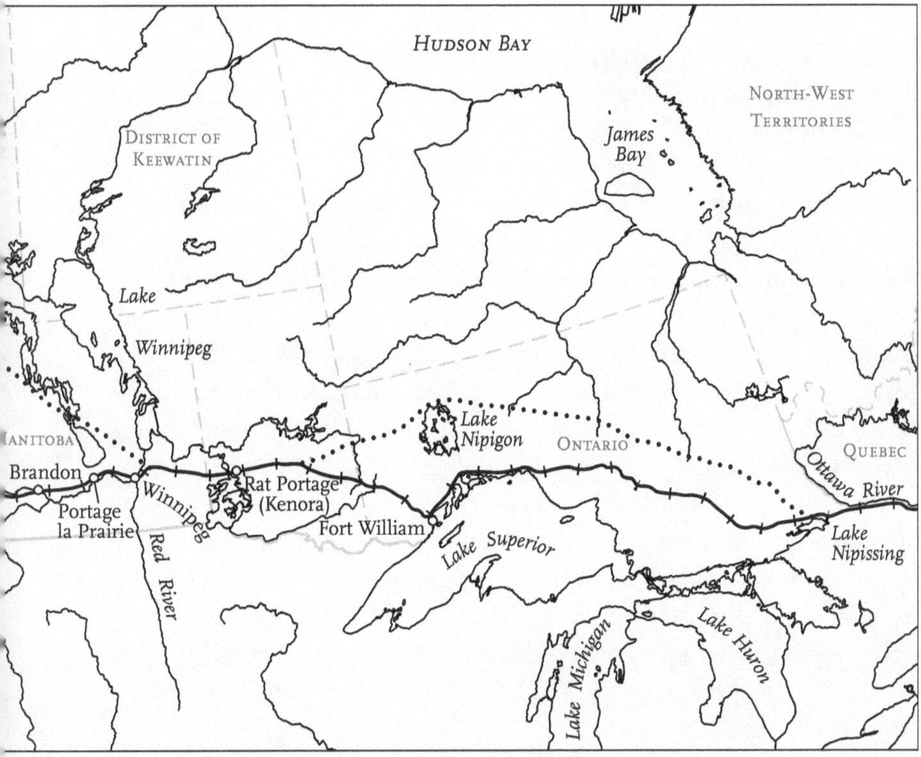

vertiginous. Similarly, his survey recommended a line much further north of Lake Superior than the shorter one eventually built along its difficult rocky shore. For much of the ensuing decade-plus of actual railroad planning and construction, there was almost no doubt that the CPR would take this northern corridor.

Finally, there was a counterpart problem north of the border as to where the railroad would terminate. The division of opinion here was, in a sense, more localized than the contest between Puget Sound and cities in California, but it was nonetheless contentious. Seven different harbours on the mainland and Vancouver Island were investigated. There were partisans for a lower mainland terminus at Burrard Inlet, north of the delta of the Fraser River, whose upland canyons provided access into the interior, where the towns of Yale and Kamloops plus other settlements would also benefit.

The Fraser/Burrard option was estimated to be the more expensive choice but would pay the biggest dividends in terms of boosting the peopling of the nascent upriver communities.

The leading contenders on Vancouver Island were Victoria, Esquimalt, and Port Alberni. These were all superb harbours, and extending the line to the island would also provide access to the Nanaimo coal fields. The question was how to get a railroad on a causeway across the Strait of Georgia. The best solution seemed to be a route down the Homathko River emerging at Bute Inlet. This was one of the fjords J.G. Kohl referenced in his hydrographic survey for Isaac Stevens. A bridge (utilizing Valdes Island near Seymour Narrows as a landing) could reach Vancouver Island halfway up the famed Inside Passage and then proceed south. Pierre Berton took Fleming to task for not doing what Stevens committed to early on, that is, waiting "five years before consulting the Admiralty about the usefulness of the various harbours at the head of the inlets his men were examining." Substantial time and money could have been saved "if these reports had been in his hands at the outset, for they made it clear that Burrard Inlet was the only really satisfactory terminus on the mainland."[12] The Liberal government of Prime Minister Alexander Mackenzie settled on the Fraser/Burrard Inlet route in July 1878.

But in the end, the dispositive factor relative to the completion of the CPR was finding a path through the mountains of British Columbia – not merely the main stem of the Rockies but the Selkirk annex plus the Coast Range. This paralleled the Rockies/Bitterroot/Cascades mountains nexus Stevens had faced, but Fleming was not as decisive as Stevens. He dilly-dallied over this question for most of a decade, ordering follow-up surveys and constantly asking for second opinions. In his account of Fleming's expedition, Grant noted dejectedly that "every attempt has been made to wrest from the mountains the secret of an easy passage to the ocean, but ... the attempts have failed. There is no direct, and no easy pass."[13] Fleming's lack of resolution in the mountains eventually cost him his job.

The first CPR segment was completed near Thunder Bay in June 1875. The citizens of Winnipeg heard their first train whistle in 1877. But as late as 1880, one year before the ostensible deadline for a connection to Pacific tidewater, CPR surveyors were still enmeshed in the mountainous labyrinth. There was talk of secession in British Columbia as early as 1876. With his crews abandoning the field to seek work with the Northern Pacific, inexorably extending westward just south of the border, Fleming was relieved of his CPR commission in 1880. Thereafter, in the midst of the recurring controversies that were common to most of the North American transcontinental railroads (scandalous contracts had previously cost Macdonald control of the government), the CPR was reconstituted as a private concern in order to insulate financing and routing decisions from political interference. In Berton's estimation, the new charter enabled Canada to break out from "the prison of the St. Lawrence lowlands."[14]

When the new company of private investors assumed their federal charter on February 15, 1881, they inherited the three relatively short segments the government had managed to construct: (1) the Pembina branch from Winnipeg south to the border, where it linked with an American concern; (2) the 433-mile-long segment from Fort William on Lake Superior to Winnipeg; and (3) the 215-mile terminal segment from Savona's Ferry on Kamloops Lake to Port Moody at the head of Burrard Inlet.

The most important decision the new directors made was to hire William Cornelius Van Horne as Fleming's successor on November 1, 1881. A native of Illinois, he completed the CPR in four years' time and guaranteed the survival of the nation as expansively conceived at Charlottetown. Thus, in historian Clark Blaise's estimation, "it was an American who saved Canada from American designs." Ironically, foremost among these threats was the Canadian-born James J. Hill, the so-called Empire Builder, at the head of the Great Northern Railway, the only American transcontinental line that never went bankrupt or missed a dividend. It was Hill who recommended Van Horne to his friends in Canada who were desperate to finish their national line. In his verdict on Fleming's tenure, Berton notes that when the CPR was reorganized as a private company, "most of his surveys were discarded and as an entirely new line was mapped out across the Shield, north of Lake Superior, through the southern prairies and across three mountain ranges."[15]

Van Horne settled the disruptive question of where to cross the Rockies when he chose Kicking Horse Pass in his first month on the job. (The Canadian National Railway, the second line to the Pacific, eventually used the route over Yellow Head Pass originally surveyed by Fleming.) Van Horne wanted a corridor closer to the Northern Pacific, fearing that a road that followed the North Saskatchewan line would eventually lose traffic across the border. (By the same token, the CPR's charter mandated that the line had to be built at least 100 miles north of the 49th parallel.) Van Horne later declared that if he had not inherited the Fraser River Canyon segment, he would never have built there because it was "one of the worst places in the world."[16] Alexander Mackenzie (the explorer who avoided the canyon, not the prime minister who built the line) would have agreed. In the end, Van Horne's CPR, measuring westward from Winnipeg, was shorter than any American line running to the Pacific from the Mississippi Valley.

Railroad Time

Fleming may have parted company with the CPR, but its operating requirements, and those of every railroad, were still on his mind. Before his separation, he had begun dabbling with the question for which he best known, one precipitated by the revolutionary capability of the railroad to compress time and space. From time immemorial, the twenty-four-hour day revolved around the local determination of noon, the time of day when the sun reached its apex in the sky. By definition, every town had its own local noon. The rare exceptions occurred only when distant towns shared the same meridian of longitude, and even then there were minor differences. More commonly, cities and towns, whether separated by a few dozen miles or a hundred miles or more, had different noontimes. As a practical matter, this meant that every minute, and upward of an hour or more of the day, differed from the time virtually everywhere else. In the United States, there were 100 different time standards. Michigan alone had 27, and Pittsburgh had 6 versions of correct time.

This proliferation created a problem for the railroads, whose operations required running on fixed schedules – not only in settling upon what times trains would leave and arrive at distant stations but also, given the fixed nature of their corridors, the avoidance of collisions. Without a standard, when it was astronomical noon in Toronto, it was 12:25 p.m. in Montreal. The solution was railroad standard time, adopted by the CPR and American railroads on November 18, 1883 (known popularly as the Sunday of Two Noons). Seventy percent of governmental bodies and commercial entities adopted the format within days. The new system quickly became known simply as "standard time." This concept entailed broad longitudinal swaths of countryside sharing the same general time zone. In Fleming's system, when it was noon at Toronto, it was noon in Montreal.

Fleming became one of the leading international advocates for this new way of organizing the twenty-four-hour day. Immediately after his dismissal from the CPR, he left for Venice to present a paper titled "The Adoption of a Prime Meridian." By dividing the world's geography into twenty-four different time zones, one for each hour of the solar day, the system could be applied globally. For this to work, each new "day" had to start somewhere, resulting in the International Date Line. Twelve hours (and time zones) in either direction, was the prime meridian, where it was noon when it was midnight at the date line. After a great deal of negotiation and angst (in Paris, principally), the meridian of Greenwich Observatory, the same one that was an essential aspect of Captain Cook's longitudinal calculations, was denominated as the prime meridian. In that sense, the acceptance of Fleming's standard time was the terrestrial analogue of John Harrison's invention of the handheld chronometer. With Harrison's timepiece or replicas thereof, sailors such as Cook could establish their longitude by glancing at a timepiece keeping Greenwich Time. They had only to calculate local noon and compare it with the time in Greenwich to establish how far west or east of the observatory their current station was. Standard time effectively achieved the same result. The concept was adopted on January 1, 1885.

Building the Northwest Passage

It is not within the scope of this book to relate the history of the first northern transcontinental railroads. A few construction highlights might be cited, starting with the CPR's initial rival south of the 49th parallel. The Northern Pacific was extended from St. Paul to Fargo, North Dakota, by June 1872, reaching Bismarck on the Missouri River a year later. At the west end of the right-of-way, construction started from Kalama, Washington Territory, on the Columbia River, toward Tacoma, which was selected as the tidewater terminus on July 14, 1873. Because of numerous depressions and bankruptcies, construction was intermittent until Henry Villard reorganized the company in 1882. The line reached Livingston, Montana, in mid-January 1883, and the company's golden spike ceremony was conducted on September 8 at Gold Creek, Montana. A bridge over the Missouri, replacing the train ferry, was completed near Bismarck on October 21, which created seamless through service from St. Paul to Tacoma. This alignment, however, followed Frederick Lander's Columbia River level water route from eastern Washington, not the mountain line that Stevens had originally idealized. A direct line from Pasco, Washington, to Tacoma through Stampede Pass (immediately south of Snoqualmie Pass) was completed on May 27, 1888.

With the decisive Van Horne in charge, the CPR charged across the prairie toward Fort Calgary in 1882, laying 417 miles of new roadbed. With Kicking Horse Pass now destined as the crossing of the Continental Divide, the decision on how to transverse British Columbia and link with the Fraser River segment could be postponed no further. The choice was between following the hairpin loop of the northern Columbia River valley or cutting across the Selkirk Mountains. Surveyor A.B. Rogers had hacked a barely workable alignment through the Selkirks and Van Horne adopted it, shaving two hours off passenger travel or four hours for freight. Rogers Pass proved an operating nightmare and was ultimately replaced in 1910 in favour of the famous spiral tunnel, the longest double-tracked tunnel in North America.

As the railroad approached completion in 1885, Van Horne made another decisive move, abandoning Port Moody in the shallow up-waters of Burrard Inlet in favour of a more expansive terminal city closer to the deeper water of English Bay. In the words of Pierre Berton, "Van Horne, the romantic, wanted to give his new metropolis a name he considered worthy of its future – that of a daring explorer who had sailed those shores long before any railway was contemplated." The fact that there was a Fort Vancouver in the neighbouring state of Washington was inconsequential. The age of river towns was fading anyway. He told an associate: "The name Vancouver strikes everybody in Ottawa and elsewhere most favourably."[17] George Vancouver had finally become associated with a workable communication to the Hudson Bay basin.

The final part of the CPR's Lake Superior section was completed on May 16, 1885. In the BC interior, the line had descended Rogers Pass to cross the Columbia River a second time at present-day Revelstoke. Via Eagle Pass, the CPR crossed over the Monashee Mountains to the South Thompson/Fraser River watershed. That fall, company executives, including Van Horne and other notables like Sandford Fleming, who had proposed a transcontinental line over twenty years before, convened near Shuswap Lake for the last spike ceremony. It was driven into a tie at Craigellachie on November 6, 1885 – four years after Macdonald's promised completion date, and two years and two months after the rival Northern Pacific's completion. Those timelines were of little concern in the moment because as of that day, Canadians could henceforth, as Berton put it, "travel the length of their nation without setting foot in a foreign land." He added, graphically, that "the nation became vertebrate." As soon as the ceremony concluded, a cry went up – "All aboard for the Pacific." The press later announced that a "New Northwest Passage" had been created; the Canadian dream of a northern corridor to the Pacific was finally realized. The first pass-through train to cross Canada left the CPR station in Montreal on June 28, 1886, and arrived at Pacific tidewater, fittingly it seems, on July 4.[18]

A Network of Iron

The Great Northern, running parallel to the Northern Pacific, was completed to Seattle in 1893. A third line, the Milwaukee Road, also extending from the Great Lakes to Puget Sound, was finished early in the twentieth century. These three American railroads, combined with the CPR and the Canadian National Railway (CN) north of the border, more than doubled the number of transcontinental linkages in more southerly latitudes, demonstrating the powerful grip of the Northwest Passage upon the continent's imagination. Asa Whitney's and Joseph Howe's visions of a Northwest Passage to Asia were finally established – as well as the dreams of Daines Barrington, Alexander Dalrymple, and John Meares, for that matter. Barry Gough observes that in the end, the Northwest Passage was made to work "by rail across North America and from the portals that lead to and out of the Strait of Juan de Fuca."[19]

Berkshire Hathaway's BNSF Railway, the twenty-first-century successor to the many "fallen flags" that preceded it in the northern US corridor, and the Union Pacific, dating to the original mid-continental link from Omaha to Sacramento in 1869, plus the CPR and CN, realize all the potential that Cook, Mackenzie, and Stevens, or their sponsors, quested after. The course of international commerce (nee the course of empire, as the eighteenth and early nineteenth trope had it), has again made its way westward with the growth of modern China. Cargoes originating there and in other Asian ports sail daily into Vancouver, Seattle, Tacoma, San Francisco Bay, and Long Beach. From those harbours, mile-long trains of container cars full of goods run eastward to Chicago and points beyond, some of it for trans-shipment to Europe. In that sense, the entire vertical expanse of North America, from southern California to the Salish Sea, has become the Northwest Passage.

The key transitional figures in this story, which ranges from the Elizabethan era to the twenty-first century, were Whitney, Stevens, and Fleming. Cook, Vancouver, and Mackenzie all viewed the Northwest Passage within the context of the British mercantilist ideal: empire as a system of connective travel routes. Even Thomas Jefferson and Meriwether Lewis worked in this idiom because their response to Great Britain did not envision occupying the interior of the continent. Stevens, more than any other individual discussed in these pages, was the great synthesizer. His effort presupposed demographic occupation and dominion, as evidenced by his concurrent offices – surveyor, governor, and superintendent of Indian affairs – which subsumed the identities of explorer, economic developer, and colonial administrator.

Walt Whitman's famous "Passage to India" was emblematic of this cultural conception. When he wrote this poem in 1871, the Union Pacific/Central Pacific combine had already bridged America; in the same year, tracks for the Northern Pacific began to be laid westward. The Suez Canal was completed, the Atlantic cable rested on the ocean floor, and for the first time in human history, it could be said that a single network linked the globe. Whitman's Romantic vision harked back to the route-foraging of the Enlightenment and anticipated Postmodernism's global village.

> *The People [are] to become brothers and sisters,*
> *The races, neighbors, to many and be given in marriage,*
> *The oceans to be cross'd, the distant brought near,*
> *The land to be welded together.*[20]

11
The Arctic Quest Redux

After George Vancouver vanquished Alexander Dalrymple's second-generation version of the Northwest Passage with the publication of his account in 1798, allowing the Pond/Mackenzie/Lewis and Clark riverine vision to mature, the Northwest Coast inevitably faded as a field of endeavour for scientific discovery. Great Britain was increasingly focused on the threats posed by revolutionary France. Napoleon Bonaparte, who became first consul in 1800, proved a brilliant and durable foe, and fully occupied Britain's attention until 1815. Ditto Russia.

But a harbinger of renewed interest in the North Pacific appeared that same year when Lieutenant Otto von Kotzebue led the first attempt to follow James Cook's track north. While Europe was preoccupied with post-revolutionary France, the Russian-American Company – a merchant coalition which secured a charter monopoly from the czar in 1799 – gradually expanded southward along the Alaska littoral. Under the leadership of regional governor Aleksandr Baranov from his new headquarters in Sitka, the company eventually extended its reach as far as Fort Ross, eighty miles north of San Francisco Bay, by 1812. Baranov's vision, however, was stressing the company's ability to transport goods and men across Siberia to an ever more distant colony running almost the entire length of North America's Pacific Slope. In this fashion, the Northwest Passage (which the Russians referred to as the *northeast* passage) through the Arctic Ocean re-emerged in Russian naval circles in St. Petersburg as the means by which the company might be resupplied more economically and the colony safeguarded from international incursion.

Departing from his Baltic port on July 17, 1815, less than a month after the battle of Waterloo, Kotzebue's *Rurik* aimed for Cape Horn, reaching Talcahuano, Chile, in February 1816. Sailing Cook-style across the Pacific (i.e., diagonally), he stopped at Kamchatka on June 7 for refreshment. This Russian Far East peninsula had been Charles Clerke's jumping-off point for a second season of northern discovery after he succeeded Cook following the great navigator's death in 1779. Kotzebue spent the next month exploring the Alaska coastline. He was able to transect the Bering Strait and cross the Arctic Circle, but he never got beyond the large inlet (not shown on Cook's charts) that now bears his name on the northwest corner of the Alaskan subcontinent, several degrees south of Icy Cape, the most northern of Cook's American landfalls. Wishing to avoid the rigours of a northern winter, Kotzebue sailed first to San Francisco Bay and then Hawai'i for overwintering. In March 1817, he sailed north to resume his search for a northern passage, but this time *Rurik* got no further than the Bering Sea's St. Lawrence Island, 170 miles shy of the Arctic Circle and the Bering Strait. The ship was damaged in a storm, leading Kotzebue to fear for its survival. He therefore turned for the South Pacific and home.

For the balance of the next half-century, the northern quest was a purely British enterprise. After Napoleon's final defeat at Waterloo and his banishment to the South Atlantic, 90 percent of the Royal Navy's officer corps became unemployed. Needing something to keep this cadre engaged, Britain chose a geographical enemy to supplant the French, and resumed the search for the original all-saltwater Northwest Passage. The difference this time was the realization that if a theoretical passage existed, as Glyn Williams put it, "it must lie far to the north, among the ice of the polar sea."[1]

In a sense, the Northwest Passage as cartographic projection returned to its roots in the high latitudes. The stakes this time were not as high as during Cook's era, or even Vancouver's. No one in an official capacity or otherwise was under the illusion of navigability, a "communication" across the continent, or any other euphemism for a shortcut like Alexander Dalrymple had once touted. All conceded that a mercantile objective was not achievable. The ensuing effort was, for the most part, Romantic in style: conquering nature for its own sake, akin to climbing a mountain. The Arctic and its mysteries became a fifty-year obsession until the British, out of emotional exhaustion more than financial, yielded the Arctic to the concern of the Americans and Norwegians.

The instructions given to the first British explorer in the field, Captain John Ross, advised that he was "to endeavor to ascertain the practicability of a Passage from the Atlantic to the Pacific Ocean along the Northern Coast of America."[2] Clearly, the new quest was to be a purely Arctic affair. This early guidance also indicates that, as was true from the origins of the search for a fully navigable passage from the sixteenth century until Cook redirected matters, the emphasis was to be on an approach from the Atlantic side. This chapter will reference the major efforts that had that directional orientation, but the emphasis will be on the approach from the Pacific side that Cook pioneered. In the end, it was an effort from that quarter that finally pieced together the Northwest Passage puzzle.

Britain's new endeavours in the North were preponderantly (though not exclusively) a maritime affair like Cook's and Vancouver's, but stylistically they bore a closer resemblance to the explorations of Alexander Mackenzie and of Meriwether Lewis and William Clark. Post-Cook, European exploration found its purpose in the delineation of continental interiors. Rivers – the Missouri, Columbia, Mackenzie, and Fraser in North America, and the Nile, Niger, and Congo in Africa – largely replaced oceans as the main theatre of operations. John Ledyard was the pioneering figure in this transition even if, in the end, he discovered nothing. The only exceptions were questions surrounding the dimensions of the Antarctic continent, which Cook had been the first to detect, and the northern boundary of North America. The labyrinthine Canadian Arctic Archipelago, which would take a century to fully discern, was usually accessed from a sea-borne vantage, but by definition the frozen state of the waters circumscribing its insular contents limited maritime opportunities. The challenge for Arctic explorers was how to establish a self-contained logistical supply, and greater reliance on Indigenous sources of geographical information than nautical explorers were accustomed to.

One aspect of the Arctic quest had not changed since Cook's time. The idea of an ice-free Arctic Ocean was as durable as the notion of the existence of the multi-variant Northwest Passage itself. This ancient concept, repopularized in the 1770s by Daines Barrington, a member of the Royal Society and an early proponent of what became Cook's third voyage, had currency until it was finally vanquished in the 1880s. The resilient premise was that the ice Cook struck north of the Bering Strait (as did any number of explorers approaching from the North Atlantic) was fastened and proximate only to the land masses that ringed the Arctic Circle. This barrier, it was supposed, walled off a temperate ocean, free of ice. If a ship could break through this barricade, as one would penetrate the interior of a continent, it was believed possible to sail directly to the North Pole. In the long history of Euro-American exploration of the globe, this theory was the epitome of wishful thinking – the geography of hope.

Coincidentally, in the post-Napoleonic era, the Little Ice Age began loosening its grip. In 1817, whalers discovered that the Greenland coast was ice-free in summer for the first time in living memory. This was an attentive community, and the most experienced among them, William Scoresby, was in regular communication with the now aged Joseph Banks (Cook's first-voyage naturalist and the patron of British exploration as the long-time president of the Royal Society) about the chemical composition of sea water and pack ice. (For that reason, Scoresby is mistakenly considered the father of polar glaciology, though, as I document elsewhere, that honour actually belongs to Cook.[3]) Scoresby knew that the idea of an ice-free pole was bunk, but the warming trend, slight as it was compared with the rate of change in modern times, provided incremental access to the northern archipelago. Nevertheless, it could still be a deadly environment, as subsequent events demonstrated.

The first addition to geographical comprehension in the Arctic came from the expedition of Captain William E. Parry (1819–20). Parry's instructions emphasized the geography of the passage. The scientific work that Scoresby (a civilian) was conducting was decidedly secondary, a slight deviation from Enlightenment-era discovery in the manner of Cook and his team of naturalists. Parry extended the cartographic high-water mark beyond the western end of Canada's Lancaster Sound off the North Atlantic. He was able to define, by proceeding through, Barrow Strait into Viscount Melville Sound. This strait adjoins Melville Island, located in the middle of the second tier of islands above the continental mainland. Subsequent exploration proved that Melville Sound was to be the crossroads of the Canadian North. From that intersection, channels provided further access to the north, south, and west. Parry's crew was the first to successfully overwinter in the Arctic (at the appropriately named Winter Harbour). No English sailor would reach as far westward as Parry did for another thirty years.

Parry, like many of his successors, was thwarted by the ice stream, fifty feet thick in places, flowing unimpeded in a southerly direction from the Beaufort Sea onto the north shore of Banks Island and into Melville Sound. The lesson other explorers took from Parry's experience was that the best prospect for a passage, as approached from the Atlantic side, lay in hugging the channels adjoining the southern tier of islands and the continental coastline. This adaptation was in the Cook third-voyage and Vancouver style of Northwest exploration (except in this case, keeping the continent to the port side). Alternatively, it would be pursued disastrously by the Arctic's greatest failure, Sir John Franklin, and the region's most successful explorer, Roald Amundsen. Parry himself was the first to apply this lesson. His follow-up expedition (1821–22) was yet another (and equally futile) search for a northwesterly outlet from Hudson Bay that might lead to the passage. This zone of putative discovery had frustrated British exploration of the Canadian North for the preceding two centuries.

The pivotal figure in the untangling of the Arctic labyrinth was Sir John Franklin. In the popular imagination, Franklin is typically viewed through the frame of his voyage to the central Arctic when, infamously, his expedition became marooned in the ice, leading to the death of every man sailing with him. But his introduction to the Far North actually came by following the track of overland explorers. His 1819–22 expedition was the first to follow Samuel Hearne's track to the mouth of the Coppermine River in the central Arctic a half-century earlier. After Franklin reached Coronation Gulf, which Hearne had seen but not named, he hugged the Arctic coast in an easterly direction, extending the northern chart to Bathurst Inlet. Franklin was able to confirm Dalrymple's suspicion that Hearne had positioned the mouth of the Coppermine River four degrees too far north.

Four years later, Franklin led another expedition (1826–27) to the Arctic Ocean. Like the first, it emanated from a base camp on Great Slave Lake. Similarly, he was the first European to voyage down Mackenzie's river since Sir Alexander had completed the feat in 1789. From its massive delta, his chief naturalist, Dr. John Richardson, led a detachment eastward to the mouth of the Coppermine to connect with the earlier expedition. Franklin was prepared to proceed west to Cook's Icy Cape in order to meet an ocean-going vessel sent to pick him up. Assisted by George Back of the Hudson's Bay Company (HBC), Franklin led the main party along this previously unexplored shore. They reached Prudhoe Bay, 150 miles short of Point

Barrow, which Captain Frederick Beechey, sailing north from Kotzebue Sound in the *Blossom,* had sighted in 1826. (Beechey named this feature after John Barrow, the Admiralty's second secretary. A latter-day Barrington, Barrow was confident that the North Pole was accessible by sail.)

Franklin's two overland expeditions charted 550 miles of Arctic coastline. From Cook's Icy Cape to the east end of Coronation Gulf (which Franklin named and into which the Coppermine flows), only the 150-mile gap from Prudhoe Bay to Point Barrow remained. At the east end of the passage, the gap to the nearest known point on Hudson Bay had also been considerably lessened. Franklin and Richardson were knighted for their efforts. Franklin also honoured Richardson by naming the spectacular mountain range that encloses the Mackenzie River Delta's western flank after him. The Richardsons are the northernmost segment of the Rocky Mountain cordillera, essentially forming the border between present-day Yukon and the Northwest Territories. In 1837, the HBC's Thomas Simpson (cousin of the company's governor, George Simpson) filled the gap between Franklin's Return Islands at Prudhoe Bay and Point Barrow.

Through the efforts of John Ross and James Clark Ross on expeditions in the early 1830s, the continent's northernmost mainland extension, the Boothia Peninsula, was defined cartographically and the magnetic north pole was fixed on its southwest coast. George Back re-entered the story by discerning Queen Maud Gulf east of Bathurst Inlet via the Great Fish River (later named after him). In 1839, two other fur traders, Simpson again, with Peter Warren Dease, teamed up to delineate the waterways at the southern end of Victoria Island. This created a continuous line of geographical comprehension running from Alaska to Rae Strait, near the foot of the Boothia Peninsula. For the first time, Arctic exploration coming from the west had run a cartographic line eastward to the meridian on which the west coast of Hudson Bay lay, though hundreds of miles north of it.

At this juncture, with many relevant pieces of the northern labyrinth having been puzzled out, the Royal Navy in 1845 launched Sir John Franklin's two-ship flotilla in what was presumed would be the definitive voyage establishing the Northwest Passage. Franklin's third (and last) voyage is one of the most sensational stories in the history of European exploration. At the urging of the Royal Geographical Society, Franklin was dispatched in the hope that he would be able to run a line from Lancaster Sound off Baffin Bay to Simpson and Dease's station at the western base of the Boothia Peninsula, thus completing the Northwest Passage cartographically.

Characteristically, Franklin discovered nothing. Worse, his voyage's complete fatality resulted in the most disastrous expedition in Arctic history. That story has been amply told elsewhere and there is no need to account for it here other than to stipulate that Franklin's expedition got stuck in the ice at the northwestern tip of King William Island, west of the Boothia Peninsula, in September 1846. Franklin died the following June, and in April 1848, after the ice still refused to loosen its grip on the *Erebus* and the *Terror,* by which time two dozen other officers and crewmen had died, the 105 survivors abandoned the ships in a futile attempt to reach the Back River outfall. Only material culture was left behind to tell the tale.

Franklin was last seen alive by whalers near the entrance to Lancaster Sound. When he failed to turn up after two years, the first in a series of fifty rescue missions was sent to look for him. Ironically, these expeditions, taking place between 1848 and 1859, turned out to be far more important than anything Franklin's itinerary might have proved. It might be said that an "Arctic fever" took hold in Great Britain. (Coincidentally, this occurred at the same time that railroad mania was dominating public concern in the lower latitudes of North America.) The net effect of the search for Franklin was the most concentrated attempt at delineating the complexity of the Canadian Arctic Archipelago in history, including the ultimate discernment that there was not *one* Northwest Passage but *several* of them. (See Figure 11.1) Since most of the searches were of British origin, a collateral consequence is that Franklin inadvertently cemented Canada's claim to sovereignty over the upper tiers of islands in the archipelago.

The first of the two most important of the Franklin-inspired investigations was James Clark Ross's 1848 expedition, with Robert McClure and Leopold M'Clintock as his senior officers. This introduced the latter two to the Arctic. Both would figure prominently in the eventual demystification of the Arctic labyrinth and of Franklin's fate therein, respectively. Ross proceeded in the traditional pre-Cook fashion, that is, from the Atlantic through Lancaster Sound. He was unable to penetrate the ice in Barrow Strait, north of Somerset Island. In a replay from the age of Cook, the Admiralty concluded that if ships could not breach the ice from the east, perhaps it was time to try again from the west, the same route through the Bering Strait that the great navigator himself had attempted in 1778. Ross's ships were refitted: the flagship *Enterprise* under Captain Richard Collinson, and the *Investigator* under McClure, Ross's former second-in-command. They left Plymouth in January 1850 – two of the nine ships that left England that year in search of Franklin.

Collinson and McClure personified the opposing characteristics of the exploratory temperament that Cook famously described during his first-voyage tribulation off Australia's Great Barrier Reef. On that occasion, he referenced "the *Timorousness* and want of Perseverance" of "an excuse for a man" who leaves "a Coast unexplored" and is therefore "the unfitest man in the world to be employ'd as a discoverer," versus the sailor who "boldly incounters all the dangers and obstacles he meets and is unfortunate enough not to succeed" who is then "charged with *Temerity* and want of conduct."[4] Bold barely begins to describe McClure's deportment on this expedition, and it was he who first pieced together the system of channels in the Canadian North that defined a version of the Northwest Passage.

The Admiralty's plan called for Collinson and McClure to meet up in Hawai'i in July 1850. They missed each other, the *Enterprise* leaving the day before McClure steered the *Investigator* into Honolulu. Collinson left instructions that if McClure did not reach him by the time the *Enterprise* met up with the supply ship *Plover* (already anchored in Kotzebue Sound, the only sheltered water north of the Bering Strait), it was his intention to sail in search of Franklin with the *Plover* as his consort. McClure, notwithstanding his subsidiary status, was challenged and energized by this presumed slight. He left Hawai'i determined not merely to catch up with Collinson but to beat him to the Arctic ice. He knew that Collinson was taking the long course north, the one pursued by Charles Clerke after succeeding to the command of Cook's third expedition, along the Kamchatka Peninsula. McClure determined upon the more direct path that Cook pursued in 1778 – due north from Hawai'i, taking a shortcut to the Bering Strait through the Aleutians. He was also keenly aware of the same prize money that had lured the great navigator out of retirement and into the icy latitudes – the financial reward for discovering the Northwest Passage.

McClure and the *Investigator* reached Kotzebue Sound on July 29 – predictably, several weeks ahead of Collinson's more circular route. There he was hailed by Captain Henry Kellet's *Herald,* a frigate whose role was to keep the anchored supply ship supplied. McClure, in dialogue with Kellet, feigned the impression that he was running behind Collinson and, equipped with three years' worth of provisions, decided to press on alone. No explorer bound for glory had jumped ahead quite like this since Meriwether Lewis did so habitually. (On four different occasions – reaching for the Yellowstone River, the Great Falls of the Missouri, the Continental Divide, and the Pacific

Ocean – Lewis left his partner, William Clark, behind to attend to nitty-gritty logistical details.) Kellett, a more senior officer, and aware of the fact that the approach from the Pacific was to be a two-ship venture and that Collinson had yet to arrive, played along with McClure's pretence and allowed the *Investigator* to proceed. In a profoundly ironic turn of events, Kellett, aboard a different ship that probed the Arctic several years later from the *Atlantic* side of North America, would supervise the effort that rescued McClure from his rash decision. Collinson eventually reached Alaska's Cape Lisburne in August, but considering the lateness of the season, he determined to spend the winter in Hong Kong.

Charging ahead, McClure breached the Arctic pack ice at Icy Cape, where even the great Cook had to call on his legendary navigational skills to avoid being marooned. The *Investigator*'s small boats later had to haul her through the ice around Point Barrow, Alaska's northern tip. Though this part of the North American coastline had been mapped previously, McClure was the first to *sail* into the adjoining waters. With the ancient pack ice visible to his north ninety miles away, the *Investigator* sailed breezily along the continent's Arctic shore. On August 7, McClure sighted and named Nelson Head at the southwest corner of what was later discerned to be an island named for Cook's famous first-voyage naturalist, Banks. Rounding the island's southeast corner, the *Investigator* was caught within the grip of an ice stream flowing northward, up the east coast of Banks Island. Though an ominous development on its face, it was fortuitous because, with the charts of Parry and other early explorers of the region in front of him, McClure knew he was being carried along in a channel leading directly toward Viscount Melville Sound, and, if he was truly lucky, through to Barrow Strait. By September 9, he had closed all but sixty miles of the westernmost extent of the sound – the only remaining gap in the fabled Northwest Passage. But here, the young ice of the new winter season frustrated any further travel, leaving the last link unexplored.

McClure's good fortune faded when the ice began to carry him southward, away from his goal. Worse, the *Investigator* was at risk of being crushed. On September 26, 1850, McClure made preparations to abandon ship. The crew, fearing it was doomed, lost all discipline. The incident proved the truth of Banks's astute observation made at the moment of greatest risk for Cook's *Endeavour,* when it was impaled on the Great Barrier Reef, namely, that the commander's coolness and precision stabilized his

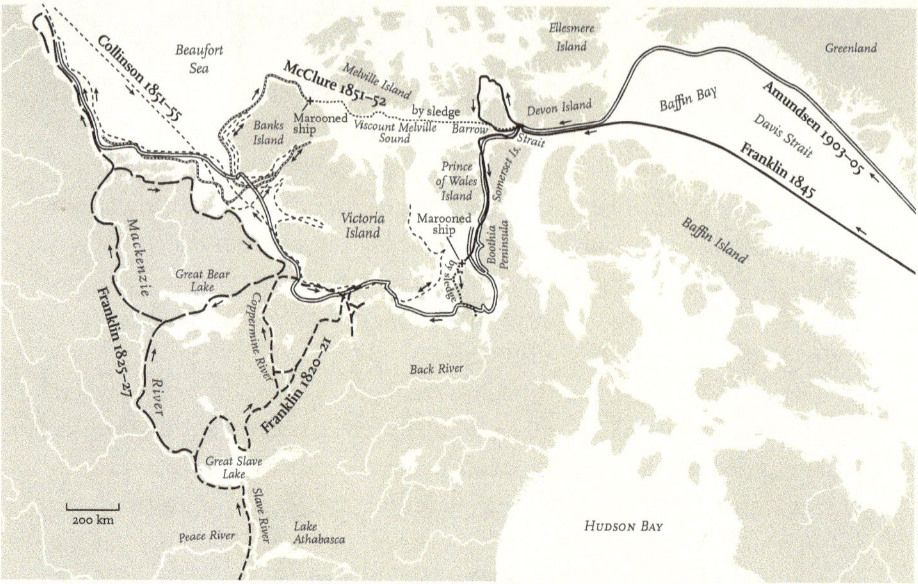

Figure 11.1 The Northwest Passage was revealed to contain seven alternate channels once the quest returned to the icy latitudes. | Based on William H. Goetzmann and Glyndwr Williams, *The Atlas of North American Exploration*, 186–87, 191, 192–93; redrawn by Eric Leinberger

crew's deportment, giving them a fair chance of being able to save the ship. For McClure, disaster was also averted, though not by effective seamanship as on the *Endeavour* but rather when Nature eased its pressure. When the ice finally froze fast, the risk of a floe crushing the ship's hull passed.

By October 1850, the *Investigator* was locked in the ice between Banks and Victoria Islands. Oddly, McClure was still in a position to assess whether the channel in which he was embedded (Prince of Wales Strait) might lead to Barrow Strait, the Northwest Passage, and personal glory. On October 26, after a five-day journey by sledge, he was able to confirm this fact. Of course, this version of the passage was fully blocked by ice, and for all he knew, perpetually so. But the substance that thwarted his ship's route was not land, and therefore, in a technical sense, from a height of land on the northeast corner of Banks Island, he identified the Northwest Passage's missing link, even if, in the moment, he could not traverse it in person.

The ice in Prince of Wales Strait did not break until July 14, 1851. Unable to sail north against the ice stream that worked like a headwind against his progress, McClure reversed course, determined to turn Banks Island in a clockwise direction to see if he could approach Viscount Melville Sound from the west instead of the south. Initially, the *Investigator* made great progress on this course, reaching the island's northeast corner in less than a week. But once again the ice stream pressing down from the northwest retarded his forward progress. The *Investigator* anchored in what would later be called McClure Strait, the commander thought temporarily, near an inlet he would later call Mercy Bay. His aim was to wait for an eastward track through the ice to clear. Lancaster Sound, the principal Atlantic gateway to the Northwest Passage, was straight ahead in that direction.

Mercy Bay would hold McClure and his crew tight for the next two years, and the *Investigator* would never leave. Some of McClure's men thought that (using the Cook matrix) their commander had been timorous – that is, he may have missed his main chance late that July to press ahead by sailing perhaps even to Parry's Winter Harbour, fully effectuating the Northwest Passage. Instead, in September 1851, with his second Arctic winter coming on fast, McClure pulled into Mercy Bay, not wanting to test his ship's fate by exposing it to the broken ice chunks in mid-channel as he had the previous year in Prince of Wales Strait. McClure the bold may very well have suffered a failure of nerve, but on the other hand, and in the end, he and most of his crew survived Mercy Bay.

In Pierre Berton's words, "in his haste to find a secluded harbour, McClure had chosen a trap."[5] He wisely rationed the food supply, fearing that the ice might not release him until the ensuing summer (of 1852). In April of that year, he set off by sledge eastward, laterally across the strait that now bears his name. (McClure Strait, running between Banks and Melville Islands, completes the northern deep-channel version of the Northwest Passage.) He was hoping to meet up with one of the many ships that he knew would be in the vicinity of Barrow Strait searching for Franklin via Lancaster Sound. By this means, he reached Parry's Winter Harbour at the east end of Melville Island, where he found a note that the ships he had expected to see had left in June 1851. Disconsolate, McClure returned to Mercy Bay on Banks Island in May 1852, fully cognizant of the other kind of strait he was now in. He feared that if there was no break in the ice by September, his expedition was going to have to spend a second winter at Mercy Bay, and his third in the vicinity of Banks Island. During the winter of 1852–53, the cold and lack of nutrition finally began to take a toll on his crew.

On March 2, 1853, the very same day the American Congress authorized the Pacific Railroad Surveys of the American West looking for a route for a steam-driven Northwest Passage, McClure began to implement a plan mapped out over the preceding winter. He hoped to save his crew by having only the twenty healthiest men stay with the ship, thereby extending the supply of rations, while parties made up of weaker members would attempt dashes in several directions hoping to find caches left by exploring parties over the previous three and half decades.

On April 6, just as the sledges were being loaded for departure, one of McClure's men in a burial detail spotted something dark-coloured moving toward them in the distance. Thought at first glance to be a muskox, it was discerned to be a sledge coming from the direction of Barrow Strait. By this point, the *Investigator*'s men had not seen someone other than themselves since July 1850. The sledge-men were sailors from another British ship, the *Resolute,* anchored in Viscount Melville Sound off the south coast of Melville Island. It was captained by the same Henry Kellett who had seen McClure off into the Arctic mists almost three years earlier. This was a turn of events that even the boldest fiction writer would not dare script. Since he last saw McClure near Alaska's Cape Lisburne, Kellett had returned to England, been given a new command on a new ship, with directions to proceed through Lancaster Sound and set up a rescue station in the crossroads of the Arctic for sailors who were looking for Franklin. Which is to say, the Royal Navy was having to send out searchers for those searching for Franklin.

It was Lieutenant (later Admiral) Bedford Pim who commanded the sledge team that rescued McClure and his men. As Berton put it: "Thus was averted a second grievous Arctic tragedy, one that could have matched that of John Franklin."[6] Accordingly, on April 6, 1853, sailors coming from the Atlantic side of the Northwest Passage miraculously met those sailing from its Pacific gateway. Kellett had reached the coast of Melville Island in the very centre of the labyrinth on September 1, 1852. From there, he sent out search parties, one of which stopped at the "Arctic telegraph" station (a cairn that Parry had constructed at Winter Harbour) and found a note McClure had left there in April that year. McClure, not knowing that he would survive, announced in this document that he had discovered the Northwest Passage.

Now armed with certain knowledge of where the *Investigator* was marooned, Kellett detached Pim with ten men aboard two sledges (one man-hauled, the other pulled by dogs) on March 10, 1853, just as McClure was readying his last-ditch attempt to reach civilization or supply stations. After twenty-eight days of travel, in weather often reaching –50°F, Pim reached Mercy Bay, where McClure's men mistook him for a muskox. Pim started ferrying the sick and near-dead survivors back to Kellett's station, but McClure, insisting on his prerogatives as a captain, tried to cajole Pim into allowing him to sail the *Investigator* out of the bay when the ice broke so that he could complete the transect of the Northwest Passage. Kellett had anticipated this. He gave Pim the latitude to indulge McClure, provided he could convince enough of the *Investigator*'s crew to wait out the ice. Only four men responded positively to this proposition, and so Kellett's order as a superior officer took effect and the *Investigator* was abandoned on May 19, 1853.

Thus, McClure completed his transect of the Northwest Passage by sledge. In that sense, Ian MacLaren quipped, McClure *"walked"* across the last link "in a most pedestrian, inglorious manner." He was knighted upon his return to England, and Parliament generously awarded him and his surviving crew members a £10,000 prize for discovering the passage, such as it was. MacLaren notes: "The passage existed; only ice got in the way of sailing it." As the *Times* of London put it, the passage "may be assumed as open once or twice a century in favourable circumstances for short periods."[7] Even that proved an overestimation. Conversely, it was impossible to understate McClure's ingratitude. He resolutely refused to share discovery credit with Franklin, Kellett, or Pim.

Inevitably, Franklin fatigue took hold in Great Britain. The Crimean War, which broke out in October 1853, expedited this because the Royal Navy could no longer afford the post-Napoleonic luxury of Arctic Romanticism. With the passage of time since Franklin was last heard of, it became increasingly evident that his crews never holed up in some quarter of the Arctic awaiting rescue. Indeed, in August 1853, just a few months after Pim found McClure, John Rae from the Hudson's Bay Company, near the foot of the Boothia Peninsula, encountered an Inuit man who told him a story of dozens of white men starving to death. Within a year, Rae secured from the Inuit a raft of materials that were unmistakably linked to Franklin's expedition. News of his findings appeared in the London press on October 23, 1854, including the sensational charge that some of Franklin's men had cannibalized their mates. From forensic evidence and Inuit oral history, it was determined that Franklin's ships had gotten stuck in the ice of Victoria Strait west of the Boothia Peninsula in September 1846. The commander died the following June, and the last survivors of the original 128-member crew died at Starvation Cove in 1848 during an ill-fated attempt to reach the continental mainland. In death, they had in fact reached the proximate location of Simpson and Dease's furthest east from their 1838–39 survey of the Arctic coast east of the Coppermine River.

By 1855, Arctic fever had reached its anticlimactic phase. Franklin's fate was known. The churlish McClure claimed his prize as the discoverer of the passage. Even Richard Collinson had turned up. After refusing to follow McClure into the labyrinth, Collinson tried to find a suitable harbour in the vicinity of Icy Cape, where Cook had nearly foundered in 1778. He was forced to spend the winter in Hong Kong, and when the *Enterprise* returned to the Arctic, it was slowed by the ice at Point Barrow in July 1851. Collinson reached Banks Island on August 26. Desperate to catch up with his presumed junior officer McClure, Collinson might have reached the *Investigator* at Mercy Bay, but instead he timidly pulled up in Walker Bay on Victoria Island facing the Prince of Wales Strait. There he spent the winter of 1851–52. (In 1910, Vilhjálmur Stefánsson, a latter-day Arctic explorer, encountered an Inuit man who remembered seeing Collinson as a youth. Stefánsson also suspected that most of the spears he saw in Indigenous use near Banks and Victoria islands were adapted from the metal works on McClure's ship.) When the ice broke, the *Enterprise* proceeded south, away from McClure's station, into Coronation Gulf, which, as noted, several HBC explorers had charted, starting with Hearne in 1771 and concluding with Simpson and Dease in 1839. Nevertheless, Collinson proved that a large ship could safely navigate the shallow waters of the southern fork of the Northwest Passage, the one successfully transected by Amundsen early in the twentieth century.

The *Enterprise* spent its second Artic winter (1852–53) at Cambridge Bay at the eastern end of Victoria Island, not far from King William Island off the west coast of the Boothia Peninsula where Franklin had been marooned. Having already missed McClure on Banks Island, Collinson was now in a position to learn what Rae subsequently discovered about Franklin's fate. But unlike Rae, who as a fur man was familiar with and reliant upon Indigenous People for geographical information and polar survival, Collinson had no one who could converse with the Inuit. After overwintering on Alaska's north coast (1853–54), he returned to England. With a timid disposition and a staff riven with dissension, Collinson had thus missed out on sharing the discovery of the passage with McClure or finding Franklin. Amundsen, however, later paid close attention to the *Enterprise*'s voyage and settled on Collinson's southern route, not McClure's northern one, for his conclusive voyage of discovery (1903–06).

Inadvertently, indirectly, infamously, Franklin's ill-fated expedition led to the discernment of the channels and insular structure of the southern half of the Canadian Arctic Archipelago. The Rosses, Parry, Collinson, Rae, McClure, and others had combined to reveal that there was more than one Northwest Passage, though only the last of these made the trip all the way through, and even then only with an assist by sledge from Pim and Kellett. None of them, however, eclipsed Franklin in contemporary popular imagination or in legend since. This proves once again that dying in the course or immediate aftermath of an expedition can be a good career move, as witness Cook and Lewis.

After Franklin, the Arctic quest moved away from the question of the Northwest Passage to the North Pole itself. This phase, however, was dominated by the Norwegians and Americans, not the British, though even here there was an echo from Cook's time. The notion (popularized prior to Cook's final voyage by Daines Barrington) still lingered that the Arctic Ocean was perhaps ice-free once a ship got far enough from Siberian or North American land masses. Great Britain's last major effort in the north – George Strong Nares's expedition (1875–76) – satisfied the Royal Navy's curiosity on this point. His effort was mirrored by the equally naive voyage of George W. DeLong, captain of the *USS Jeanette*.

DeLong sailed through the Bering Strait in 1879 hoping to replicate Cook's short-lived attempt at a passage above Siberia in 1778, after his run toward the Atlantic had been stalled at Alaska's Icy Cape. DeLong became embedded in the ice northwest of Wrangel Island in the fall of 1879. Over the course of the next twenty-two months, the ice pack carried the *Jeanette* as far west as the Lena Delta before crushing it. Like Franklin, DeLong was killed, but unlike Franklin's crew, some of his men survived to tell the tale. Also like Franklin, DeLong inadvertently spawned a legacy pursued by other explorers. In 1884, pieces of his ship washed up thousands of miles away on the east coast of Greenland. This stirred the imagination of Fridtjof Nansen, helping him to conceive the concept of Arctic drift. After sailing from Norway into the Arctic above Russia, he purposely embedded the *Fram* in the pack ice off the same Lena River where it reaches the Arctic on the north central coast of Siberia in September 1893. Slowly drifting westward, the *Fram* emerged from the ice two and a half years later.

Nansen's experimental voyage proved the concept later employed by Amundsen in command of the *Gjøa*. Filling the vacuum created by Anglo fatigue, Amundsen sailed from Christiana, Norway, on June 16, 1903. Accompanied by just six other men, he pursued the classic route that had long been a British obsession by cruising into Lancaster Sound and through Barrow Strait into the channels separating the Boothia Peninsula from Prince of Wales and King William Islands. During this transect, he proved that the magnetic north pole, which he re-established near the northern tip of the Boothia Peninsula, was mobile. Amundsen was locked in the ice for two winters at Gjoa Haven, on the southeast corner of King William Island, southwest of the peninsula, kitty-corner from where Franklin had been embedded and died.

Amundsen was now in a position to push into the southern fork of the passage, largely delineated by Franklin's overland expeditions, a series of HBC explorers, and Collinson. He was the first to sail the final 150-mile gap in Queen Maud Gulf, connecting the southwest corner of King William Island with Collinson's Cambridge Bay anchorage on the south coast of Victoria Island. Like Nansen's expedition, Amundsen's voyage was experimental. Unlike some stiff Englishmen, he was quite adaptable to Indigenous ways. Among other things, he learned about the reliability of dogs for pulling sledges. He also came to understand that in frigid climates sweat kills, and it had to be combatted through loose, fast-drying, windproof clothing. These lessons he re-employed to beat Robert Falcon Scott to the South Pole in 1911, breaking the hearts of Englishmen a second time.

In another way, however, Amundsen was a throwback to the greatest British explorer of all time. On his first voyage, Cook famously stated that Australia's Aboriginal people "may appear to some to be the most wretched people upon Earth, but in reality they are far more happier than we Europeans; being wholy unacquainted not only with the superfluous but the necessary Conveniences so much sought after in Europe, they are happy in not knowing the use of them." On his second voyage, after observing the cultural damage from his crew's interaction with the Maori, Cook observed that

> such are the concequences of a commerce with Europeans and what is still more to our Shame [as] civilized Christians, we debauch their Morals ... and we interduce among them wants and perhaps diseases which they never before knew and which serves only to disturb that happy tranquillity they and their fore Fathers had injoy'd. If any one denies the truth of this assertion let him tell me what the Natives of the whole extent of America have gained by the commerce they have had with Europeans.[8]

Echoing Cook's vein of thinking, Amundsen wrote:

> During the voyage of the *Gjöa* we came into contact with ten different Eskimo tribes in all, and we had good opportunities of observing the influence of civilization on them, as we were able to compare those Eskimo who had come into contact with civilisation with those who had not. And I must state it as my firm conviction that the latter, the Eskimo living absolutely isolated from civilisation of any kind, are undoubtedly the happiest, healthiest, and most honourable and most contented among them.[9]

The *Gjøa* transected the last stretch of previously unnavigated waters off the southern coast of Victoria Island on August 17, 1905. Amundsen's Northwest Passage was navigable, barely (in places only a few fathoms deep) and not truly practicable in the sense that all the early visionaries imagined possible. (McClure Strait, impassable in Amundsen's time, let alone the late phase of the Little Ice Age faced by Cook and Franklin, is a much wider, deeper, and more direct corridor, and with the passage of time and warming climate, is a better prospect for a workable passage.) On August 26, Amundsen met a vessel that came from the Pacific, a schooner from San Francisco. Blocked by the ice on September 9 off the Mackenzie River Delta (along with a dozen other vessels) he was forced to spend a third winter in the north. Again like Cook, who chose to stay busy when circumstances put the work of discovery on hold, Amundsen left Herschel Island on October 24 for a trek overland to Eagle, Alaska, where the closest telegraph station was located. He dispatched a thousand-word message recounting his success to Nansen and the wider world.

Amundsen rejoined the *Gjøa* in March 1906. The ice broke in July, and on August 10 he sailed for San Francisco and the end of the voyage. In a sense, two centuries' worth of effort had come to naught for Great Britain. As Pierre Berton put it, "Roald Amundsen had snatched the prize of centuries from the greatest navy in the world."[10]

Amundsen, like Cook, viewed the work of discovery as a scientific and ethnographic endeavour. With the Northwest Passage conquered, the quest for the North Pole became "a sporting event."[11] The exploratory process was now all about goal, effort for its own sake, like climbing a mountain to say you had done it – admirable perhaps, but not particularly edifying. This mode of "exploration" held a particular appeal for the American public. At the turn of the twentieth century, which the United States was to make its own, the country was hungry for heroes with international stature. Robert Peary, with his disputed claim to being the first to reach the North Pole in 1909, fit that bill, but it was a nationalistic age and even Amundsen got caught up in this paradigm in his quest for the South Pole.

Epilogue
The New North

In the post-exploratory phase of Northwest Passage history, sovereign control of the corridor replaced geographical discernment as the driving force of international interest. This phase commenced during the Second World War when Constable Henry Larson of the Royal Canadian Mounted Police transited the passage in command of the *St. Roch* with a nine-man crew. In 1940, Larson (like Amundsen, a Norwegian by birth) was tasked with piloting his 104-foot-long motorized schooner from Vancouver to Halifax. Essentially following Amundsen's route in reverse, his was the first west-to-east transit of the passage and took two winters to complete. The voyage of the *St. Roch* was nominally a war-related mission but its principal aim was to reinforce Canadian claims to jurisdiction over the waterway at a time when Great Britain and its Commonwealth allies stood alone against Nazi Germany. The *St. Roch* returned to Vancouver in 1944, taking the more northerly route above Victoria Island, the insular hub of the western Canadian Arctic around which each of the several channels of the Northwest Passage run. With McClure Strait north of Banks Island still blocked by ice, Larson headed south through Prince of Wales Strait between it and Victoria Island to access Amundsen Gulf and the Beaufort Sea.

The post-exploratory era inaugurated by the *St. Roch* has also been accentuated of late by the pace of global warming and concomitantly the ever-increasing navigability of the Northwest Passage and the Northern Sea Route atop Russia. Not so long ago, Barry Lopez could write about the physical, psychological, and policy remoteness of the Arctic from "temperate zone people." This would no longer seem to be the case. Lopez tells us that the Arctic's "expanses of tundra and ice" were typically viewed as "wastelands." As a consequence, the northern tip of North America and its adjoining seas, he stated, were "the last undiscovered complex ecosystem on the planet."[1] Today, polar ice science, and events in the Arctic in particular, are of greater global concern than ever, certainly since the eras of Franklin and Amundsen. To cite just one example from the natural world, the Far North's iconic animals – polar bears and walruses – have joined the endangered species in Africa as the world's most imperilled creatures.

There is widespread scientific consensus regarding climate trends in the polar regions. The warmest years since at least the 1860s, if not recorded human history, have occurred during the last decade and a half. Ice scientist Willie Weeks writes: "One does not have to be a statistician to realize that the probability of this being a random sequence is miniscule." He adds that, based on data reconstructions going back ever further in time, "current surface temperatures are warmer than they have been at any time during at least the last thousand years." In 2021, NASA and the European Union's Copernicus Climate Change Service determined that the average global surface temperature of 2020 matched 2016 as the warmest year on record. Weeks concedes that the episodic occurrence of warming phases in global history means that "questions still remain concerning the separation of naturally occurring thermal cycles from temperature changes resulting from the increasingly modified state of the Earth's surface and the rising level of greenhouse gases resulting from human activities." At a minimum, informed opinion, in his view, properly concludes that the anthropogenic factor is "significant." In addition to the great warming's effect on ice, Weeks notes that the Arctic polar vortex, the circumpolar wind in the stratosphere, "has both strengthened and become more persistent." Cyclonic storms are now penetrating the Arctic, and thunderstorms, once literally unheard of, now occur there. Melting ice may even have more distant effects by perhaps contributing to the extreme fall fire weather in the western United States and Canada. Under this theory, the absence of ice is thought to be nudging the jet stream out of its normative patterns. Certainly it is now known that some vegetative species have moved further north to locations where heretofore they were not able to survive.[2]

The Arctic's sea-ice cover is decreasing in both extent and duration, at the rate of an 8 percent loss per decade since the 1970s. Indeed, the rate of loss has accelerated, the meltwater having raised the sea level worldwide by more than an inch since 1994. According to Weeks, "the oldest and presumably thickest ice types have essentially disappeared." In 2007, ice five years and older was just 7 percent of the total pack volume. These and other changes, he states, are "not only surprisingly large but real."[3] The Chukchi Sea, into which Cook sailed in 1778, has seen the most dramatic sea-ice declines in the Arctic zone, losing 70 percent of its summer ice concentration south of the 80th parallel. In 2017, Arctic sea-ice coverage reached its lowest maximum seasonal coverage: 5.57 million square miles. This rate of diminution has led some scientists to suggest that the basin will be largely ice-free seasonally by 2035, or by the mid-2040s at the latest.

In effect, the Arctic summer now arrives more than two months earlier than it has historically, the melt season is longer, and less ice survives to thicken the following winter. Among other natural consequences, polar bears are increasingly forced to seek food on land, where they are not nearly as effective as hunters as on the ice, and given current trends, killer whales are likely to replace them as the top marine predator. Indeed, the polar bears are now spending so much time on land that there is some evidence they are hybridizing with grizzlies. The fate of the bears and other northern species also poses obvious correlative risks to the Indigenous Peoples of the circumpolar region, whose legendary adaptability will be tested once again.

There are other dimensions to this denouement with historical echoes. Cook's last voyage precipitated a rush to the Northwest Coast to extract furs for trade in China. The recession of the Arctic ice pack has occasioned a new commercial boom in the Far North. A diamond mine has opened in Nunavut, and iron ore is being quarried on Baffin Island. In the last decade, diamonds have also been discovered in Greenland, where an Australian company has expressed interest in mining what is thought to be a forty-year supply of uranium. Alcoa has discussed smelting aluminum there as well, and in 2009 a London mining company announced a plan to open an iron ore mine north of Nuuk for shipment to serve the Chinese market.

Stories about the search for oil and gas in the Arctic, both above ground and below the sea, are ubiquitous in modern news media. Russia, which has the world's greatest natural gas reserves in Siberia, is partnering with China, which wants to wean itself off coal by linking its northern industrial hubs to Russia's Arctic gas fields. This pattern has accelerated since the outbreak of the Russia-Ukraine War in February 2022 and the concomitant Western embargo and price caps on the international sale of Russia's carbon-based energy supply. Shipments of both liquefied natural gas in massive ice-breaking ships and crude oil in large tankers, many of the latter without ice classification (reinforced hulls), are surging to Chinese ports via Russia's Northern Sea Route and Bering Strait. This trend could be long-lasting. The US Geological Survey estimates that 13 percent of the world's undiscovered conventional oil, and approximately a third of the world's undiscovered natural gas, is located above the Arctic Circle. Of the sixty largest oil and natural gas fields discovered in the Arctic, forty-three are in Russia, eleven in Canada, six in Alaska, and one in Norway.

To extend the post-Cook fur trade analogy, natural resource wealth now accessible in the Arctic has similarly heightened international tensions. Shelagh Grant writes that climate change is "the key driver in a frantic race to claim ownership of the Arctic's mineral-rich seabed and freedom of passage through northern sea routes." Sovereignty issues in the north were an arcane fascination for diplomats in the two centuries after Cook because the Arctic ice cover restricted commercial development, including shipping lanes, but "rapid ice melt changed all that."[4]

To cite some examples since Grant's book was published in 2010: in 2020, a state-run Chinese company attempted to purchase a Canadian gold mine near Hope Bay in northern Nunavut, 120 miles north of the Arctic Circle. The government of Prime Minister Justin Trudeau was rattled by this quest for mineral resources so deep in this sensitive region. Using a provision in law that triggers a review of any prospective foreign state-owned enterprises within Canada, the government vetoed the deal because the mine, located on tidewater adjacent to Bathurst Inlet and the main southern channel of the Northwest Passage, would give China too much ready access to the Arctic. This was of a piece from an episode in 2018, when Finland blocked an attempt by China to buy an old air base in Lapland, allegedly for scientific research. Similarly, Denmark stopped a Chinese plan to build three airports in Greenland, its semi-autonomous territory. Such are the diplomatic suspicions that are now endemic to the basin.

To further its sovereign aims in the Arctic, Russia has developed the world's largest fleet of major icebreakers (thirty-six), many of them nuclear-powered and capable of cutting through ice ten feet thick. These craft escort commercial traffic through Russia's 3,400-mile Northern Sea Route, which sees an ever-increasing number of ships and tonnage. (In June 2021, the Russian government reported that traffic was running 11 percent ahead of the record 1,014 trips conveying 33 million tons of cargo, oil, and liquefied natural gas the previous year.) Some of these icebreakers have the potential of being armed with cruise missiles, reflecting Russia's desire to press its strategic advantage in a region it sees not only as vital to its economic interests but as one that has the prospect of putting it astride the shortest trade route between the Eastern and Western Hemispheres, the centuries-old dream.

China has committed to building an icebreaker fleet of its own, even though its northernmost point is 900 miles from the Arctic Circle. Currently it has four, including two research ships that can navigate in the icy latitudes. (American officials suspect the latter have strategic purposes though China maintains its interest lies solely with the quest for scientific information.) In some respects, this is a calculated backup plan for China, because of the increasingly congested Strait of Malacca between Malaysia and Indonesia or more pointedly, in case of any disruption of its trade routes in the South China Sea by the United States and its allies in the region: Australia, the United Kingdom, and India.

Canada has been debating, seemingly without end, whether to build an Arctic ice-breaking fleet. The United States sails a single heavy icebreaker into the Arctic, the Coast Guard's aging *Healy*, though in 2020 President Donald Trump ordered the procurement of several more. (The equally venerable *Polar Star*, nearing the end of its serviceability, is assigned to the Antarctic. The home base for both is Seattle.) Given the lead time necessary to build such ships (delivery on the first new American heavy ice-breaker since 1976 has slipped to 2028), in November 2022 the administration of President Joseph R. Biden committed to the purchase of the *Aiviq* icebreaker for the Coast Guard's use from a private firm, Edison Chouest Offshore. To buy and equip this ship (which is slated to operate out of Juneau, Alaska) will cost an estimated $125 million to $150 million. Otherwise, the United States relies on satellites, aircraft, and its submarine fleet to patrol access to Arctic waters.

The Northwest Coast fur trade of the late eighteenth century unexpectedly altered the geopolitical importance of a region, nearly to the point of hostilities between Great Britain and Spain over Nootka Sound, and today climate change is likewise affecting international stability in the Arctic. Indeed, militarization of this "New North" is already well underway as a result of ice dissipation. Vladimir Putin, president of the Russian Federation, welcomes global warming as a strategic benefit, a view that, one might say, is the polar opposite of conventional opinion in North America, where fighting the circumstances accelerating climate change has taken on the trappings of a moral crusade. As a function of this new priority, Russia is reopening old Soviet naval stations and air bases across the top of Siberia, and developing new ones. In June 2021, Russia created a new military district out of this pool of assets under the command of its Northern Fleet, the country's most formidable naval force, headquartered at Kola Bay near Murmansk. Russian special forces have increased their presence in the Arctic by 30 percent in the last few years.

Correspondingly, the United States is expanding its presence in the Arctic, including the development of plans in 2019 for reopening a naval station in Adak, at the western end of the Aleutian chain, where a base was first built during the Second World War. In 2020, the US's Alaska command detected more Russian military aircraft in the American Air Defense Identification Zone than at any time since the heyday of the Cold War. In the fall of 2022, the US Coast Guard, out of its base at Dutch Harbor in the Aleutian Islands, detected a seven-ship Russian and Chinese fleet, including the latter's newest class of cruiser-destroyers, on a joint exercise. Warned by radio that they had entered US regulated waters in the Bering Sea, the ships left. That same year, the US Department of Defense announced that it was replacing its heavy armoured Stryker Brigade in Alaska with a more mobile infantry unit better suited for combat in the Arctic zone. The rationale, which had been developed even before the Russian invasion of Ukraine, was that the large Strykers were fundamentally ill-suited for potential combat in the ice and snow, whereas an airborne division aided by lighter vehicles would be more useful.

It was this kind of strategic thought that prompted the United States to foil a plan China had to expand its influence in Greenland, a territory of Denmark's that won the right to limited self-rule in 2008. To fill that vacuum in Greenland's development, while at the same time countering the prospect of Russian and Chinese influence in the Arctic, the United States granted the Danish protectorate a $12.1 billion aid package in 2020. With Denmark's approval, Greenland had opened a diplomatic office in Washington, DC, in 2014. Within the context of growing international rivalry in the Far North, the United States finally reciprocated by opening a consulate in Nuuk, the capital of Greenland, over the winter of 2020–21.

The Arctic does not share Antarctica's jurisdictional status, and none of the several northern nations with contested sovereignty claims – Russia, Norway, Denmark, Canada, and the United States – has renounced territorial prerogatives regarding the definition of international waters or rights to the polar seabed. China has declared itself a "near-Arctic state" due to its emerging interests in the "New North," which it styles as a "Polar Silk Road," part of Premier Xi Jinping's larger Belt and Road initiative. Given its economic heft, China has recently gained "observer status" with the Arctic Council, which also includes Finland, Iceland, and Sweden. (Russia had once opposed China's membership but the war in Ukraine changed its calculus.) In August 2022, the Biden administration announced that it was going to upgrade American engagement with this body by appointing an ambassador-at-large for Arctic affairs. This position, an elevation from the previous Arctic coordinator, is of sufficient stature to require Senate confirmation.

In this fashion, global warming is facilitating an international pivot to the Far North. The modern surge in national security, environmental concerns, and economic development issues in the region because of climate change has taken many forms. Most recently, telecommunication companies in Finland, Japan, and the United States have formed a joint venture to lay fiber-optic cable through the Northwest Passage in Canada and above Alaska from Europe to Asia. The region has even seen the reintroduction of sovereignty rituals, of the kind last seen during the age of Cook. In 2007, Russia planted a flag on the Arctic floor. This kind of neo-imperial unilateral extension of state interests stands in stark contrast to the post-imperial multilateral governance system that seemed to be the dominant international paradigm after the end of the Cold War. The prospective supply of vital oil, gas, and minerals is a powerful incentive for such behaviour. The Arctic powers have overlapping and largely unresolved claims regarding how far their continental shelves – and by extension their resource rights – stretch toward the North Pole. Even Canada and Denmark have been sparring since the 1970s over the precise boundary between Greenland and Ellesmere Island.

Separately, Canada and the United States are negotiating over rights to a 6,250-square-mile area in the Beaufort Sea north of Alaska and Yukon Territory. In terms of state security, a part of the world that was once largely inaccessible is now vulnerable, Grant says, to "smuggling, poaching, terrorist activities," or even invasion.[5] This has led to the odd situation wherein the two countries, though continuing to disagree about the international waters status of the Northwest Passage (addressed below), committed in 2021 to a strengthened joint defence posture in the Arctic. This strategy includes an upgrade of the old NORAD radar facilities ringing the Arctic that were previously considered Cold War relics.

Similarly, even Sweden, prior to the Russian invasion of Ukraine in February 2022, became so alarmed by Russia's more assertive stance in the Arctic that it reinstituted the military draft and developed a joint regional defence plan with Finland. Heretofore, the biggest advocate in Scandinavia for enhanced security in the Arctic was Norway. Both Sweden and Finland were finally provoked into applying to join Norway as NATO members in May 2022, after the war in Ukraine broke out, but it is nonetheless true that global warming and the concomitant increase in air and naval activity in the Arctic basin laid the basis for this fundamental shift in outlook. Finland has significant icebreaker shipbuilding capability and Sweden an effective submarine fleet for use in polar defence. Relatedly, the Baltic Sea – Russia's shortest route to the Atlantic – has taken on a strategic importance not seen since the eighteenth century. In June 2022, there was speculation that NATO might hold a future meeting of the alliance north of the Arctic Circle, perhaps at Tromsø, to demonstrate the increased relevance of northern security.

In one last parallel to the post-Cook rush to the Pacific Northwest in the eighteenth and early nineteenth centuries, it is important to note that the rights and concerns of the Arctic's regional Indigenous inhabitants are often the last factor to be considered.[6] This trend may be changing, however. In April 2021, the pro-environment Inuit Ataqatigitt party in Greenland emerged victorious and formed a coalition government that cast doubt on the future of a rare-earths mine on the island's southern coast. The Shenghe Resources Holding Company of China holds the largest stake in Greenland Minerals, an Australian-based firm that secured the project's original licence. If allowed to operate the mine would be the largest source of rare-earth deposits outside China.

Whither the Far North?

The future of economic development in the Arctic is an increasingly dense topic, with a host of issues destined to capture the attention of the world for the balance of this century and likely beyond. It is outside the scope of this study to fully explicate these complexities except for the one story that occupies our attention: the effect of the dissipation of Arctic sea ice on the resurrection of the ancient dream of a Northwest Passage, and its corollary, the Northeast Passage, also known as the Northern Sea Route, atop Russia.

This is a stunning development to be living through. With the melting of the Arctic ice pack, we are witnessing an event within a human lifetime or two that is functionally equivalent to a geological occurrence that normally plays out over the course of many millions of years, such as the growth of a mountain range or the scouring of a river gorge.[7]

In terms of the practical adaptation to this circumstance, it seems as if the clock has been turned back to the last third of the eighteenth century and the era of James Cook and Alexander Dalrymple – that is, assessing prospects for reduced shipping distances between Europe and Asia is once again in vogue, with a test phase already well underway.[8] It began in 2012 when the Russian gas firm Gazprom chartered one of the few existing ice-class tankers, the *Ob River,* from the Greek shipper Dynagas, to transport deep-chilled liquefied natural gas from Norway's fields to Japan. The *Ob River's* month-long northern voyage shaved one-third off the route distance in three weeks less time than if it had followed a more conventional course through the Mediterranean Sea and the Suez Canal and around South Asia, without regard to savings in fuel costs. (The most common route of transport to and from East Asia and Europe, via the Suez Canal, is approximately 13,000 miles long. It is thought that a conventional northern passage would cut 5,000 miles and typically ten to fifteen days of travel off that run.) In 2013, the China Ocean Shipping Company (COSCO) conducted a successful trial run with the *Yong Sheng,* its only cargo ship with an icebreaking hull, and in 2015 announced plans for regular sailings through the Arctic to Europe during the warm-weather travel window. Only a decade and a half ago, this window opened in July and closed in November. In 2020, it was eight months long. The 2021 the shipping season started earlier than ever, in February, when the *Christophe de Margerie,* a natural gas carrier, sailed from China to Russia's Yamal Peninsula via the Northern Sea Route.

Among the diplomatic complications at issue in the north is the lack of agreement between Canada and the United States over how to categorize the several threads of the Northwest Passage that course through the Arctic Archipelago, the world's second-largest aggregation of insular massing (after Indonesia). The Canadian position is that these several straits, bays, gulfs, and sounds are internal territorial waters over which Canada has sole jurisdiction, arguing that surface and submarine vessels from other nations must ask its permission to pass. The long-standing position of the United States, as crystallized during the Cold War with the Soviet Union, is that these corridors are international waters through which the vessels of any nation may pass. Here we find yet another echo from the late eighteenth century. The American "freedom of the seas" argument brings to mind Great Britain's earlier view vis-à-vis Spain in the Nootka Sound controversy, reflecting the innate conceit of nations with the most powerful navy. The United States maintains that its position relative to the Northwest Passage is consistent with its stance on access through the Strait of Hormuz exiting the oil-rich Persian Gulf and Qiongzhou Strait between Hainan Island and mainland China.

The discussion between the United States and Canada over the Northwest Passage was purely an intramural dispute of very modest proportions until, as Shelagh Grant writes, rapidly melting ice in the passage offered "a shorter and safer route to Asia"; now "many western European countries and China ... support the American position." But this development comes with its own problems, equally disadvantageous to both the United States and Canada. Neither country would look favourably upon potential adversaries, such as Russia, sailing a warship through North America in the fashion that Daines Barrington once envisioned on behalf of Cook. Grant, citing Arctic sovereignty expert Donat Pharand, stipulates that "there is still insufficient use of the Northwest Passage to classify it as an international strait, but admits that should such a degree of use develop, Canada should take immediate steps to ensure it has the ability to protect the environment and general security of the area." The issue may become moot, however. Drawing on historian Jack Granatstein, Grant suggests that "the status of the Northwest Passage will become redundant if scientists are correct in estimating that within fifteen years the waters around the North Pole will be ice-free, thus opening a much shorter route across the Arctic Ocean." Scott Borgerson and Michael Byers have argued for a bilateral US/Canada treaty clarifying the status of the Northwest Passage using the model of the St. Lawrence Seaway project.[9]

Though the timeline for the emergence of an ice-free North Pole, at least seasonally, is in dispute, the fact remains that in such an eventuality the Northwest Passage will have evolved yet again, for the fifth time in modern (post-Enlightenment) history. However, an Arctic Ocean without ice would transcend the historical definition of the Northwest Passage as consisting of mere channels and corridors through North America, and become a "new Mediterranean" – the kind that George Vancouver once mocked. More broadly, if the Arctic Ocean ever becomes completely ice-free, it will be the first time that has occurred in at least 13 million years,[10] the climatic epoch during which the great apes – humans, chimpanzees, gorillas, and orangutans – first appeared on the evolutionary path.

Appendix

An Exploration Chronology for the Pacific Portal to the Northwest Passage

Note: The chronological sequence of maritime fur traders calling at Nootka Sound on the Northwest Coast or harbours in Alaska are enclosed in square brackets.

1768 August 26	James Cook sails from Plymouth on the *Endeavour*
1769 June	Cook observes the transit of Venus in Tahiti; William Wales is at the Hudson's Bay Company (HBC) post at Fort Churchill for the transit of Venus and meets Samuel Hearne
1771 July 17	Hearne reaches the mouth of the Coppermine River on the Arctic Ocean
1771 August 13	*Endeavour* voyage concludes
1772 January	Upon returning from the Arctic, Hearne crosses Great Slave Lake
1772 June 30	Hearne returns to Churchill
1772 July 14	Cook leaves England on the *Resolution* on a voyage toward the South Pole
1774	British Admiralty learns that Hearne has reached the Arctic
1774 March	John Ledyard sails to Great Britain and joins the Royal Navy

1775 July 30	Cook returns to England from second voyage
1776 July 1	Alexander Henry the elder discloses knowledge of a northwestern river connection to salt water
1776 July 12	Cook sails from England on Northwest Passage voyage
1778 March–September	Cook explores the Northwest Coast, the Alaska Coast, and the Arctic
1778 Summer	Peter Pond discovers Methye Portage, a new gateway to the Athabasca country
1778–79	Pond winters in the Athabasca country
1779 February 14	Cook is killed in Hawai'i
1779–80	Pond winters in the Athabasca country
1780 September 30	John Gore leads the ships of Cook's third voyage back to England
1780–81	Pond winters in the Athabasca country
1781 October	Ledyard sails back to America as sergeant of marines in the Royal Navy
1781 October 18	Pond's news of a new transcontinental route reaches Montreal, then Joseph Banks
1782	Ledyard deserts the Royal Navy
1782–83 Winter/Spring	Ledyard writes his journal in Hartford
1783	Pond's success leads to a formal North West Company partnership with headquarters in Montreal
1783 Spring	Ledyard starts commercial canvas in New York City, New London, and Philadelphia for a trading voyage to the Northwest Coast
1783 Summer	Ledyard's journal is published
1783 September	Thomas Jefferson assesses George Rogers Clark's interest in conducting a trans-continental expedition
1783–84 Winter	Pond winters on the Athabasca River; queries Natives about waters flowing to the northwest and receives first intimation of Great Slave Lake

1784	Russians reach Kodiak Island; Pond leaves the West and arrives at Grand Portage
1784 February	Ledyard visits Ezra Stiles at Yale University in New Haven
1784 June	Cook's third-voyage account is published
1784 Summer	Ledyard sails for Europe
1784 August	Thomas Jefferson arrives in Paris
1784–85 Winter	Pond is in Montreal and the United States
1785 March 1	Pond presents his map to the Continental Congress in New York
1785 April 15	James Hanna sails for the Northwest Coast from Macao
1785 April 18	Pond meets with Henry Hamilton, lieutenant governor of Canada, and shows his second map, with hints of "Cook's River" added
1785 Spring	Ledyard arrives in Paris; Pond leaves Quebec for the Athabasca country intending to reach Russia
1785 Summer	Ledyard meets Jefferson in Paris
1785 August	Nathaniel Portlock and George Dixon sail to the Northwest Coast from England for Etches and Company
1785 August 18	Hanna becomes the first to arrive at Nootka in the *Sea Otter* [1]
1785 December	East India Company merchant James Strange sails from Bombay as supercargo for an expedition to the Northwest Coast consisting of the *Captain Cook* (Henry Laurie) and the *Experiment* (Henry Guise)
1786 February	Ledyard corresponds with Jefferson in Paris, and determines on Siberian sojourn (delayed); Jefferson describes Ledyard's intent to the Marquis de Lafayette
1786 March	John Meares sails from India
1786 June 25	Strange arrives at Nootka with the *Captain Cook* [2] and the *Experiment* [3]

1786 July	Portlock (*King George*) [4] and Dixon (*Queen Charlotte*) [5] arrive at Cook Inlet; Ledyard corresponds with Jefferson about magnetic variation
1786 August	Ledyard leaves for London hoping to join a fur trade venture to the Northwest Coast; Hanna arrives at Nootka for the second time [6], where he meets Strange and his two ships (first competitive meet-up)
1786 Summer	Meares arrives on the Alaskan Coast from Calcutta via the *Nootka* [7] and the *Sea Otter* [8] (Tipping)
1786 September 5	Strange has his second competitive meet-up when he encounters Meares in Cook Inlet
1786 December	Ledyard leaves London on Siberian quest
1786–87 Winter	Meares winters at Prince William Sound; Dixon and Portlock retreat to Hawai'i
1787	The North West Company is reorganized; Alexander Mackenzie is sent to Lake Athabasca as Pond's deputy
1787 March	Jefferson receives letter from Ledyard in St. Petersburg, who is about to venture into Siberia
1787 April 2	Empress Catherine accepts James Trevenen's proposal to explore Japan and China coasts, missed by Cook's successor (John Gore), to establish new maritime fur trade route via Alaska and the Northwest Coast
1787 May	Portlock and Dixon find Meares under duress in Prince William Sound
1787 June	Ledyard leaves St. Petersburg for Siberia; William Barkley in the *Imperial Eagle* [9] reaches Nootka, having sighted de Fuca's strait
1787 July	Pond prepares his third map, showing northern lake outflow directly to "Cook's River"
1787 August 8	Portlock and Dixon reach Nootka and find Charles Duncan (*Princess Royal*) [10] and James Colnett (*Prince of Wales*) [11]

1787 September	Ledyard arrives in Yakutsk, his furthest point; Jefferson tells others that Ledyard is on his way to Kamchatka
1787 October	Ledyard meets Cook veteran Joseph Billings in Yakutsk, and returns to Irkutsk
1787–88 Winter	Pond's last season in Athabasca, Mackenzie's first
1788 February	Ledyard is arrested in Irkutsk; Mackenzie determines to follow up on Pond's cartographic vision
1788 March	Colnett (*Prince of Wales*) [12] and Duncan (*Princess Royal*) [13] return to the Northwest Coast; Duncan pursues Barkley's lead about de Fuca's strait
1788 May	Pond leaves Athabasca permanently; Meares reaches Nootka (*Felice Adventurer*) [14] via Hawai'i, joined by William Douglas (*Iphigenia Nubiana*) [15]; Ledyard returns to London
1788 July	Etches proposes to Banks a coast survey voyage involving a complement of prisoners and soldiers
1788 August	Duncan explores and charts the western end of de Fuca's strait (published in 1790)
1788 Summer	Roderic Mackenzie builds Fort Chipewyan
1788 September	Meares launches the *North West America* at Nootka
1788 September 16	Robert Gray (*Lady Washington*) [16] and John Kendrick (*Columbia Rediviva*) (September 23) [17] arrive at Nootka; they find Meares, who leaves Robert Funter in command of the station and the *North West America*
1789 February 15	Mackenzie tells North West Company partners that Fort Chipewyan will become his "principal post"
1789 March 18	Alexander Dalrymple publishes his seminal essay, *Plan for Promoting the Fur-Trade,* on the need for a new official expedition to the Northwest Coast to consolidate British fur trade interests

1789 June 9	The first English ships are seized by Esteban José Martínez at Nootka Sound
1789 June 21	José María Narváez commences the first Spanish exploration of de Fuca's strait, and returns to Nootka (July 5) with a report of open water to the southeast
1789 Summer	George Vancouver returns to Great Britain after a five-year duty in the Caribbean; Colnett's ship is seized by Spain at Nootka
1789 July	Portlock and Dixon write Evan Nepean, British undersecretary in charge of colonial affairs quoting Dalrymple's *Plan* on the need for British forts and settlement because of Russian moves to "Cook's River" and Prince William Sound
1789 July 14	Mackenzie reaches the "Northern Ocean"
1789 August	Duncan arrives in England with a chart of de Fuca's strait
1789 August 10	Mackenzie attempts to climb a mountain near Camsell Bend, looking for westward-flowing river spoken of by Natives
1789 September	Mackenzie returns to Fort Chipewyan; Jefferson leaves Paris to become US secretary of state
1789 November	Pond shows his latest map to Isaac Ogden in Quebec City
1789 November 4	Dr. John Nooth writes to Banks about Pond's return from the West
1789 November 7	Isaac Ogden writes to David Ogden in England about Pond's travels
1789–90 Winter	Mackenzie winters at Fort Chipewyan
	Washington administration considers a covert exploratory mission to the trans-Mississippi west.
1789 December 19	The *Discovery* is launched with Henry Roberts as captain and George Vancouver as first lieutenant, for an expedition to West Africa

1790 January	David Ogden receives Isaac Ogden's letter
	David Ogden presents Pond's information to Nepean; Roberts and Vancouver board the *Discovery* for a South Atlantic voyage; Dalrymple publishes Duncan's chart of de Fuca's strait
1790 January 23	Nepean receives the Ogden correspondence
1790 Late January	First news of the Colnett/Martínez fracas at Nootka reaches England
1790 February	John Frederick Holland tenders plan for cross-country expedition to the mouth of "Cook's River"
1790 March	Pond's fourth and final map appears in *Gentleman's Magazine* showing Pacific and Arctic outlets; Pond retires to Connecticut
1790 March 15	William Bligh returns to London having survived a mutiny on the *Bounty*
1790 Late March	Roberts is notified of new "more distant" mission; station in New South Wales is notified that the *Discovery* is to head to the Northwest Coast with a companion ship to establish a fort to sustain the fur trade and redress Colnett's losses, to establish a possible penal colony, and to survey the coast from latitude 51° N to "Cook's River" (old Etches/Banks plan of July 1788)
1790 April	Meares arrives in London with more detail of Nootka fracas
1790 April 1	The first Aaron Arrowsmith map is published, showing two outlets from Great Slave Lake (per Pond's fourth map)
1790 April 30	Meares publishes a memorial demanding redress for the Spanish action at Nootka
1790 May	Vancouver oversees stand-down of the *Discovery* as a function of the Spanish Armament, after which he is posted to the *Courageous,* a ship of the line, for maneuvers through the fall
1790 May 13	Meares's memorial is presented to Parliament
1790 June	Bligh's account of the mutiny is published

1790 June 24	Mackenzie meets HBC's Philip Turnor at Cumberland House
1790 July 13	Mackenzie reaches Grand Portage; the voyage is ignored by North West Company partners.
1790 Late August	Mackenzie meets Turnor again at Cumberland on his way back to the Athabasca country
1790 October 28	The Nootka Convention is signed
1790 November	The legend of Ferrer Maldonado is resurrected by Jean-Nicolas Buache de la Neuville, affecting the course of Alejandro Malaspina's voyage to the Pacific for Spain
1790 November 10	Holland arrives in Quebec, receives/records news of Mackenzie, and cancels his overland expedition
1790 November 17	Plans for a voyage by the *Discovery* and the *Chatham* to "remote parts" resume
1790 November 20	Vancouver is called to a meeting at the Admiralty
1790 November/ December	Meares's account is published
1790 December	The first sketchy news of Mackenzie's voyage to the north reaches London
1790 December 13	Vancouver is named captain of the *Discovery*; Roberts is reassigned to the original Africa mission
1790 December 16	Vancouver assumes command
1790 December 18	Holland's letter with (sketchy) news of Mackenzie's first voyage and possible new "western river" opposite the Mackenzie River is received in London
1791 February 11	The draft of Vancouver's instructions is reviewed, reflecting the Pond/Mackenzie river agenda
1791 March 2	Alexander Mackenzie writes Roderic Mackenzie at Great Slave Lake post from Fort Chipewyan asking him to inquire of Natives about the new "Great River" west of his "River Disappointment" that reached the Arctic
1791 March 8	Vancouver receives final instructions

1791 April 1	Vancouver departs
1791 June 1	Mackenzie meets Turnor at Grand Portage; Turnor heads to the Athabasca country to "settle some dubious points" about Hearne/Pond
1791 July 26	Turnor writes HBC officials about Mackenzie's voyage "to the Sea"
1791–92 Winter	Mackenzie winters in England and learns of the Meares/Vancouver voyages
1792 April	Vancouver begins his survey of the Northwest Coast
1792 May 10	Vancouver detects Puget Sound
1792 May 11	Gray crosses the bar of the Columbia River
1793 April	Jefferson, for the American Philosophical Society, inscribes Michaux's instructions for a western expedition
1793 May 8	Mackenzie sets out on his second expedition, "to traffick with the Russians"
1793 May 31	Mackenzie changes course and takes the southern fork of the Peace River
1793 June 10	Vancouver's report from Nootka, via Zachary Mudge, arrives at the Admiralty
1793 August	News of Gray's discovery of the Columbia River reaches the Atlantic seaboard
1794 April 12	Vancouver sails into Cook Inlet
1794 May 6	Vancouver sees the termination of Cook Inlet
1794 August 17	Vancouver concludes his survey of the Northwest Coast
1794 September 8	Vancouver writes the Admiralty that he was "unable to penetrate beyond the barrier"
1794 September 10	From Toronto, Mackenzie reports to J.G. Simcoe about "Tacoutch Tesse," which, because of its southerly course, he takes to be a branch if not the whole of the "River of the West"
1794 November 17	Mackenzie tells Lord Dorchester, governor of Canada, that the 1789 voyage had followed Pond's vision of "Cook's River" but instead led him to the "Northern Ocean"

1795 February	News of Mackenzie reaching the Pacific is published in New York
1795 September	Vancouver returns to England
1798 August/September	The account of Vancouver's voyage appears in print
1799	The *Annual Register* concludes "that it is by no means yet proved, that a N.W. passage does not exist"
1801 December	Mackenzie publishes *Voyages from Montreal,* considers the classic saltwater form of the Northwest Passage "settled"
1802 January	Jefferson learns of the publication of Mackenzie's account
1802 October	Spain suspends right of American "deposit" in New Orleans
1803 January	Jefferson sends message to Congress calling for an official expedition to the Pacific
1803 March	Jefferson drafts instructions for Meriwether Lewis and the search for the Northwest Passage
1803 April	France offers to sell all of Louisiana to the US
1803 July	News of the Louisiana Purchase agreement reaches Washington, DC
1803 October	Treaty approving the Louisiana Purchase adopted by the US Senate
1803 December	US officials, including Lewis, take possession of Louisiana
1804 May	Lewis and William Clark venture west up the Missouri from St. Louis
1806 September	Lewis and Clark complete the voyage to the Pacific with their return to St. Louis, profess to have found a practicable Northwest Passage
1814	Nicholas Biddle's *History of the Expedition under the Command of Captains Lewis and Clark* appears
1842	Charles Wilkes explores the Columbia River and Puget Sound
1842–43	John C. Frémont explores the Oregon Trail and Willamette Valley

1845	Asa Whitney proposes the creation of a land grant program to underwrite the construction of a transcontinental railroad from Lake Michigan to Oregon
1846 June	The Treaty of Washington ends joint British and American "occupancy" of Oregon and divides the territory at the 49th parallel
1846 September	Sir John Franklin is marooned in the Arctic ice
1849	Whitney publishes *A Project for a Railroad to the Pacific*
1850 June	Richard Collinson and Robert McClure leave Plymouth, England, in a search for Franklin using Cook's old approach to the Arctic from the Pacific
1850 October	McClure is locked in the ice between Banks and Victoria Islands
1851	In Canada, Joseph Howe echoes Whitney with a proposal for an all-Canadian transcontinental railroad
1851 September	McClure's ship is permanently locked in the ice in Mercy Bay, off Banks Island, where he remains for two years
1852 April	McClure reaches Winter Harbour by sledge, linking his Pacific voyage to the westernmost extent of an Atlantic-based exploration
1853 March 2	The US Congress passes the Pacific Railroad Survey act
1853 April 6	McClure is rescued, avoiding Franklin's fate
1853 April 8	Isaac Stevens is appointed to head of the northern Pacific railroad survey
1853 June 6	Stevens departs Minnesota for the Washington Territory
1853 August	The first evidence of Franklin's fate is detected
1853 October 18	Stevens and George McClellan meet on the Columbia Plateau
1853 November 26	Stevens reaches Puget Sound, completing his survey

1854 October 23	News of Franklin's fate reaches England
1855 November	The US War Department endorses the southern railroad corridor
1855–58	Congress issues the first eleven volumes of the *Pacific Railroad Reports*
1857–60	Robert Palliser surveys the Saskatchewan River Basin
1860 May	Stevens tenders his final report (volume twelve of the *Pacific Railroad Reports*) to the War Department
1863 July 5	Sandford Fleming proposes an all-Canadian overland passage to the Pacific
1867 July 1	Canadian Confederation
1871	British Columbia joins Confederation; the Canadian government promises a transcontinental rail link within ten years
1872 April	Fleming is appointed chief engineer of the Canadian Pacific Railway; he commences his survey of the proposed route westward later that year
1872 June	Northern Pacific Railway reaches Bismarck, North Dakota
1873 July 14	Tacoma is designated as the western terminus for the Northern Pacific
1875 June	The first segment of the Canadian Pacific Railway (CPR) is completed near Thunder Bay, Ontario
1881 November 1	William Cornelius Van Horne succeeds Fleming as chief engineer for CPR construction
1882	The CPR reaches Calgary
1883 September 8	The Northern Pacific "golden spike" ceremony is held at Gold Creek, Montana
1883 November 18	Railroad Standard Time is adopted by North American railroads
1885	Van Horne establishes Vancouver as CPR terminus, replacing Port Moody

1885 January 1	Greenwich is adopted as the prime meridian for calculation of Universal Standard Time
1885 May 16	The Lake Superior section of the CPR is finished
1885 November 6	The CPR "golden spike" ceremony is held at Craigellachie, BC
1888 May 27	The Northern Pacific's Stampede Pass shortcut to Puget Sound is completed
1893	The Great Northern Railroad is completed to Seattle
1903 June 16	Roald Amundsen sails from Norway
1905 August 17	Amundsen completes the first transit of the Northwest Passage

Notes

PROLOGUE | The Remarkable Tale of John Ledyard's Attempt to Copy Cook

1 James Munford, ed., *John Ledyard's Journal of Captain Cook's Last Voyage* (Corvallis: Oregon State University Press, 1963), xxiv.
2 J.C. Beaglehole, ed., *The Journals of Captain James Cook on His Voyages of Discovery,* 3 vols. (Cambridge: Cambridge University Press/Hakluyt Society, 1955–67), 3: 1564–65; Munford, *Ledyard's Journal,* xxxii.
3 Munford, *Ledyard's Journal,* xxxi. I introduced the concept of the palm-tree paradigm as the controlling idea for the study of Cook in *Captain Cook Rediscovered: Voyaging to the Icy Latitudes* (Vancouver: UBC Press, 2020).
4 Munford, *Ledyard's Journal,* 70.
5 Ibid., xlvi, xlviii.
6 Ibid., xlvii.
7 Ibid., xxxii–xxxiii.
8 For Jefferson's quotation, see Lee Alan Dugatin, "André Michaux's 'Almost Expedition' to the West in 1793," *We Proceeded On* 47, 1 (February 2021): 5, 11; Munford, *Ledyard's Journal,* xxxiii.
9 James Zug, *American Traveler: The Life and Adventures of John Ledyard* (New York: Basic Books, 2005), 167.
10 James Zug, ed., *The Last Voyage of Captain Cook: The Collected Writings of John Ledyard* (Washington, DC: National Geographic Society, 2005), 143; Howard T. Fry, *Alexander Dalrymple (1737–1808) and the Expansion of British Trade* (Toronto: University of Toronto Press, 1970), 199.
11 Zug, *Last Voyage of Captain Cook,* 145–46.
12 Ibid., 151, 153, 160, 167.
13 Ibid., 163.
14 Ibid., 167–68, 171.
15 Ibid., 175.
16 Ibid., 176, 191.
17 Ibid., 204.

18 Ibid., 233, 236.
19 James Gibson, "A Notable Absence: The Lateness and Lameness of Russian Discovery and Exploration in the North Pacific, 1639–1803," in *From Maps to Metaphors,* ed. Robin Fisher and Hugh Johnston Vancouver: UBC Press, 1993), 103.
20 James Trevenen, *A Memoir by James Trevenen,* ed. Christopher Lloyd and R.C. Anderson (London: Navy Records Society, 1959), 93, 96.
21 Kenneth Owens, with Alexander Yu. Petrov, *Empire Maker: Aleksandr Baranov and Russian Colonial Expansion into Alaska and Northern California* (Seattle: University of Washington Press, 2015), 42–44, 46.
22 Zug, *Last Voyage of Captain Cook,* 224.
23 Ibid.
24 Ibid., 225.
25 Ibid., xvii, 241, 243; Zug, *American Traveler,* 217, 219, 220.
26 Zug, *Last Voyage of Captain Cook,* 244.
27 Ibid., 247.
28 Ibid., 250.

CHAPTER 1 | Reinventing the Northwest Passage

1 Johann Reinhold Forster, *History of Discoveries Made in the North* (Dublin: Luke White and Pat. Burn, 1786), xvi, 451, 455, 462-63; Howard T. Fry, *Alexander Dalrymple (1737–1808) and the Expansion of British Trade* (Toronto: University of Toronto Press, 1970), 207.
2 Glyn Williams, *Arctic Labyrinth: The Quest for the Northwest Passage* (London: Penguin, 2009), 151.
3 David L. Nicandri, *Captain Cook Rediscovered: Voyaging to the Icy Latitudes* (Vancouver: UBC Press, 2020), 9–10, 70, 88, 115, 163, 198, 204, 208, 273, 281, 304, 321, 356; Barry Gough, *Juan de Fuca's Strait: Voyages in the Waterway of Forgotten Dreams* (Madeira Park, BC: Harbour, 2012), 85–86.
4 Richard A. Van Orman, *The Explorers: Nineteenth-Century Expeditions in Africa and the American West* (Albuquerque: University of New Mexico Press, 1984), 34, 48.
5 Felipe Fernández-Armesto, *Pathfinders: A Global History of Exploration* (New York: Norton, 2006), 310.
6 Barry M. Gough, *The Northwest Coast: British Navigation, Trade, and Discoveries to 1812* (Vancouver: UBC Press, 1992), 70, 77; Peter Stark, *Astoria: John Jacob Astor and Thomas Jefferson's Lost Pacific Empire* (New York: HarperCollins, 2014), 71.
7 William Beresford, *A Voyage Round the World… to the North-West Coast of America: Performed in 1785, 1786, 1787 and 1788, in the* King George *and* Queen Charlotte, *Captains Portlock and Dixon* (London: Geo. Goulding, 1789), xvi, 95, 262, 315.
8 John Meares, *Voyages Made in the Years 1788 and 1789, from China to the North West Coast of America* (London: Logographic Press, 1790), xxiii.
9 Gough, *Northwest Coast,* 77.
10 Beresford, *A Voyage Round the World,* 216, 236.
11 Gough, *Northwest Coast,* 89.
12 Ibid., 94.
13 Derek Hayes, *First Crossing: Alexander Mackenzie, His Expedition across North America, and the Opening of the Continent* (Seattle: Sasquatch, 2001), 35.
14 Barry Gough, *First across the Continent: Sir Alexander Mackenzie* (Norman: University of Oklahoma Press, 1997), 62; David Chapin, *Freshwater Passages: The Trade and Travels of Peter Pond* (Lincoln: University of Nebraska Press, 2014), 163; Barry Gough, *The Elusive Mr. Pond:*

The Fur Trader and Explorer Who Opened the Northwest (Madeira Park, BC: Douglas and McIntyre, 2014), 119.

15 Chapin, *Freshwater Passages,* 207–8; Hayes, *First Crossing,* map, 60.

16 Chapin, *Freshwater Passages,* 232; Hayes, *First Crossing,* map, 60.

17 Chapin, *Freshwater Passages,* 244.

18 Pond's second map is found in Hayes, *First Crossing,* 56–57; Gough, *The Elusive Mr. Pond,* 123.

19 Hayes, *First Crossing,* 58.

20 Chapin, *Freshwater Passages,* 245.

21 Hayes, *First Crossing,* map, 48–49; *Chapin, Freshwater Passages,* 279.

22 Gough, *First across the Continent,* 74, 105; Chapin, *Freshwater Passages,* 288.

23 Chapin, *Freshwater Passages,* 278; Gough, *The Elusive Mr. Pond,* 134; W. Kaye Lamb, ed., *The Letters and Journals of Sir Alexander Mackenzie* (London: Cambridge University Press, 1970), 13; Richard Dillon, "Peter Pond and the Overland Route to Cook's Inlet," *Pacific Northwest Quarterly* 42, 4 (October 1951): 326–27.

24 Dillon, "Peter Pond," 327–28; Lamb, *Letters and Journals,* 13. Isaac Ogden's letter was eventually published in *Gentlemen's Magazine* and later reprinted in *New-York Magazine.* James Madison brought this account of Pond's travels to the attention of Thomas Jefferson. Chapin, *Freshwater Passages,* 295, 340 n. 15.

25 Gough, *Northwest Coast,* 178; Chapin, *Freshwater Passages,* 278; Gough, *The Elusive Mr. Pond,* 130–31; Dillon, "Peter Pond," 329.

26 Dillon, "Peter Pond," 328; Chapin, *Freshwater Passages,* 278. Ironically, Alexander Mackenzie descended what would later be called the Bentinck Arm of Dean Channel when he reached Pacific tidewater in 1793. "Jesuits Harbor" is a mysterious place name that was probably a garbled version of Three Saints Bay on Kodiak Island, the first permanent Russian colony in North America, founded in 1784.

27 Chapin, *Freshwater Passages,* 282; Dillon, "Peter Pond," 327–28.

28 Gough, *Northwest Coast,* 180.

29 Glyn Williams, "Myth and Reality: The Theoretical Geography of Northwest America from Cook to Vancouver," in *From Maps to Metaphors: The Pacific World of George Vancouver,* ed. Robin Fisher and Hugh Johnston (Vancouver: UBC Press, 1993), 37.

CHAPTER 2 | Dalrymple's Revenge?

1 Glyn Williams, *Voyages of Delusion: The Search for the Northwest Passage in the Age of Reason* (London: HarperCollins, 2002), 357.

2 Howard T. Fry, "Alexander Dalrymple and Captain James Cook: The Creative Interplay of Two Careers," in *Captain Cook and His Times,* ed. Robin Fisher and Hugh Johnston (Seattle: University of Washington Press, 1979), 54.

3 Alexander Dalrymple, *Plan for Promoting the Fur-Trade, and Securing It to This Country, by Uniting the Operations of The East-India and Hudson's-Bay Companys* (London: George Bigg, 1789), iv.

4 Fry, "Alexander Dalrymple and Captain Cook," 53; Howard T. Fry, *Alexander Dalrymple (1737–1808) and the Expansion of British Trade* (Toronto: University of Toronto Press, 1970), 145, 192.

5 Barbara Belyea, "Just the Facts: Vancouver and Theoretical Geography," paper presented at the Vancouver Conference on Exploration and Discovery, Simon Fraser University, Burnaby, BC, April 23–26, 1992, 8.

6 Dalrymple, *Plan,* 1–2 (emphasis original).

7 Ibid., 5–6 (emphasis original).

8 Ibid., 6 (emphasis original).
9 Gordon Speck, *Samuel Hearne and the Northwest Passage* (Caldwell, ID: Caxton, 1963), 64.
10 Samuel Hearne, *A Journey from Prince of Wales Fort in Hudson's Bay to the Northern Ocean,* ed. and intro. Richard Glover (Toronto: Macmillan, 1958), xlix, 195; Derek Hayes, *First Crossing: Alexander Mackenzie, His Expedition across North America, and the Opening of the Continent* (Seattle: Sasquatch, 2001), 32.
11 Dalrymple, *Plan,* 6–7 (emphasis original).
12 Ibid., 8 (emphasis original).
13 Ibid., 1, 9 (emphasis original).
14 George Vancouver, *A Voyage of Discovery to the North Pacific Ocean and Round the World, 1791–95,* ed. W. Kaye Lamb, 4 vols. (Cambridge: Cambridge University Press/Hakluyt Society, 1984), 3: 1554–55.
15 Dalrymple, *Plan,* 14, 16–17.
16 Ibid., 17–18, 20.
17 Barry Gough, *Juan de Fuca's Strait: Voyages in the Waterway of Forgotten Dreams* (Madeira Park, BC: Harbour, 2012), 63.
18 Speck, *Samuel Hearne,* 213.
19 Dalrymple, *Plan,* 22.
20 Ibid., 22–25 (emphasis original).
21 Ibid., 25–26 (emphasis original).
22 Ibid., 26, 28, 30.
23 Ibid., 2, 30–32.

CHAPTER 3 | Mackenzie Solves Pond's Puzzle

1 Derek Hayes, *First Crossing: Alexander Mackenzie, His Expedition across North America, and the Opening of the Continent* (Seattle: Sasquatch, 2001), 65–66.
2 Barry Gough, *The Elusive Mr. Pond: The Fur Trader and Explorer Who Opened the Northwest* (Madeira Park, BC: Douglas and McIntyre, 2014), 143; Hayes, *First Crossing,* 67, 73. Dalrymple, in another of his many pamphlets from this period, and no doubt informed by Pond's dalliance with the American Congress, questioned whether he was "attached to this country" and wondered whether "encouraging him we may be fostering a viper in our bosom." David Chapin, *Freshwater Passages: The Trade and Travels of Peter Pond* (Lincoln: University of Nebraska Press, 2014), 301.
3 W. Kaye Lamb, ed., *The Letters and Journals of Sir Alexander Mackenzie* (London: Cambridge University Press, 1970), 437.
4 Ibid., 58, 529.
5 Ibid., 196–97.
6 Ibid., 212.
7 Ibid., 213. Two weeks later, near Camsell Bend, where the Mackenzie River separates from the Mackenzie Mountains that had directed his course north toward the Arctic instead of westward on what Pond had intimated was Cook's River, Mackenzie made a last-ditch effort to catch a glimpse of this new watershed that the visiting Natives had told him about. On August 10, 1789, he started out with two guides, but several hours into the trek through thick woods, his companions wanted to return to the Mackenzie River camp because their leggings and shoes were worn out. The Nor'Wester insisted they continue, but soon he too "found it impossible to proceed in a Straight Line & the Marsh extended as far as I could see." So the party made its way back to the river camp at the bend, arriving at midnight "much fatigued." Ibid., 222.
8 Hayes, *First Crossing,* 141.

9 Barry Gough, *First across the Continent: Sir Alexander Mackenzie* (Norman: University of Oklahoma Press, 1997), 66-67; Gough, *The Elusive Mr. Pond*, 170.

CHAPTER 4 | John Meares and the Nootka Affair

1 I am indebted, as in so many other ways, to Robin Inglis for sorting out the specific voyages and general timeline of the Russian-Spanish nexus in the North Pacific both pre- and post-Cook.
2 J. Richard Nokes, *Almost a Hero: The Voyages of John Meares, R.N., to China, Hawaii, and the Northwest Coast* (Pullman: Washington State University Press, 1998), 125.
3 W. Kaye Lamb and Tomás Bartroli, "James Hanna and John Henry Cox: The First Maritime Fur Trader and His Sponsor," *BC Studies* 84 (Winter 1989): 31.
4 Nokes, *Almost a Hero*, 145.
5 John Meares, *Voyages Made in the Years 1788 and 1789, from China to the North West Coast of America* (London: Logographic Press, 1790), opposite i.
6 Stephen R. Bown, *Madness, Betrayal and the Lash: The Epic Voyage of Captain George Vancouver* (Vancouver: Douglas and McIntyre, 2008), 35; Glyn Williams, *Arctic Labyrinth: The Quest for the Northwest Passage* (London: Penguin, 2009), 153; Barbara Belyea, "Just the Facts: Vancouver and Theoretical Geography," paper presented at the Vancouver Conference on Exploration and Discovery, Simon Fraser University, Burnaby, BC, April 23–26, 1992, 14.
7 Meares, *Voyages*, vi-vii.
8 Ibid., xliii.
9 Ibid., xliii–xlv.
10 Ibid., xlv–xlvii. Meares later contradicted himself on this point, asserting that "in no part of the year is ice found in Cook's River," a bewildering notion given his overwintering in adjoining Prince William Sound (lviii).
11 Ibid., xlvii–xlviii, lxiv.
12 Ibid., l–li.
13 Ibid., lvi-lvii, lxiii.
14 Ibid., lvii, lxii.
15 Jim McDowell, *José Narváez: The Forgotten Explorer* (Spokane: Arthur H. Clark Company, 1998), 39.
16 Ibid., 42.
17 Meares, *Voyages*, lvii–lix.
18 Ibid., lxix. For the Cook-Meares-Mackenzie connection, see David Nicandri, "The Rhyme of the Great Navigator: The Literature of Captain Cook and Its Influence on the Journals of Lewis and Clark. Part 1: A Canoe's Teeth," *We Proceeded On* 42, 1 (February 2016): 22–28.
19 Meares, *Voyages*, 156, 179.
20 Williams, *Arctic Labyrinth*, 154.
21 Alexander Dalrymple, *Spanish Pretensions Fairly Discussed* (London: George Bigg, 1790), 15–16.
22 Barry Gough, *Juan de Fuca's Strait: Voyages in the Waterway of Forgotten Dreams* (Madeira Park, BC: Harbour, 2012), 104; Belyea, "Just the Facts," 10.

CHAPTER 5 | The Complicated Origins of the Vancouver Expedition

1 W. Kaye Lamb, ed., *The Letters and Journals of Sir Alexander Mackenzie* (London: Cambridge University Press, 1970), 19; Robert J. King, "George Vancouver and the Contemplated Settlement at Nootka Sound," *The Great Circle: Journal of the Australian Association for Maritime History* 32, 1 (2010): 10; Robin Fisher and Hugh Johnston, eds., *From Maps to*

Metaphors: The Pacific World of George Vancouver (Vancouver: UBC Press, 1993), 7; Barbara Belyea, "Just the Facts: Vancouver and Theoretical Geography," paper presented at the Vancouver Conference on Exploration and Discovery, Simon Fraser University, Burnaby, BC, April 23–26, 1992, 2; Barry Gough, *Juan de Fuca's Strait: Voyages in the Waterway of Forgotten Dreams* (Madeira Park, BC: Harbour, 2012), 24; Freeman Tovell, trans. and ed., and Robin Inglis and Iris H.W. Engstrand, eds., *Voyage to the Northwest Coast of America, 1792: Juan Francisco de la Bodega y Quadra and the Nootka Sound Controversy* (Norman, OK: Arthur H. Clark Company, 2012), 82.

2 David L. Nicandri, *Captain Cook Rediscovered: Voyaging to the Icy Latitudes* (Vancouver: UBC Press, 2020); Gough, *Juan de Fuca's Strait*, 193.

3 Glyn Williams, *Arctic Labyrinth: The Quest for the Northwest Passage* (London: Penguin, 2009), 165; J.C. Beaglehole, ed., *The Journals of Captain James Cook on His Voyages of Discovery* (Cambridge: Cambridge University Press/Hakluyt Society, 1955–67), 1: cxxii.

4 Felipe Fernández-Armesto, *Pathfinders: A Global History of Exploration* (New York: Norton, 2006), 308.

5 King, "George Vancouver," 3-4.

6 Ibid., 4–5.

7 Ibid., 6–8.

8 Ibid., 8. For an exposition of the doctrine of discovery, see Robert J. Miller, *Native America, Discovered and Conquered: Thomas Jefferson, Lewis and Clark, and Manifest Destiny* (Westport, CT: Praeger, 2006), 9–24.

9 King, "George Vancouver," 10 (emphasis added).

10 George Vancouver, *A Voyage of Discovery to the North Pacific Ocean and Round the World, 1791–95*, ed. W. Kaye Lamb, 4 vols. (Cambridge: Cambridge University Press/Hakluyt Society, 1984), 1: 41, 283–84.

11 Ibid., 1: 284.

12 Ibid.

13 Derek Hayes, *First Crossing: Alexander Mackenzie, His Expedition across North America, and the Opening of the Continent* (Seattle: Sasquatch, 2001), 48–49, 56–57, 60, 65, 147.

14 Vancouver, *Voyage of Discovery*, 1: 284–85.

15 Ibid., 1: 41.

16 Hayes, *First Crossing*, 122–25.

17 Lamb, *Letters and Journals of Sir Alexander Mackenzie*, 163; Hayes, *First Crossing*, 128.

18 Ibid., 139, 142; Howard T. Fry, *Alexander Dalrymple (1737–1808) and the Expansion of British Trade* (Toronto: University of Toronto Press, 1970), 212.

19 Lamb, *Letters and Journals of Sir Alexander Mackenzie*, 453.

20 Hayes, *First Crossing*, 173.

21 Lamb, *Letters and Journals of Sir Alexander Mackenzie*, 455–56.

22 Vancouver, *Voyage of Discovery*, 1: 42.

23 Ibid., 1: 284.

CHAPTER 6 | Vancouver's Maze

1 George Vancouver, *A Voyage of Discovery to the North Pacific Ocean and Round the World, 1791–95*, ed. W. Kaye Lamb, 4 vols. (Cambridge: Cambridge University Press/Hakluyt Society, 1984), 1: 317.

2 Ibid., 2: 500.

3 Ibid., 2: 511.

4 Ibid., 2: 500–1, 503.

5 Ibid., 2: 511–12.
6 Ibid., 2: 525–26, 536, 540.
7 Ibid., 2: 541, 543.
8 Ibid., 2: 558, 561. Leaving Admiralty Inlet on his way north, Vancouver also briefly saw, but did not name, another mountain to the south of Rainier but distinct from St. Helens. This peak, later named Mount Adams by American explorers after the second president of the United States, is very elusive, and not just from today's western Washington. Later, when off the bar of the Columbia River, Vancouver could not distinguish it from St. Helens as they are orthogonal to each other from that vantage. Ibid., 2: 572.
9 Ibid., 2: 549–50, 563, 572–73.
10 Ibid., 2: 568-69.
11 Ibid., 2: 584.
12 Glyn Williams, *Arctic Labyrinth: The Quest for the Northwest Passage* (London: Penguin, 2009), 157–58, 160. Again, I express my appreciation to Robin Inglis for clarifying the chronology in this and adjacent paragraphs.
13 Vancouver, *Voyage of Discovery*, 2: 616; Williams, *Arctic Labyrinth*, 160. Vancouver eventually turned for Nootka, and the scheduled rendezvous with his diplomatic counterpart Bodega y Quadra, at Point Menzies near latitude 52°N.
14 Vancouver, *Voyage of Discovery*, 2: 628–29, 631.
15 Ibid., 2: 642.
16 Ibid., 2: 657.
17 Ibid., 2: 743.
18 Ibid., 2: 683.
19 Ibid., 2: 689, 692.
20 Ibid., 2: 692.
21 Ibid., 2: 693; J.C. Beaglehole, ed., *The Journals of Captain James Cook on His Voyages of Discovery*, 3 vols. (Cambridge: Cambridge University Press/Hakluyt Society, 1955–67), 1: 380.
22 Vancouver, *Voyage of Discovery*, 2: 694.
23 Ibid., 2: 747, 751.
24 Gary E. Moulton, ed., *The Journals of the Lewis and Clark Expedition*, 13 vols. (Lincoln: University of Nebraska Press, 1983–2001), 6: 33 (emphasis original).
25 Vancouver, *Voyage of Discovery*, 2: 757, 759.
26 Ibid., 2: 760.
27 Ibid; 762 n. 5 (emphasis original).
28 Ibid., 2: 761, 768.
29 Ibid., 2: 767.

CHAPTER 7 | Defining the Labyrinth, Cook's Legacy, and the Delusion of the Northwest Passage

1 George Vancouver, *A Voyage of Discovery to the North Pacific Ocean and Round the World, 1791–95*, ed. W. Kaye Lamb, 4 vols. (Cambridge: Cambridge University Press/Hakluyt Society, 1984), 3: 920, 926.
2 Ibid., 3: 933, 938, 954, 969, 972.
3 Ibid., 3: 990, 1005.
4 Ibid., 3: 1024, 1062–63 (emphasis original).
5 Ibid., 3: 1203.
6 Ibid., 3: 1203.
7 Ibid., 3: 1211, 1214.

8 Ibid., 3: 1223, 1227; 4: 1234.
9 Kenneth Owens, with Alexander Yu. Petrov, *Empire Maker: Aleksandr Baranov and Russian Colonial Expansion into Alaska and Northern California* (Seattle: University of Washington Press, 2015), 95.
10 Vancouver, *Voyage of Discovery*, 4: 1235.
11 Ibid., 3: 1228.
12 Ibid., 3: 1229-30.
13 Ibid., 4: 1238–39, 1242–43.
14 Ibid., 4: 1243 (emphasis original).
15 Robin Inglis, personal communication with author, May 6, 2016.
16 Vancouver, *Voyage of Discovery*, 4: 1244.
17 Again, I am indebted to Robin Inglis for this last insight.
18 Vancouver, *Voyage of Discovery*, 4: 1244.
19 Ibid., 4: 1260–61.
20 Ibid., 4: 1261, 1266, 1297 (emphasis original).
21 Ibid., 4: 1285.
22 Owens, *Empire Maker*, 57, 104–5.
23 Vancouver, *Voyage of Discovery*, 4: 1301-02, 1308.
24 Ibid., 4: 1297, 1312.
25 Owens, *Empire Maker*, 101, 302 n. 53. In July 1794, Vancouver met merchant-captain William Brown in Yakutat Bay and "from him we received the latest accounts of the state of Europe that had appeared in China before his sailing. These contained not only the melancholy intelligence of the death of Louis XVI and of the anarchy which existed in France, but likewise her declaration of war against England." Vancouver, *Voyage of Discovery*, 4: 1314. In this way, Vancouver's voyage mirrored Cook's third and final one, in that it took place against the backdrop of conflict with France and other countries.
26 Vancouver, *Voyage of Discovery*, 4: 1317, 1361.
27 Ibid., 4: 1322.
28 Vancouver, *Voyage of Discovery*, 4: 1319, 1346; Brian Fagan, The *Little Ice Age: How Climate Made History, 1300–1850* (New York: Basic Books, 2000), 126.
29 Vancouver, *Voyage of Discovery*, 4: 1348–50. The engraved chart of the world included in Cook's third-voyage atlas did not contain this lake because Cook's cartographer, Henry Roberts, lost a battle with Banks and Dalrymple over its inclusion. However, Cook lieutenant James King and Roberts immediately published their own chart later in the year, and it is this one that contains all interior lakes and waterways. See Andrew David, ed., *The Charts and Coastal Views of Captain Cook's Voyages*, vol. 3, *The Voyage of Resolution and Discovery, 1776–1780* (London, Hakluyt Society, 1997), charts 3.95A and 3.95B.
30 Vancouver, *Voyage of Discovery*, 4: 1350–51.
31 Ibid., 4: 1371.
32 Ibid., 4: 1375, 1381.
33 Ibid., 4: 1382.
34 Ibid., 4: 1382, 1390 (emphasis original).
35 Ibid., 4: 1391 (emphasis original).
36 Ibid., 4: 1598.
37 Ibid., 4: 1601.
38 Barry M. Gough, *The Northwest Coast: British Navigation, Trade, and Discoveries to 1812* (Vancouver: UBC Press, 1992), 160.
39 Vancouver, *Voyage of Discovery*, 4: 1627.
40 Barry Gough, *Juan de Fuca's Strait: Voyages in the Waterway of Forgotten Dreams* (Madeira Park, BC: Harbour, 2012), 211.

41 Vancouver, *Voyage of Discovery,* 1: 271 (emphasis original).
42 Ibid., 1: 274–75 (emphasis original).
43 Ibid., 1: 275 (emphasis original).
44 Ibid., 1: 275–78.
45 Ibid., 4: 1542 (emphasis original).
46 Glyn Williams, "Myth and Reality: The Theoretical Geography of Northwest America from Cook to Vancouver," in *From Maps to Metaphors: The Pacific World of George Vancouver,* ed. Robin Fisher and Hugh Johnston (Vancouver: UBC Press, 1993), 50.
47 Vancouver, *Voyage of Discovery,* 4: 1556.
48 Stephen R. Bown, *Madness, Betrayal and the Lash: The Epic Voyage of Captain George Vancouver* (Vancouver: Douglas and McIntyre, 2008), 225.
49 David Mackay, "The Burden of *Terra Australis:* Experiences of Real and Imagined Lands," in *From Maps to Metaphors: The Pacific World of George Vancouver,* ed. Robin Fisher and Hugh J.M. Johnston (Vancouver: UBC Press, 1993), 266–67.
50 J.C. Beaglehole, ed., *The Journals of Captain James Cook on His Voyages of Discovery,* 3 vols. (Cambridge: Cambridge University Press/Hakluyt Society, 1955–67), 1: cxx.
51 Gough, *The Northwest Coast,* 182.
52 Donald Jackson, ed., *Letters of the Lewis and Clark Expedition, with Related Documents: 1803–1854,* 2 vols. (Urbana: University of Illinois Press, 1978), 1: 320.
53 Glyn Williams, *Voyages of Delusion: The Search for the Northwest Passage in the Age of Reason* (London: HarperCollins, 2002), 408.
54 For these last insights, I am indebted to Robin Inglis plus James P. Delgado, "The Continuing Quest: The Lure of the Northwest Passage in the Nineteenth and Twentieth Centuries," and Lawson W. Brigham, "Marine Navigation in the Arctic Ocean and the Northwest Passage," in *Arctic Ambitions: Cook and the Northwest Passage,* ed. James K. Barnett and David L. Nicandri (Seattle: University of Washington Press, 2015), 328, 371.

CHAPTER 8 | The American Response, or, the Lewis and Clark Expedition as Diplomatic History

1 Donald Jackson, ed., *Letters of the Lewis and Clark Expedition, with Related Documents: 1803–1854,* 2 vols. (Urbana: University of Illinois Press, 1978), 2: 654–55.
2 Ibid., 2: 655–56. Clark was the oldest brother of the future explorer William Clark.
3 Ibid., 2: 588.
4 Ibid., 2: 588–89.
5 Ibid., 2: 661.
6 Ibid., 2: 661–62.
7 Ibid., 2: 662–63, 665.
8 Ibid., 2: 665–66.
9 Ibid., 2: 589.
10 Ibid., 2: 667–68.
11 Ibid., 2: 669–70.
12 Ibid., 2: 670.
13 Ibid., 2: 671.
14 Ibid., 2: 589, 670.
15 Henry Nash Smith, *Virgin Land: The American West as Symbol and Myth* (New York: Vintage Books, 1950), 21.
16 Donald Jackson, *Thomas Jefferson and the Rocky Mountains: Exploring the West from Monticello* (Norman: University of Oklahoma Press, 2002), 121.
17 Jackson, *Letters of the Lewis and Clark Expedition,* 1: 4–5.

18 Ibid., 1: 5.
19 Ibid.
20 Ibid., 1: 11.
21 Ibid., 1: 11–12; 2: 589.
22 Ibid. 1: 12; 2: 589.
23 Ibid., 1: 12–13.
24 Ibid., 1: 13.
25 Ibid.
26 Ibid., 1: 14–15.
27 Ibid., 1: 15, 18.
28 Ibid., 1: 17, 19, 21-22.
29 Ibid., 1: 22–23.
30 Ibid., 1: 25–26.
31 Ibid., 1: 26–27.
32 Ibid., 1: 34.
33 Ibid., 1: 28, 32.
34 Ibid., 1: 33.
35 Ibid., 1: 42, 44.
36 Ibid., 1: 53, 56.
37 Ibid., 1: 57–58.
38 Ibid., 1: 58–59 (emphasis added).
39 Ibid., 1: 59; 2: 591.
40 Ibid., 1: 61, 63.
41 Ibid., 1: 65.
42 Ibid., 1: 65.
43 Ibid., 1: 270.
44 Ibid., 1: 46–47 (emphasis original).
45 Ibid., 1: 47; Henry Nash Smith, *Virgin Land: The American West as Symbol and Myth* (New York: Vintage Books, 1950), 9, 11, 22.
46 Jackson, *Letters of the Lewis and Clark Expedition,* 1: 109–10; Smith, *Virgin Land,* 11, 22.
47 Jackson, *Letters of the Lewis and Clark Expedition,* 1: 105.
48 Ibid., 1: 173–74.
49 David L. Nicandri, *River of Promise: Lewis and Clark on the Columbia* (Bismarck, ND: Dakota Institute Press, 2009), and *Lewis and Clark Reframed: Examining Ties to Cook, Vancouver, and Mackenzie* (Pullman: Washington State University Press, 2020).
50 Jackson, *Letters of the Lewis and Clark Expedition,* 1: 320 (emphasis added).
51 Ibid., 1: 321.
52 Ibid.
53 Ibid., 1: 321–22.
54 Ibid., 1: 331.
55 Ibid., 1: 352.
56 Smith, *Virgin Land,* 18.
57 Jackson, *Letters of the Lewis and Clark Expedition,* 2: 385, 395.
58 Ibid., 2: 401–2 (emphasis original).
59 Ibid., 2: 403–4.
60 Ibid., 2: 405–7.
61 Ibid., 2: 406–7 (emphasis original).
62 Ibid., 2: 469, 555.
63 Ibid., 2: 411.
64 Ibid., 2: 547.

CHAPTER 9 | Following in the Footsteps of Lewis and Clark

1 Henry Nash Smith, *Virgin Land: The American West as Symbol and Myth* (New York: Vintage Books, 1950), 24. Emphasis in the original.
2 Ibid., 23.
3 William H. Goetzmann, *Army Exploration of the American West, 1803–1863* (Lincoln: University of Nebraska Press, 1979), 274.
4 Isaac I. Stevens, *Narrative and Final Report of Explorations for a Route for a Pacific Railroad, near the Forty-Seventh and Forty-Ninth Parallels of North Latitude, from St. Paul to Puget Sound* (Washington, DC: Thomas H. Ford, 1860), 31. Stevens's narrative formed the concluding volume (12) of the *Pacific Railroad Reports,* commonly cited as *Reports of Explorations and Surveys to Ascertain the Most Practicable and Economical Route for a Railroad from the Mississippi River to the Pacific Ocean.*
5 Ibid., 32.
6 Ibid., 33.
7 Kent D. Richards, *Isaac I. Stevens: Young Man in a Hurry* (Provo, UT: Brigham Young University Press, 1979), 139.
8 Stevens, *Narrative,* 33.
9 Richards, *Isaac I. Stevens,* 102, 139.
10 Stevens, *Narrative,* 34–35.
11 Ibid., 35.
12 Ibid., 37–38.
13 Ibid., 42.
14 Ibid., 44, 49.
15 Ibid., 59.
16 Ibid., 60.
17 Ibid.
18 Ibid., 78, 83, 87.
19 Ibid., 86.
20 Ibid., 95–96.
21 Ibid., 98.
22 Ibid., 103–4.
23 Ibid., 104.
24 Ibid., 110.
25 Ibid., 119.
26 Ibid., 121–22.
27 Ibid., 151.
28 Ibid., 158.
29 *Reports of Explorations and Surveys to Ascertain the Most Practicable and Economical Route for a Railroad from the Mississippi River to the Pacific Ocean,* 12 vols. (Washington, DC: Beverley Tucker, Printer, 1855–60), 1: 52.
30 Ibid., 1: 72.
31 Stevens, *Narrative,* 163–64.
32 Ibid., 165, 249.
33 Ibid., 166–67.
34 Ibid., 167.
35 *Reports of Explorations,* 1: 183.
36 Ibid., 1: 621.
37 Ibid., 1: 113.
38 Ibid., 1: 114.

39 Ibid., 1: 114–15.
40 Ibid., 1: 115–16.
41 Stevens, *Narrative,* 18.
42 Ibid., 189.
43 Ibid., 257.
44 Ibid., 275.
45 Ibid., 280–81.
46 Ibid., 281.
47 Ibid., 284–86.
48 Ibid., 288–90.
49 Ibid., 275.
50 Ibid., 346–47.
51 Richards, *Isaac I. Stevens,* 141.
52 Goetzmann, *Army Exploration,* 282.
53 Ibid., 282–83.
54 *Reports of Explorations,* 2: 5.
55 Ibid., 2: 7–8, 13, 18.
56 Ibid., 2: 18–19.
57 Ibid., 2: 25–27.
58 Ibid., 2: 25.
59 Ibid., 2: 26.
60 Ibid., 2: 27.
61 Ibid., 2: 29, 33, 43.
62 Ibid., 2: 36, 38.
63 Ibid., 2: 41, 43.
64 Stevens, *Narrative,* 197–98.
65 Ibid., 331.
66 Ibid., 307–9.
67 Ibid., 313, 331.
68 *Reports of Explorations,* 7: 7.
69 Ibid., 7: 11.
70 Ibid., 7: 8.

CHAPTER 10 | The Canadian Dream

1 Clark Blaise, *Time Lord: Sir Sandford Fleming and the Creation of Standard Time* (New York: Vintage, 2000), 46.
2 Ibid., 58–59.
3 George M. Grant, *Ocean to Ocean: Sandford Fleming's Expedition through Canada in 1872* (Toronto: Rose Belford Publishing, 1879), 21.
4 Pierre Berton, *The National Dream: The Great Railway, 1871–1881* (Toronto: Anchor Canada, 2001), 10; Grant, *Ocean to Ocean,* 23.
5 Grant, *Ocean to Ocean,* 17; Berton, *The National Dream,* 11.
6 Grant, *Ocean to Ocean,* 22.
7 Blaise, *Time Lord,* 79.
8 Berton, *The National Dream,* 33.
9 Ibid.; Grant, *Ocean to Ocean,* 19.
10 Grant, *Ocean to Ocean,* 18, 22–23.
11 Ibid., 21.
12 Berton, *The National Dream,* 273.

13 Grant, *Ocean to Ocean,* 372–73.
14 Berton, *The National Dream,* 354.
15 Blaise, *Time Lord,* 125; Berton, *The National Dream,* 278.
16 Pierre Berton, *The Last Spike: The Great Railway, 1881–1885* (Toronto: Anchor Canada, 2001), 309.
17 Ibid., 305.
18 Ibid., 328, 330, 416.
19 Barry Gough, *Juan de Fuca's Strait: Voyages in the Waterway of Forgotten Dreams* (Madeira Park, BC: Harbour, 2012), 235.
20 Henry Nash Smith, *Virgin Land: The American West as Symbol and Myth* (New York: Vintage Books, 1950), 51.

CHAPTER 11 | The Arctic Quest Redux

1 Glyn Williams, *Arctic Labyrinth: The Quest for the Northwest Passage* (London: Penguin, 2009), 169.
2 Pierre Berton, *Arctic Grail: The Quest for the Northwest Passage and the North Pole, 1818–1909* (Toronto: Lyons Press, 2000), 16.
3 David L. Nicandri, *Captain Cook Rediscovered: Voyaging to the Icy Latitudes* (Vancouver: UBC Press, 2020).
4 J.C. Beaglehole, ed., *The Journals of Captain James Cook on His Voyages of Discovery,* 3 vols. (Cambridge: Cambridge University Press/Hakluyt Society, 1955–67), 1: 380.
5 Berton, *Arctic Grail,* 229.
6 Ibid., 247.
7 I.S. MacLaren, "Bones of Empire: Cook and Franklin Reaching to Alaska for a Northwest Passage," in *Imagining Anchorage: The Making of America's Northernmost Metropolis,* ed. James K. Barnett and Ian C. Hartman (Fairbanks: University of Alaska Press, 2018), 118–19; Williams, *Arctic Labyrinth,* 309.
8 Beaglehole, *Journals of Captain James Cook,* 1: 399; 2: 175.
9 Berton, *Arctic Grail,* 542.
10 Ibid., 548.
11 Ibid., 551.

EPILOGUE | The New North

1 Barry Lopez, *Arctic Dreams: Imagination and Desire in a Northern Landscape* (New York: Vintage, 2001), 8, 12.
2 W.F. Weeks, *On Sea Ice* (Fairbanks: University of Alaska Press, 2010), 520–21.
3 Ibid., 530, 537.
4 Shelagh Grant, *Polar Imperative: A History of Arctic Sovereignty in North America* (Vancouver: Douglas and McIntyre, 2010), 429, 435.
5 Ibid., 445.
6 For a longer discussion of Arctic sovereignty issues, including environmental regulatory controls, international treaty status, and the Inuit nexus, see Gudrun Bucher and Robin Inglis, "The Arctic in Focus: National Interests and International Cooperation," in *Arctic Ambitions: Captain Cook and the Northwest Passage,* ed. James K. Barnett and David L. Nicandri (Seattle: University of Washington Press, 2015), 375–89.
7 The same phenomenon, in general terms, is transpiring at the South Pole, where Antarctic ice sheets sitting over land and the glacier ice shelves, tongue-like extensions from the continental land mass into the sea, are losing volume (some from twenty-four to sixty-three feet

per decade) through melting and the calving of icebergs. There are seasonal variations and natural factors during the southern winter, such as changes in the strength and motion of the winds, which result in intermittent rebounds to the northerly extent of the saltwater ice pack surrounding Antarctica. But unlike the Southern Ocean, which surrounds Antarctica, the Arctic Ocean is landlocked, enclosed by North America, Greenland, and Eurasia, whose surfaces are inherently warmer than water, and the oceanic currents in the Northern Hemisphere are warmer as well, factors that combine to result in more vigorous melting.

8 For a deeper analysis of this topic, see Lawson W. Brigham, "Marine Navigation in the Arctic Ocean and the Northwest Passage," in *Arctic Ambitions: Cook and the Northwest Passage,* ed. James K. Barnett and David L. Nicandri (Seattle: University of Washington Press, 2015), 359–74.

9 Grant, *Polar Imperative,* 450–52. Granatstein's timeline for early arrival of an ice-free North Pole model differs from most models, which predict 2040 to mid-century as the earliest prospect for open-water voyages across the North Pole. Scott Borgerson and Michael Byers, "The Arctic Front in the Battle to Contain Russia," *Wall Street Journal,* March 9, 2016, A15.

10 Weeks, *On Sea Ice,* 540.

Bibliography

Beaglehole, J.C., ed. *The Journals of Captain James Cook on His Voyages of Discovery.* 3 vols. Cambridge: Cambridge University Press/Hakluyt Society, 1955–67.

Belyea, Barbara. "Just the Facts: Vancouver and Theoretical Geography." Paper presented at the Vancouver Conference on Exploration and Discovery, Simon Fraser University, Burnaby, BC, April 23–26, 1992.

Beresford, William. *A Voyage Round the World … to the North-West Coast of America: Performed in 1785, 1786, 1787 and 1788, in the* King George *and* Queen Charlotte, *Captains Portlock and Dixon.* London: Geo. Goulding, 1789.

Berton, Pierre. *Arctic Grail: The Quest for the Northwest Passage and the North Pole, 1818–1909.* Toronto: Lyons Press, 2000.

–. *The Last Spike: The Great Railway, 1881–1885.* Toronto: Anchor Canada, 2001.

–. *The National Dream: The Great Railway, 1871–1881.* Toronto: Anchor Canada, 2001.

Blaise, Clark. *Time Lord: Sir Sandford Fleming and the Creation of Standard Time.* New York: Vintage, 2000.

Borgerson, Scott, and Michael Byers, "The Arctic Front in the Battle to Contain Russia," *Wall Street Journal,* March 9, 2016, A15.

Bown, Stephen R. *Madness, Betrayal and the Lash: The Epic Voyage of Captain George Vancouver.* Vancouver: Douglas and McIntyre, 2008.

Brigham, Lawson W. "Marine Navigation in the Arctic Ocean and the Northwest Passage." In *Arctic Ambitions: Captain Cook and the Northwest Passage,* edited by James K. Barnett and David L. Nicandri, 359–73. Seattle: University of Washington Press, 2015.

Bucher, Gudrun, and Robin Inglis. "The Arctic in Focus: National Interests and International Cooperation." In *Arctic Ambitions: Captain Cook and the Northwest Passage,* edited by James K. Barnett and David L. Nicandri, 375–91. Seattle: University of Washington Press, 2015.

Chapin, David. *Freshwater Passages: The Trade and Travels of Peter Pond.* Lincoln: University of Nebraska Press, 2014.

Dalrymple, Alexander. *Plan for Promoting the Fur-Trade, and Securing It to This Country, by Uniting the Operations of the East-India and Hudson's Bay Companys.* London: George Bigg, 1789.

–. *Spanish Pretensions, Fairly Discussed.* London: George Bigg, 1790.
David, Andrew, ed. *The Charts and Coastal Views of Captain Cook's Voyages*, vol. 3, *The Voyage of Resolution and Discovery, 1776–1780*. London: Hakluyt Society, 1997.
Delgado, James P. "The Continuing Quest: The Lure of the Northwest Passage in the Nineteenth and Twentieth Centuries." In *Arctic Ambitions: Captain Cook and the Northwest Passage,* edited by James K. Barnett and David L. Nicandri, 313–29. Seattle: University of Washington Press, 2015.
Dillon, Richard. "Peter Pond and the Overland Route to Cook's Inlet." *Pacific Northwest Quarterly* 42, 4 (October 1951): 324–29.
Dugatin, Lee Alan. "André Michaux's 'Almost Expedition' to the West in 1793." *We Proceeded On,* 47,1 (February 2021): 5–11.
Fagan, Brian. *The Little Ice Age: How Climate Made History, 1300–1850*. New York: Basic Books, 2000.
Fernández-Armesto, Felipe. *Pathfinders: A Global History of Exploration*. New York: Norton, 2006.
Fisher, Robin, and Hugh Johnston. *From Maps to Metaphors: The Pacific World of George Vancouver.* Vancouver: UBC Press, 1993.
Forster, Johann Reinhold. *History of Voyages of Discovery Made in the North.* Dublin: Luke White and Pat. Byrne, 1786.
Fry, Howard T. *Alexander Dalrymple (1737–1808) and the Expansion of British Trade.* Toronto: University of Toronto Press, 1970.
–. "Alexander Dalrymple and Captain Cook: The Creative Interplay of Two Careers." In *Captain Cook and His Times,* edited by Robin Fisher and Hugh Johnston, 41–57. Seattle: University of Washington Press, 1979.
Gibson, James. "A Notable Absence: The Lateness and Lameness of Russian Discovery and Exploration in the North Pacific, 1639–1803." In *From Maps to Metaphors: The Pacific World of George Vancouver,* edited by Robin Fisher and Hugh Johnston, 85–103. Vancouver: UBC Press, 1993.
Goetzmann, William H. *Army Exploration of the American West, 1803–1863*. Lincoln: University of Nebraska Press, 1979.
Gough, Barry M. *The Elusive Mr. Pond: The Fur Trader and Explorer Who Opened the Northwest.* Madeira Park, BC: Douglas and McIntyre, 2014.
–. *First across the Continent: Sir Alexander Mackenzie.* Norman: University of Oklahoma Press, 1997.
–. *Juan de Fuca's Strait: Voyages in the Waterway of Forgotten Dreams.* Madeira Park, BC: Harbour, 2012.
–. *The Northwest Coast: British Navigation, Trade, and Discoveries to 1812*. Vancouver: UBC Press, 1992.
Grant, George M. *Ocean to Ocean: Sandford Fleming's Expedition through Canada in 1872.* Toronto: Rose Belford Publishing, 1879.
Grant, Shelagh. *Polar Imperative: A History of Arctic Sovereignty in North America.* Vancouver: Douglas and McIntyre, 2010.
Hayes, Derek. *First Crossing: Alexander Mackenzie, His Exploration across North America, and the Opening of the Continent.* Seattle: Sasquatch, 2001.
Hearne, Samuel. *A Journey from Prince of Wales Fort in Hudson's Bay to the Northern Ocean,* edited and introduced by Richard Glover. Toronto: Macmillan, 1958.
Jackson, Donald, ed. *Letters of the Lewis and Clark Expedition, with Related Documents: 1803–1854*. 2 vols. Urbana: University of Illinois Press, 1978.
–. *Thomas Jefferson and the Rocky Mountains: Exploring the West from Monticello.* Norman: University of Oklahoma Press, 2002.

King, Robert J. "George Vancouver and the Contemplated Settlement at Nootka Sound." *The Great Circle: Journal of the Australian Association for Maritime History* 32, 1 (2010): 1–16.

Lamb, W. Kaye, ed. *The Letters and Journals of Sir Alexander Mackenzie.* London: Cambridge University Press, 1970.

Lamb, W. Kaye, and Tomás Bartoli. "James Hanna and John Henry Cox: The First Maritime Fur Trader and His Sponsor." *BC Studies* 84 (Winter 1989): 3–36.

Lopez, Barry. *Arctic Dreams: Imagination and Desire in a Northern Landscape.* New York: Vintage, 2001.

Mackay, David. "The Burden of *Terra Australis:* Experiences of Real and Imagined Lands." In *From Maps to Metaphors: The Pacific World of George Vancouver,* edited by Robin Fisher and Hugh J.M. Johnston, 263–89. Vancouver: UBC Press, 1993.

MacLaren, I.S. "Bones of Empire: Cook and Franklin Reaching to Alaska for a Northwest Passage." In *Imagining Anchorage: The Making of America's Northernmost Metropolis,* edited by James K. Barnett and Ian C. Hartman, 105–25. Fairbanks: University of Alaska Press, 2018.

McDowell, Jim. *José Narváez: The Forgotten Explorer.* Spokane: Arthur H. Clark Company, 1998.

Meares, John. *Voyages Made in the Years 1788 and 1789, from China to the North West Coast of America.* London: Logographic Press, 1790.

Miller, Robert J. *Native America, Discovered and Conquered: Thomas Jefferson, Lewis and Clark, and Manifest Destiny.* Westport, CT: Praeger, 2006.

Moulton, Gary E., ed. *The Journals of the Lewis and Clark Expedition.* 13 vols. Lincoln: University of Nebraska Press, 1983–2001.

Munford, James, ed. *John Ledyard's Journal of Captain Cook's Last Voyage.* Corvallis: Oregon State University Press, 1963.

Nicandri, David L. *Captain Cook Rediscovered: Voyaging to the Icy Latitudes.* Vancouver: UBC Press, 2020.

–. *Lewis and Clark Reframed: Examining Ties to Cook, Vancouver, and Mackenzie.* Pullman: Washington State University Press, 2020.

–. "The Rhyme of the Great Navigator: The Literature of Captain Cook and Its Influence on the Journals of Lewis and Clark. Part 1: A Canoe's Teeth." *We Proceeded On* 42, 1 (February 2016): 22–28.

–. *River of Promise: Lewis and Clark on the Columbia.* Bismarck, ND: Dakota Institute Press, 2009.

Nokes, J. Richard. *Almost a Hero: The Voyages of John Meares, R.N., to China, Hawaii, and the Northwest Coast.* Pullman: Washington State University Press, 1998.

Owens, Kenneth, with Alexander Yu. Petrov. *Empire Maker: Aleksandr Baranov and Russian Colonial Expansion into Alaska and Northern California.* Seattle: University of Washington Press, 2015.

Richards, Kent D. *Isaac I. Stevens: Young Man in a Hurry.* Provo, UT: Brigham Young University Press, 1979.

Smith, Henry Nash. *Virgin Land: The American West as Symbol and Myth.* New York: Vintage, 1950.

Speck, Gordon. *Samuel Hearne and the Northwest Passage.* Caldwell, ID: Caxton, 1963.

Stark, Peter. *Astoria: John Jacob Astor and Thomas Jefferson's Lost Pacific Empire.* New York: HarperCollins, 2014.

Stevens, Isaac I. *Narrative and Final Report of Explorations for a Route for a Pacific Railroad, near the Forty-Seventh and Forty-Ninth Parallels of North Latitude, from St. Paul to Puget Sound.* Washington, DC: Thomas H. Ford, 1860.

Tovell, Freeman, trans. and ed., and Robin Inglis and Iris H.W. Engstrand, eds. *Voyage to the Northwest of America, 1792: Juan Francisco de la Bodega y Quadra and the Nootka Sound Controversy.* Norman, OK: Arthur H. Clark Company, 2012.

Trevenen, James. *A Memoir by James Trevenen,* edited by Christopher Lloyd and R.C. Anderson. London: Navy Records Society, 1959.

Van Orman, Richard A. *The Explorers: Nineteenth-Century Expeditions in Africa and the American West.* Albuquerque: University of New Mexico Press, 1984.

Vancouver, George. *A Voyage of Discovery to the North Pacific Ocean and Round the World, 1791–95,* edited by W. Kaye Lamb. 4 vols. Cambridge: Cambridge University Press/Hakluyt Society, 1984.

Weeks, W.F. *On Sea Ice.* Fairbanks: University of Alaska Press, 2010.

Williams, Glyn. *Arctic Labyrinth: The Quest for the Northwest Passage.* London: Penguin, 2009.

–. "Myth and Reality: The Theoretical Geography of Northwest America from Cook to Vancouver." In *From Maps to Metaphors: The Pacific World of George Vancouver,* edited by Robin Fisher and Hugh Johnston, 35–50. Vancouver: UBC Press, 1993.

–. *Naturalists at Sea: Scientific Travelers from Dampier to Darwin.* New Haven, CT: Yale University Press, 2013.

–. *Voyages of Delusion: The Search for the Northwest Passage in the Age of Reason.* London: HarperCollins, 2002.

Zug, James. *American Traveler: The Life and Adventures of John Ledyard.* New York: Basic Books, 2005.

–, ed. *The Last Voyage of Captain Cook: The Collected Writings of John Ledyard.* Washington, DC: National Geographic Society, 2005.

Index

Printed in the United Kingdom
Set in Myriad and Garamond by Artegraphica Design Co. Ltd.
Copy editor: Frank Chow
Indexer: Rachel Robinson
Cartographer: Eric Leinberger
Cover designer: George Kirkpatrick
Cover image: *The Discovery on the Rocks in Queen Charlotte's Sound.* Delineated by William Alexander from the original watercolour sketch by Zachary Mudge, engraved by B.T. Pouncey. Courtesy Washington State Historical Society, 1911.5.4.39

Authorized Representative:
Easy Access System Europe -
Mustamäe tee 50, 10621 Tallinn, Estonia,
gpsr.requests@easproject.com

Papier | Fördert gute Waldnutzung
FSC® C083411

Kolibri 360 GmbH
Ferdinand-Jühlke-Straße 7
99095 Erfurt, Deutschland
produktsicherheit@kolibri360.de